AF822024

PRAISE FOR
THE NEUROBIOLOGY OF CONNECTION

"A marvelous text, full of brilliant transdisciplinary insights!"

–Darcia Narvaez, PhD
Professor Emerita of Psychology,
Developer of the Evolved Nest

"A perfect bridge between neuroscience and the wisdom traditions."

–JD Daniels, MD, MPH
Professor of Family & Community Medicine
Assistant Dean of Students for SIU Medicine

"The origin story of human connectedness is not frozen in deep time. It is here, alive, in our present-now, with pulse and with resonance so deep that in somehow coming into contact with this book, I am simultaneously coming more into contact with this pulse. Gabriel's ability to dissolve limiting paradigms allows the reader to dissolve them, too - reconfiguring our own origin story. This is a book for any human interested in vitality, interested in decoding wellbeing, and interested in coming home."

–Jordhynn Guy, CCC
Developer of the Elite Resilience Framework

"Gabriel has brought together some of the cutting-edge work in autonomic physiology, neurobiology and neuroscience within the broader context of late-stage capitalism. Healing in absence of context can amplify trauma. This book attempts to address the decontextualization of modern science, and in the process, engages the reader in deep neural cartography for troubled times. This work is pioneering the emerging field of what Gabriel calls connection phenomenology. It is timely and necessary."

–Alnoor Ladha
co-author of *Post Capitalist Philanthropy: Healing Wealth in the Time of Collapse*

"Neurobiology of Connection is a book we all need—a user's manual for our nervous systems. Brilliantly insightful and practical, we can learn to understand why we are the way we are moment-to-moment, meet our needs in ways that heal old

THE NEUROBIOLOGY OF CONNECTION

traumas, and enhance deep connection with other humans and the world around us. Gabriel brings together material from indigenous cultures, neuroscience, neurochemistry, mindfulness, somatically-informed trauma work, Polyvagal Theory, and literature, combined with attention to his inner world, in order to create an updated model of the Autonomic Nervous System. I have made so many changes to the way I teach, and my students are loving it!"

–Marcia Miller
Certified Yoga Therapist
Reiki Master Teacher
senior teacher for UZIT

"My first nervous system health mentor introduced me to both Polyvagal Theory and the concept of 'cellular safely'. As a bodyworker practicing Visceral Manipulation through a trauma and nervous system lens, I use the sensitivity of my hands to introduce cellular safety to decontextualized stress responses (also known as tension) in the body. From the release of tension anchored in safety, systemic well-being emerges and continues to emerge as the body continues to be approached and supported in this way. It has been a challenge to find my work accurately represented through a lens of neurobiology. I am beyond grateful for this eloquent expansion of Polyvagal Theory that brings in the knowing of my hands and the wisdom that flows from our bodies to our heart, then our brain.

I also deeply appreciate Gabriel for naming the domination mindset that is imbedded into the fabric of modernity and for challenging the status quo around it. It is bold and necessary."

-Kelley Curtis
Bodyworker and Nervous System Health Practitioner

"Moving forward from fixed systems, this book provides us a fluid, confluence template and teaches us that conscious navigation of our internal rivers is what affords transformation and sovereign embodiment."

–Holly Bridges
Creator of Anxiety Reframe Technique
and Author of 'Reframe Your Thinking Around Autism'

"Gabriel has waved a poetic wand over off-putting medical-speak and pulled the curtain back on the divine delighting herself in her magic shop. In this monumental work that is being constructed through him, I get to press my nose against that glass. I tell people the work Gabriel is conducting is the science behind safety. Bravo maestro! A Slam Dunk."

–Jimbo Graves,
Author of *Tillage*

"As a facilitator working with groups to develop organizational strategy for more than 25 years, I've been a student of many interdisciplinary methods that offer insight on how to better understand human behavior. When I stumbled upon Hearth Science's work with the neurobiology of connection, the pathway opened wide and a whole new sense of potential began to emerge—a path that feels accessible to everyday people and rooted in experience.

Gabriel's work with the autonomic nervous system offers powerful keys to understanding how to participate in relational spaces that consciously nurture trusting relationships and invite creativity. These are key ingredients for the kind of ongoing attunement we need to generate a culture of care, belonging, and collective wellbeing. The neurobiology of connection is a wide open gateway to inform the awareness of human connection that is essential to create contexts for diverse people to come together and navigate the many complex, intersecting challenges we face as a species on this planet. If you recognize the importance of lived experience on our ability to form healthy relationships with ourselves and others, build community and collaborate for social and environmental change, this body of work offers an embodied, highly functional map to actively heal the fractured social territory of our times."

–Carri Munn
Founding Partner, Circle Generation

BY NATUREZA GABRIEL

Hearth Science

Restorative Practices of Wellbeing

Keywords: A Field Guide to the Missing Words

So, About that Death Cult you Joined?

The Archeology of Shadows

Sport of Kings

Can't Get Home (Stories)

Destoryer of Empire (Collected Poems 2020-2022)

Origin Stories for Children

THE NEUROBIOLOGY OF CONNECTION

RE-WILDING THE DEEP NERVOUS SYSTEM FOR WELLBEING

NATUREZA GABRIEL

FOUNDER OF HEARTH SCIENCE

THE NEUROBIOLOGY OF CONNECTION

NATUREZA GABRIEL

Published by:
Jaguar Imprints
PO Box 567, Nicasio, CA 94946, USA

A CIP record for this book is available from the Library of Congress Cataloging-in-Publication Data

Printed in USA

As the fever of day calms towards twilight
May all that is strained in us come to ease.
We pray for all who suffered violence today,
May an unexpected serenity surprise them.
For those who risk their lives each day for peace,
May their hearts glimpse providence at the heart of history.
That those who make riches from violence and war
Might hear in their dreams the cries of the lost.
That we might see through our fear of each other
A new vision to heal our fatal attraction to aggression.
That those who enjoy the privilege of peace
Might not forget their tormented brothers and sisters.
That the wolf might lie down with the lamb,
That our swords be beaten into ploughshares
And no hurt or harm be done
Anywhere along the holy mountain.

~ John O'Donohue

FOR ALL THE CHILDREN OF THE LIVING EARTH

CONTENTS

Fractal (noun)-
Self-similar at varying levels of scale.

THE NEUROBIOLOGY OF CONNECTION: A FRACTAL IN THREE MOVEMENTS

FIRST MOVEMENT: MANDALA

COVER The Dust Jacket of the Book

SECOND MOVEMENT: AUTONOMICS

LEGEND Legend 18

BACKGROUND Background 25

AUTONOMICS Principles of Autonomics 33

MEDITATIONS Meditations on the Autonomic Wheel 46

3 X 3 X 3 Three by Three by Three 62

AUTONOMIC STATE Autonomic State 69

KEYS We've Lost the Housekeys 75

THIRD MOVEMENT: HEALTH CREATION MANUAL

GL Glossary (Specialist Terms, Non-English words) 87

GETTING STARTED

01 How we Got Here 93

02 Setting the Stage 103

03 Your Guide 119

A Note about Supplies & Cadence 132

04 Evolutionary Context 136

05 Soil You Were Born In 153

06 The Altar of the Spirit 170

FOREST

07 Listening with the Eyes 189

08 Inward Forest 204

09 Attention is the Light 209

NEUROLOGY OF THE THREE AUTONOMIC SYSTEMS

10 I Feel Myself, Therefore I am 212

11 I Touch the Ground, Therefore I am 224

12 I Move, Therefore I am 235

13 Epiphenomena of Autonomic State 241

CHEMISTRY

14 Neurochemistry 247

DOORWAY OF SAFETY

15 Safety is a Sovereign Decision 265

16 Safety in a World of Existential Threat 293

DOORWAY OF CONNECTION

17 Connection is the Food 299

A Note on Attention & Autonomic State 304

18 The Body You are Wearing 307

PORTALS TO CONNECTION

19 Rhythm & Regulation 326

20 Gravity as Dance Partner 335

21 The Great Mystery of Breath 341

22 Face & Voice 353

23 Singing in the Shower 359

24 Language as Doorway 365

25 I See You 372

26 The Errant Eyebrow 376

27 I Hear You 382

28 Hands 393

29 Craft Is Language of the Hands 405

30 Your Hands are Radar 416

31 Touch the Ground 423

32 Your Feet Begin in Your Back 432

33 Movement 442

34 Gesture as Doorway 458

35 Posture as Doorway 477

36 Enter the Heart Directly 486

INTEROCEPTION

37 Settled Interoception 495

DOORWAY OF INTUITION

38 Intuition 509

DOORWAY OF RELATEDNESS

39 The Doorway to Relatedness 520

DOORWAY OF ENDURING WELLBEING

40 The Doorway to Enduring Wellbeing 530

POST TEXT

Acknowledgements 542

FOR INTERACTIVE VIDEO- AND TEXT-BASED INSTRUCTION IN THIS BOOK

http://www.hearthscience.io

LEGEND

The focus of this book is on activating the root drivers of well-being, through the deepest neural architecture of your body: the Autonomic Nervous System. This system contains the most powerful levers that govern how your body processes experiences – responsible for ensuring survival on the one hand, and creating thriving on the other.

Although a good part of the knowledge about how the Autonomic Nervous System works is available in specialist circles, it is often so deeply neurologically technical that even most wellness professionals do not understand it fluently enough to grasp its full implications or make good use of it. This is a shame, because if autonomic physiology was centralized in the diagnosis and treatment of disease it would completely transform medicine and mental health. As for people who are not wellness professionals, most everyday people have no idea what the Autonomic Nervous System is, or what it does.

Yet what we are talking about in this book is what it means to be human, refracted through the lens of your deep neurobiology. Everything here is actually very close to your experience. You already know what I'm talking about, and will be talking about, you probably just don't realize this because it hasn't been organized and presented to you in such a way that you can understand and digest it. You've never taken a class on it because one has never been offered, and so no one has *brought it home for you* in a pattern language you can recognize. No one has made it so relevant that you cannot ignore it.

Until now, no one has helped you use the lens of the Autonomic Nervous System to bring your life into focus.

My goal in this text, therefore, is to hold a particular kind of magnifier up to our shared human experience, and look through it with you in language and metaphors that you can see and feel and taste and digest, so that they can begin to re-organize you. I'm not simply interested in you *understanding* what I am saying. I'm not interested in you having an academic experience. I want it to transform you.

The simplicity that we are aiming for is such that you could turn around and teach this to your six-year-old, because even a six-year old has experienced most of the things I'll be talking about. I have taught this stuff to kindergartners.

When you encounter a map, often in a lower corner of it, you find a legend. The legend teaches you how to read the map. It helps you decipher signs and symbols and colors and shapes so you understand what you are looking at.

This entire book is a map. The image on the cover of the book is an archetypal depiction of it, a symbolic fractal, a holism that stands in for it all, and the text will teach you to read the map of your life using the symbol as a kind of compass. On this page two paragraphs up– there it is in miniature.

The symbol depicts twelve autonomic states arranged in a sacred geometry around an open center. An autonomic state is an energy-processing template. It is an energy-processing template because, like the graphic equalizer on a stereo, it governs the way experience flows through you. Like the tuning present on a stereo, it electro-magnetically governs how your speaker vibrates with the flow of life.

It is an energy-processing template surfaced, moment-to-moment, by the deepest layers of your nervous system, whose two primary functions are to regulate your physiological interior and tune it in response to your experiences.

As you wake up, get out of bed, stretch, take a shower, get dressed, tell the kids to get up, read your first emails, make breakfast, yell to the kids that you mean it, crawl to school through traffic, greet the crossing guard, and head to work, you are having experiences that shape these templates and are shaped by them. The first time you wake the kids, fresh from bed, stretched, showered, sipping your coffee with enjoyment, you are wearing a different energy-processing template than when you yell up to them fifteen minutes before you need to leave as breakfast rapidly cools on the table. As you negotiate traffic (*Why is this street so busy?*) you are wearing a different template than when you exchange pleasantries with the crossing-guard moments later (*Alright then. Good to see ya.*).

Underneath the thinking layer of your experience there is a river of energy: the quality of the way it feels to be you, moment-to-moment, in your essence, in your body this morning, the landscape of your emotions, of thoughts, perceptions about the world around you, and then your behavior in the midst of it all.

Autonomic state is what is shaping the banks of this river of energy.

Forcing the river into class 5 rapids at high velocity as you negotiate a complex business agreement that your yearly commission depends upon, slowing out into a broad slow pool so placid that it is almost unmoving as you sit with your feet kicked back on vacation, reading a novel with the kids at camp and no place to be.

Your Autonomic Nervous System, running it all, is the biological system with the strongest shaping force over how it feels to be you in each and every moment of your life.

Present moment autonomic state shapes the way it feels to wear your body (muscle tension or relaxation, posture, alignment, heartrate, breathing, digestion, body temperature, etc.), your emotions, thinking, perceptions of the world around you, and behaviors, according to a precise set of parameters.

Each state on the map, each station, each colored circle with an image of water at its center, represents a combination of three primary factors: 1) a determination of whether you are safe, in danger, or under lifethreat (*neuroception* is the technical word for this), 2) which autonomic systems are available to you (this is the actual neurology that is active: your Connection, Movement, and Grounding Systems), and what neurochemistries are flowing over them (there are three primary chemistries: connection, activation, and shutdown): the neurohormonal fluid flowing through your inward ocean.

In the compass on the cover, we use color and water to represent each state. Color because in the same way that any color can be represented by three variables: hue, value, and chroma, autonomic state can be represented by neuroception, autonomic system, and neurochemistry. And color because just as blue and yellow predictably yield green, the combination of Connection and Movement, in the presence of safety yields Play. Color lets us see and feel the way neuroception, neurology, and neurochemistry combine predictably into autonomic state.

And water, because this mysterious element, which is Life, also exists in three primary states. Imagine that you had never been to earth. Arriving, you encounter water wearing each of her three faces. First, as liquid. Then, as steam. Finally, as ice. If you could not watch her transform, I do not believe you would accept that she was just one single substance. Her physical and chemical properties are completely different across these three faces. And yet, she is One. One element that transforms depending entirely on energy flowing through her, in the form of temperature.

Like water, the version of you that is available to meet the world when you feel safe, in danger, or under lifethreat is unrecognizeably different. And yet, also like water, if we understand water, and understand *why* it changes form, we can change it back by changing the inputs to it. Imagine how bereft of hope you would be, not realizing that liquid water could change to ice and then change back, to discover in winter that your entire supply of drinking water had turned to ice, believing it would never change back.

You would be so distressed! *It's ruined,* you would think. *This beautiful liquid that I need to drink has turned into stone! Woe is me! Surely I will never recover.*

And yet, understanding nothing of the lifethreat response, nothing of shutdown chemistry, people's Autonomic Nervous Systems shift beneath them into defensive states and they experience themselves permanently and irrevocably changed. Stuck with autonomic glasses they cannot take off, the world irrevocably smudged, they lose hope, without realizing that we can change the lenses back!

With your water that has turned to stone, the plumber is called. The plumber, knowing that water exists in three states, that it is not *really* stone, precisely applies heat to thaw it. And, *lo and behold*, your liquid water returns.

This is the power of the map you are holding. To understand the inner workings of this deep biology, so that you can harness it in the direction of thriving.

I want to let you in on a little secret, right here at the beginning. The direction of wellbeing is liquid water. But nearly everything about our modern world is conspiring to turn you into steam and ice. Wellbeing is a revolutionary act. And the recipe that will give rise to it is not what you think.

Because much of the tilt of modern medicine and mental health is focused on *what's wrong*, mirroring the negativity bias of our minds, which neuroplasticity expert Rick Hanson notes grips onto negative experiences like they are velcro, and off which positive experiences bounce like they are sliding off Teflon, this book brings into focus how we can change the inputs to the deepest most primal parts of our nervous systems to create the conditions for thriving.

It proposes that without your consert or awareness you have been indoctrinated into a brain-centric world, a world that is actually a byproduct of alienation and empire, and taught to organize your sense of self, and your understanding of experience from the perspective of your thinking rather than your feeling. The book teaches another, more ancient way to embody your relationship with reality by tuning into certain kinds of inputs that activate your thriving. It will teach you to become as discerning about feeding your Connection System as you are about nourishing your children. In the same way that you would never forget to eat for a week, after reading this book I don't think you'll ever forget to nourish your Connection System for a week. Because by the end you will understand that it is the foundation of your wellbeing. And in the same way that you would hesitate to fill your children's bellies with the empty calories of candy instead of proper meals, you will hesitate to feed your Connection System with the connection simulations of the modern world.

By the end of the book you'll be empowered to harness a process of self-directed neuroplasticity that can literally change the way that you experience reality, and who you are in it, entirely.

Because some of you coming to this book will be doing so as interested general readers, and some of you will be doing so as wellness professionals, I would like to provide a bit of explanatory scaffolding at the outset with the attempt to best meet both of your sets of learning needs, which may be quite different. If you find this background section too technical, or not relevant, skip it.

BACKGROUND

At Hearth Science, the organization that I founded and direct, we have developed a new foundation model of autonomic physiology called *Autonomics*. If you are a generalist reader I don't expect you to understand what I'm talking about yet, but I need to address this at the outset of the book for those of you who may be coming from the study of Polyvagal Theory and/or orthodox neuroscience or neurophysiology backgrounds, so that we can be clear on the distinctions we are making in this text and you don't get confused by them. We are intentionally parsing these systems differently than both Polyvagal Theory and orthodox neurology.

While autonomic physiology has existed as a field for over a hundred years, this novel conceptual framework replaces a number of terms and dramatically updates the conceptual and functional neurological maps of how the autonomic systems inter-operate IRL (in real life).

Stephen Porges, PhD, the Developer of the Polyvagal Theory, is a genius, and a personal mentor of mine. To honor his work, and with his supervision I developed the first authorized visual presentations of the Polvyagal Theory in 2020, in the form of a poster entitled *The Official Polyvagal Poster*.

In order to create this poster, I spent nine months in dialogue with him while I was building a software platform to help wellness professionals diagnose and treat stress-related disorders. When we completed the poster, which describes seven autonomic states (Connection, Play, Fight, Flight, Freeze, Shutdown, and Intimacy), my organization built a diagnostic engine for our software that used these seven states to filter diseases into different treatment frameworks. Rather than starting with a person's symptoms, we began by trying to understand the present-moment state of their Autonomic Nervous System. Once we understood that, rather than treating the symptoms, we worked to shift that underlying state (which, if they were

AUTONOMICS

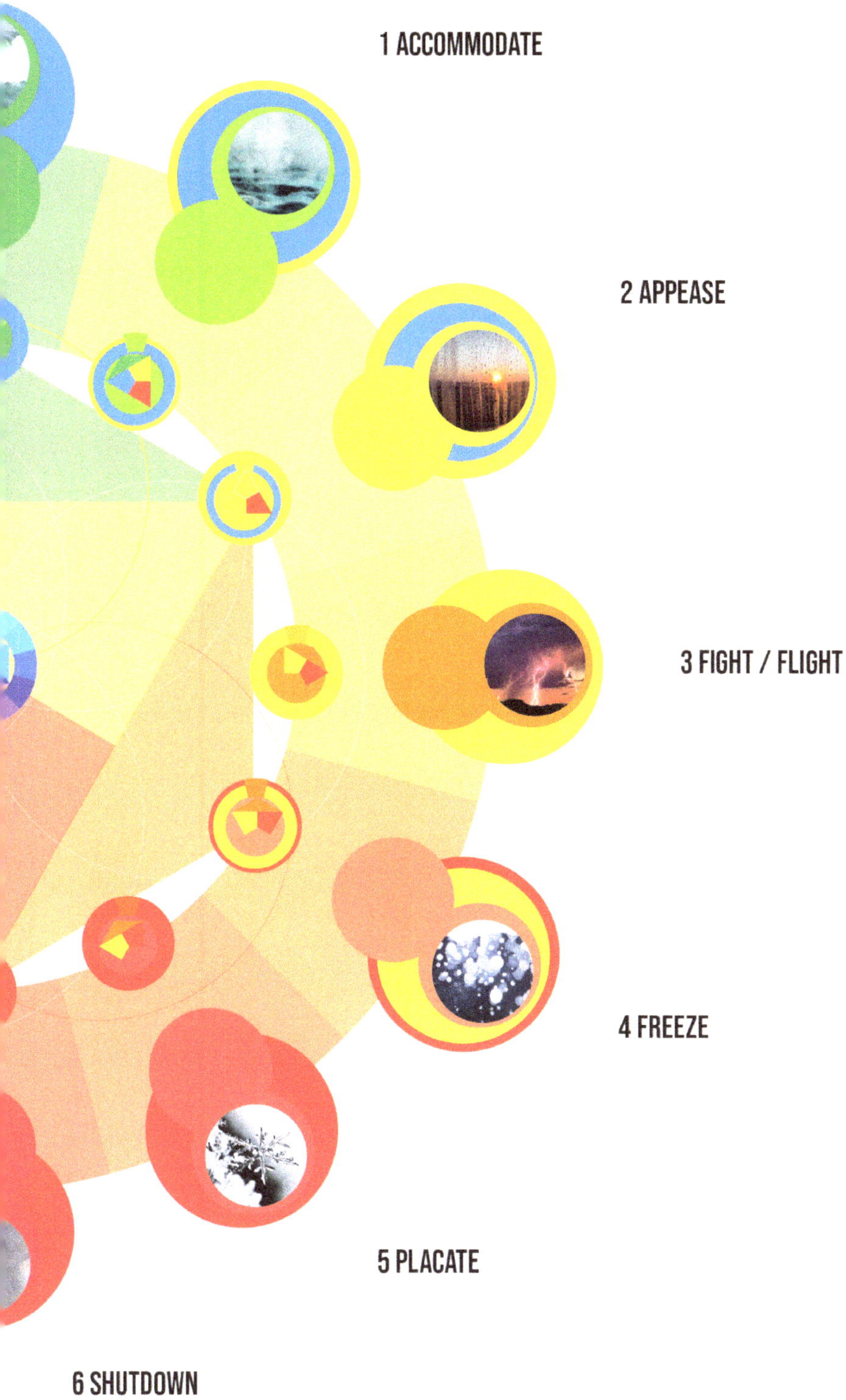
1 ACCOMMODATE
2 APPEASE
3 FIGHT / FLIGHT
4 FREEZE
5 PLACATE
6 SHUTDOWN

experiencing disease was always a defensive state) back to safety. We were shifting ice and steam back to liquid water. The results were astonishing. Clients made more progress on healing in a few weeks with our platform than they had made in the previous year of therapy. Mental health practitioners got upset with us because their clients would tell them, sometimes within a couple of weeks of starting to use our platform, that they were better and didn't need therapy anymore. When the therapists pushed back, saying surely it couldn't have happened this fast, clients would say things like, "Now I understand that my anxiety is a response to feeling threatened. I am listening for that feeling. When it arises, I trust the intelligence of it, and listen to what my body needs. I understand why it is there, and I have a set of practices I can use to address it. And guess what? I'm not anxious anymore. I'm ok."

What happened over the next four years was that we worked with thousands of people in 50 countries around the world. We worked with patients suffering from mental health issues including anxiety, depression, anger issues, sleep disorders, gastro-intestinal disease, auto-immune issues, complex chronic illness, post-traumatic stress injuries, sexism, racism, other forms of oppresion, climate anxiety, ecological alienation, and a host of other issues. Things that make people feel in danger, and under lifethreat.

We worked with wellness professionals, their patients and clients, and activists. Because we were running an incredible amount of data through the model, we saw both its explanatory power, which is immense, and places where it was inadequate to address the living texture of people's experiences. And from Day One, we saw autonomic states that were not on the map of Polyvagal Theory that Dr. Porges had drawn.

Because the way that I like to learn is in call-and-response with people, and because I have more than 100 mentors around the world in more than 25 disciplines of healing and 24 cultures, and because I train a lot of wellness professionals, I spoke with many people from many backgrounds about patterns we were seeing clinically, and in the data. And because many of these people were deeply insightful, these conversations began to lead to a novel set of understandings. We brainstormed, and collaborated, and bickered, and joked, and shared ideas and questions and challenges.

One of the first places that we saw deficits in the Polyvagal Theory model was around defensive responses that make use of social neurobiology– varieties of *appeasing*. We saw, very quickly, a clear and direct correlation between this autonomic state as a default energy processing template, and many categories of auto-immune illness. Most of the people who came to the platform seeking relief from auto-immune issues (we worked extensively with a clinic treating patients with Mast Cell Activation Syndrome, for example) utilized *appeasing* to manage threats. This autonomic pattern, more common in women than in men, more common in the United States in people of color than white folks, was one I had pointed out to Porges while we were making the poster, and that he had insisted was not physiological. But he was wrong. I can say this with confidence both because we could clearly see that it was so with hundreds of clients, and because now, four years later, he has added it to the Polyvagal map.

Not only is *Appease* a physiological state, an autonomic energy processing template, but it comes in two varieties, depending on which neurochemistries of defense are involved, and how they therefore govern the availability of autonomic states.

Gradually a view began to consolidate in our work that expanded the terrain of autonomic states to encompass not merely the seven that Stephen had described, but five more that we saw with regularity.

As our confidence in our own data grew, and as we continued to refine our understanding of the living texture of the experiences our clients were having, something deeper began to take place and I began to realize that foundational elements of the explanatory and pedagogical scaffolding of Polyvagal Theory were incomplete. I did not set out to come to this conclusion. It came from studying patterns, from my own interoceptive tracking with myself, from working with experts in early childhood development, from trans-disciplinary research, and from patients and clients with whom we were working.

This dawning realization did not come as news to Steve. One of the reasons I was able to trust him, personally, is because of his humility. In one of the private conversations I had with him, while developing the poster, I asked him why, given that Polyvagal Theory had an evidence base of more than 300 papers in many fields underneath it, he was still calling it a theory. He shrugged. *It could still be proved wrong*, he said, and our conversation moved on.

As Thomas Kuhn pointed out in his seminal work on scientific paradigms, *The Structure of Scientific Revolution*, there is a meta-pattern to advances in science. A paradigm arises as an explanatory scaffold that helps us understand why things are the way that they are. The endurance of the theory is a function of its explanatory power. It is often hard for us to see beyond a paradigm while we are operating within it, it is sometimes even difficult to recognize that we are operating within a paradigm at all: it can just seem like *how things are*. The geocentric view of the late middle Ages, which places earth at the center of the Solar system, yielded extremely reluctantly to the Copernican heliocentric model hundreds of years after the Polish astronomer Nicolaus Copernicus proposed it in 1543. Galileo, the Italian mathematician and astronomer who ground the lenses of his own telescopes so that he could independently verify Copernican assertions, was branded 'vehemently suspect of heresy' in 1616 for publicly agreeing with Copernicus, and spent the last decade of his life under house arrest.

But science, unlike religious doctrine, is empirical rather than messianic. And so theories continue to evolve, to transform, based on new evidence. About a year ago it became clear that the dissonance between Polyvagal Theory and our own work had reached a point where it either needed to be reconciled, or we needed to call our work something else.

Autonomics would not exist if Stephen had not developed Polyvagal Theory, because Polyvagal Theory is how I re-organized an understanding of my own Autonomic Nervous System. We were only able to develop Autonomics standing on the shoulders of Polyvagal Theory. But Stephen is a psycho-physiologist, and I am an Indigenously-trained interoceptive tracker and neural cartographer, and so we have perceived different things and come to different understandings.

Beyond Polyvagal Theory, in the development of Autonomics we have relied deeply on an understanding of the deep ancestral context of our Small Band Hunter Gatherer forebears, which is the manner of life in which humans have lived for 99% of our lineage history, and which is the context in which our hominid nervous systems were forged socially over the long spans of evolutionary time. If you would like to better understand this context, I direct you to the extraordinary trans-disciplinary synthesis of Darcia Narvaez, PhD. Her work on the extra-genetic scaffolding that provides the ideal context for thriving in the young of a species is called *The Evolved Nest*[1].

For thirty years I have had the good fortune to be trained in non-cognitive ways of knowing in a variety of Indigenous cultural and ancestral contexts. Darcia's work elegantly articulates commonalities of an Indigenous cosmo-vision with particular focus on the Indigenous childrearing technologies that emerge from an Animist worldview of profound relatedness: a participatory engagement with the Creation.

And in this triad of autonomic physiology and the evolved nest

1 Her most accessible version of this work is *The Evolved Nest: Nature's Way of Raising Children and Creating Connected Communities* by Darcia Narvaez, PhD and GA Bradshaw, PhD, North Atlantic Books, 2023

within the context of deep ancestral wisdom about relationship as a verb, I found the unique formulation of our work.

We have also, in our research, chosen to give primacy to autonomic signals from respiration, rather than Heart-rate variability, because these are continuous autonomic signals with greater magnitudes of information encoded about the state of the *entire* Autonomic Nervous System. It is possible, using a class of mathematical equations called Fourier transforms, to decompose the breath-wave into the component frequences of the Autonomic Nervous System, providing a window into its most nuanced manifestations in realtime. At the time Stephen's lab began studying HRV, in the early 1970s, the compute was inadequate to store and subsequently decompose these signals. And so we have been able to see most of our theses validated, empirically, and replicably, after having tracked them interoceptively in the first place. You could say that our mode of science has moved from intuition to empirical validation: the hallmark of an inner science.

I explain all of this to do honor to my mentors, Steve in particular, and all of those who have contributed to expanding the understanding of his work, as well as to respectfully disentangle ours from it. As a teacher myself, it is my happiness when my students take my work forward, if they do so with insight, rigor, and respectful attribution, because this is, after all, the point. How do we carry the flame forward? How do we expand the explanatory power of the work so that it becomes more useful? How do we use it in service to Life?

I therefore humbly present to you a general theory of *Autonomics*. This is the theoretical foundation upon which this book, *The Neurobiology of Connection*, builds.

We have changed the names of autonomic systems because the historical namings fail to differentiate the neurology from the neurochemistry that shapes the way these systems operate differentially *in vivo* in the presence of safety, danger, and lifethreat. Therefore...

AUTONOMICS

There are three autonomic systems in the human body:

in BLUE, a Connection System comprised of the ventral branch of the Vagus that unites the neural regulation of the face, voice, eyes, ears, and turning of the head and neck, with the heart and breath in newborns. As this system myelinates prior to the closing of the fontanelles around 18 months of age, it grows into the ventral surfaces of the hands, parts of the ventral surface of the feet, the skin and the genitals. Much of this book is focused on exploring the various physiological portals to this system because when this system is driving, we experience health-creating states. Previous systems and conceptualizations have termed parts of this system the 'Ventral Vagal system' and the 'Social Engagement Complex.' When Polyvagal Theory references this system it is referring to the face/voice elements of it.

in YELLOW, a Movement System organized along the entire length of the spine and into the brainstem whose central pattern generators govern rhythmic movements in the body including sucking/swallowing, breathing, the movements of the arms and legs, the movements required to pee, poop, and experience an orgasm. This system has been understood by classical neurology and Polyvagal Theory primarily with respect to undergirding the fight-flight responses (it is known in classical neurology as the Sympathetic Nervous System), yet this is only how it operates when threat is detected and activation chemistry is present. It has an entire repertoire of neurological function beyond this as it coordinates all centrally pattern generated movement (crawling, walking, running, swimming, complex coordinated arm swings, coordinated full spinal movement patterns, sucking, swallowing, vomiting, breathing, digesting, peeing, pooping, sexual thrusting/ hip rotational range, orgasm,

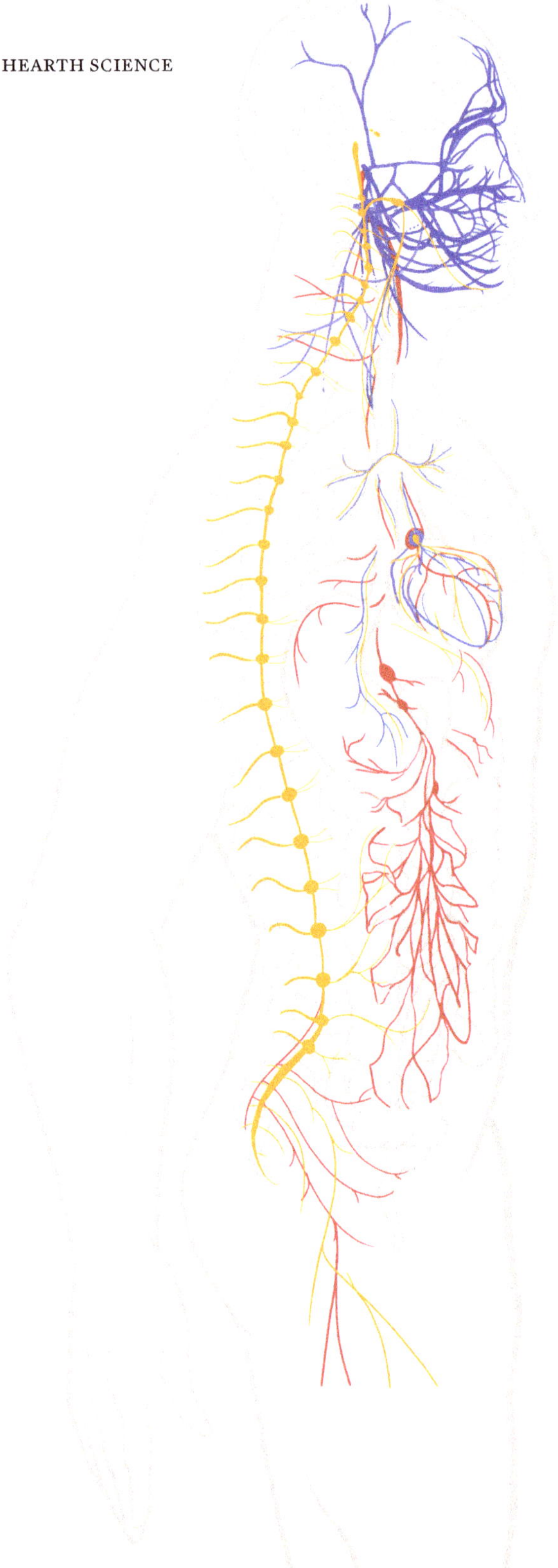

etc). When we refer to danger responses from this system we will use the terminology fight/flight response. In this configuration, it is part of our response repertoire to neuroceptions of danger.

in RED, a Grounding System comprised of the dorsal branch of the Vagus that innervates the deep belly primarily subdiaphragmatically beneath the respiratory diaphragm and includes the enteric nervous system. This system has been understood by Polyvagal Theory and the related somatically-oriented trauma healing modalities only in the configuration that it enters when we experience lifethreat, and is referred to by Polyvagal Theory as the Dorsal Vagal System. Yet it is actually the *hara* (Japanese), or *lower dantien* (Chinese), understood in eastern cultures to be the 'elixir field': the location where the elixir of life is made. This grounding system is the center of metabolism, the center of authentic movement. When we are referring to the role of this system in lifethreat responses, we will call it the Shutdown System. Symbolically simplified (and related to water in its three primary states), these systems look like this:

Prior autonomic physiology has categorically failed to distinguish neural pathways from the associated neurochemistry that activates based on our neuroception (present moment neurological detection) of safety, danger, or lifethreat. Yet neuroception governs neurochemistry, determines which autonomic systems are recruited, and how they relate to one another.
In the presence of safety, which is pulsative, connection chemistry coordinates the activities of the Connection System, Movement System, and Grounding System. (In order to coordinate sucking and swallowing with digestion, these systems must be coordinating- this is the first autonomically coordinated activity that newborns engage). (See states 8, 9, 10, 11, 12 on the Autonomics wheel)

As neuroception of danger arises, connection chemistry withdraws, and the Connection System is no longer able to rhythmically modulate and coordinate the Movement and Grounding Systems. In the presence of activation chemistry (adrenaline/cortisol) this moves us into the continuum of danger responses (accommodate/appease/flight/fight). (See 1, 2, 3 o'clock)

As neuroception of lifethreat arises, the addition of shutdown chemistry (endogenous opioids) begins to immobilize. The addition of endogenous opioids in combination with activation chemistry yields states of tonic immobility (rigid and immobilized). (See 4 and 5 o'clock)

In full lifethreat, as endogenous opioids become the primary chemistry, muscle tone is lost, the Movement System comes offline and we are left in the fully collapsed devastation of shutdown (See 6 o'clock).

In order to organize our model, we have primary colors combining into ranges, combining into a full spectrum. These neurological elements then interplay with associated neurochemistry. A symbolic simplification of our model is the logo for Autonomics, representing the ways that these three autonomic systems combine into a full spectrum of states, like this:

Throughout the text we will use this nomenclature, and the times on the clock to reference the full spectrum of energy-processing templates the autonomic nervous system can surface.

Blue and green states are, broadly speaking, health-birthing when we reside in them. Orange and red states are, broadly speaking, disease-birthing when we reside in them.

Autonomic state, in its living moment-to-moment texture, can be usefully conceptualized like color, (a blend of hue, chroma, and value), as a blend of neuroception (do we feel safe, in danger, or under lifethreat), autonomic system (Connection, Movement, Grounding), and neurochemistry (connection chemistry, activation chemistry, shutdown chemistry).

So now you have the artist's palette with which we will spend the book painting our portraits.

COORDINATION OF AUTONOMIC SYSTEMS

The most advanced current model of autonomic physiology (Polyvagal Theory) proposes a ventral vagal social engagement system that functions when we experience safety. The pervading notion of the living operation of this system understands it operating independently of other autonomic systems. This does not happen.

When we experience safety, we have access to the ventral vagal system that I refer to simply as the Connection System. This system, in the presence of the coordinating pulsation of safety, functionally organizes both the Movement system and the Grounding System. It cannot operate alone, in the absence of these other systems. Rather, when we feel safe enough, *the Connection System harnesses these other autonomic systems*. It is a **fundamental misconception** to understand these three distinct systems as operating in isolation.[1]

What happens rather, when we feel safe, is that we have access to all three of these neurological systems, and their synergistically associated felt intelligences, cohering under the auspices of a pulsation of safety. This pulsation is coordinated by the connection chemistry of oxytocin: the chemistry of love, bonding, cooperation, and collaboration. When we feel safe *enough*, these three autonomic systems work together in concert to cohere our physiological experience, and create the neural platform for our deepest connection to self, and the fullest expression of our capabilities. These capabilities extend all the way down to the soles of our feet, so that we can access what the Lakota call *anamagoptanye*, the information that the earth is constantly uploading to us. And they extend all the way up to our medial prefrontal cortex, the seat of Executive function, that gifts us with our most expansive strategic vision. With the

---

1 The only autonomic state where a single autonomic system is online is the Grounding System under lifethreat: the Shutdown response at 6 o'clock on the model. In every other autonomic state, there are two or three systems coordinating. Sometimes synergistically, sometimes antagonistically, but in all cases active.

coordinating pulse of safety, led by our hearts, our autonomics coordinate to grace us with the fullest range of our capacities.

The coordination of these three states happens on a health-creating continuum. It ranges from states where stillness is predominant (*Union* and *Restoration*) to states where sociality is predominant (*Connection*), to states where movement is predominant (*Play, Competition*). This continuum utilizes all three autonomic systems on an arousal continuum from the deeply settled to the effervescent. It includes the 8, 9, 10, 11, and 12 o'clock positions on the Autonomic Spectrum map.

If this balance has more stillness in it, we can think of it as the meditative states where deep inwardness, communion with spirit, rejuvenation and regeneration can happen. If this state has more movement in it, more vitality, we experience it as play, which deepens into experiences of enduring flow. These are the zones of elite performance, where we are both social and mobile. We have access to explosive movement, acceleration, and velocity and nuanced movement that is enabled by the Movement system, yet without any of the chemistry of adrenaline and cortisol that tightens the muscles or vasopressin that makes this system aggressive. Here we can dance, perform on stage, engage in collaborative work without friction, and engage in elite athletic performance. And between these two, when there is neither an abundance of stillness, or a high level of movement, we have the kind of social range of connection where intimate conversations take place. Where we gather around the ancestral hearth, be it the campfire or the kitchen table, to tell stories, share sorrows, build community bonds. The zone where we can lean into relating with other trust-worthy humans. This entire continuum is health-creating. Within the scaffolding of safety, the harmonization of these three systems in different balances gives us the range of the full behavioral repertoire of wellbeing.

NOTICE, in the diagram on the facing page, the three triangles– blue, yellow, and red– nested within each circle. These are the three autonomic systems, and the diagram is showing us that in the salugenic states all three of them are available.

When we lose safety, when we lose the orienting rhythmicity that gives us access to the Connection System, it can no longer organize and harness these other two autonomic systems. When we experience threat, we move sometimes to danger, and sometimes directly to life threat, depending on our deep nervous system's assessment of the present moment danger in light of our neural conditioning. In the danger responses, we lose the nuanced regulation of the face and the voice and the heart, but we retain the ability to move and mobilize. Activated by danger, the chemistry of adrenaline and cortisol work with the vasopressin / oxytocin balance to transform these spinal movements into aggressive or escape physiology. If we move into an aggressive stance, our hands and arms are flooded with blood so that we can fight something off. Fear physiology, the continuum of flight, powers up the legs and feet so that we can run away. In both cases, we lose access to the highest range of our faculties, as adrenaline and cortisol shut down the medial pre-frontal cortex and our executive function.

When escape is inhibited, and our systems move fully into life threat, we also eventually lose access to the Movement system. Our bodies, in addition to adrenaline and cortisol are flooded with endogenous opioids that target interneurons in the sympathetic ganglia that immobilize us. And here in this realm, commonly experienced during terror, we enter the zone of tonic immobility (the *Freeze* state), and eventually deep collapse (*Shutdown*). We become like the deer in headlights, rigid and frozen, or we simply pass out, losing all muscle tone.

Safety provides the coordinating pulse that organizes and harmonizes these three crucial and distinct autonomic systems. In the absence of safety, we cannot cohere our autonomic systems.

NOTICE NOW, in the diagram on the facing page, the three triangles– blue, yellow, and red– nested within each circle. In pathogenic (defensive) states, notice that only two of the three systems are available, until we reach shutdown, where there is but one.

MEDITATIONS ON THE AUTONOMIC WHEEL

For consistency in helping you orient, each of these meditations begins at the same place on the Autonomics wheel, with the state called *Union*. If the wheel is a clock, this is 7 o'clock.

As you read these, please try to sense the energies of each state. Our naming of states is provisional, an attempt to basket some of the key attributes of the state into a few words. But word choices often carry associational baggage, and so the name we have chosen may not resonate with you. Pay more attention to the felt quality we are describing, because the language of the Autonomic Nervous System is sensation, emotion, and memory. We pin words to it *after the fact* to help us make sense of the energies. We can call them state names, or months of the year, or hours on the clock dial– really they are *templates for energies in motion*. States are *verbs*. Try to *feel* them.

MEDITATION 01

TWELVE PRIMAL FLAVORS OF SELF

Begin in the stillness of meditation. The body you are wearing is nearly immobile, but without fear. On the meditation cushion, attention inward, you ride your breath as if surfing a wave, attention riveted to it, feeling each contour. There is almost no desire to move. Rather, what you sense is open inwardness and expansion. In the stillness, in the dark, it is difficult to tell where the body ends– in fact, in pure awareness it would be difficult to say that the body is here at all, really. There is awareness, it can focus to a single point or expand into a field, there is this sense of merging with everything, a sense of opening such that we can touch and be touched by everything. We do nothing, everything is done through us. Mind is vast as sky when we reach up, solid as earth when we reach down. This is the state of *Union*. We are available to what the Universe requires of us, permeable to her instructions.

Delighted by our experience of contact with All That Is, a tickle of vibration (*is it joy?*) hums through this primordial stillness. Birthed from a sense of belonging, a sense of deep home, something inward releases, settles, and some deeper part of the body gives us permission to rest. Suddenly we are drawn inward upon the breath, drop deeper into the basement of the body, and find it congealing around us. Deep in the guts, in our bowels, a gurgling. Hand on the deep belly, the body appears around us, we are no longer formless, but the primary contours of this state are our merging with rest. We slip off the meditation cushion and lie down, allow ourselves to sink into the floor. Gravity is experienced, and although we are not asleep, we hear the body engaged in inward metabolic process: the swoosh of blood through the arteries and veins, the gurgle of the stomach and intestines processing not simply food but the depths of our experience: composting, rejuvenating. Temperature rises for a few moments, then falls. Our immune system just performed some miraculous and everyday housecleaning. At some point we drift into sleep. This is the state of *Restoration.*

Waking, we find ourselves refreshed. Who knows how long we have been asleep. Like the cat or dog on waking, the body moves of its own accord, into a slow stretch. We tense until we shiver, and then release. We arch forward, feeling the full extension of the spine, opening a tiny space at each vertebrae, feeling the expansion of inward bodily space. Our senses awaken like winter animals coming out of hiberation. We sniff at the breeze. *What is that smell?* Some flower is blooming.

Opening the chest, we allow our arms to float to the sides, and then raise them up. We roll the head on the neck, loosening it. None of this movement is planned. We follow no routine. Simply listening to the body it tells us how it wants to move. The primary emotional texture this gives rise to is a sense of ease and enjoyment. Although having a body is not always pleasurable, in this state of *Enjoyment*, the felt simplicity of embodiment is what stands out. We are slow, yet the embodied nature of this state is clear. Our form has congealed: we stretch and move to feel it around us, this body that we are wearing. There are hints of sociality. We are wanting to share this experience with others. The cat and dog seek us out unbidden. Cat is on us, too close, rubbing feline whiskers across our face, *yuck* and also how deeply sweet, this proximity of feline affection, he senses our availability, wants to be this close to the emanation of us known as our face.

Now we want to connect. Pets are wonderful, the garden beckons, but in this state of connection we are wanting the kind of relational contact for which an appropriately selected human is best. We want to lean into conversation with someone over coffee, to share story, to listen with our attention trained on their face, the sound of their voice, the micro-expressions that rise and fall across their face, the way they gesture with their hands, the way their posture shifts as they grow more animated in the telling. Who from amongst our friends? We yearn for community, libations, a dinner party, gathering around the hearth. We want to meet our friends at a restaurant at nine pm, hug them all close, share food and drink, walk through the warm spring night laughing, feeling part of a community. This is the state of *Connection*.

In truth, we want to dance. And perhaps there is a particular person with whom we wish to dance. Someone with whom we would like to play, in call-and-response, the art of movement, hold and be held by. Or maybe no, it's not necessary, maybe it is simply the freedom of movement of our own bodies, wanting to move now, needing to, wanting to exert a bit more, wanting to extend into their full range of motion, to jump, to twirl, to race, to accelerate. As we continue to warm, we find ourselves moving. Music playlists call to us. The dog, sensing our increasing vibrancy, begins to wag, begins to jump up and down. He is sure, somehow, that we are getting ready to take him for a run.

Ok, ok, you say, and head upstairs to get your running shoes.

Dog, downstairs, begins to bark. *Hurry up hurry up*, he says in dog. But you, looking for your running shoes, have passed the easel on which your most recent painting sits half-complete and in this state of *Play,* passing it you pause, the brushes beckoning you, the smock inviting you to put it on and fall into creative reverie, into flow, to dance on canvas instead of in the street.

Your running partner texts. Can you meet them in 30 minutes for a run? Something inside you swells. *Yes*, you text back. Out with the dog now, you are jogging with him, stretching, you are already running, but secretly you are preparing. Secretly, some part of you is expanding with the knowledge that when you meet your running partner in half an hour, the moment you get out on the trail, you want to already be stretched out, because you want to let loose, to really run. This is not even necessarily something that has to do with beating your running partner, with being faster than them. It has to do with you wanting to find the elite edge of your capabilities. Is is a state that you have learned to access, after thousands of hours of practice, when you perform on stage. It is also a state you can enter, a domain of flow, in collaborative projects with your team. You simply want to be able to let loose, to push yourself to the emergent edge where something else takes over. In this moment, the form it wants to take is to really sprint. You want to hit the trail and just take off. You can feel it in your heart, the anticipation. A trilling of velocity. You want to test yourself against something. You want to perform. You want people to watch you cross the finish line. When you think of finishing first in a race, some tiny muscle in your jaw flickers tight. You want to explore yourself against the state of *Competition.*

Out there, on the trail with your running partner, unbeckoned, as your heartrate crosses north of 120 beats-per-minute, you find yourself rehashing, mentally, your salary negotiation from the week before. You can see now, from this vantage on the trail, where your running partner is at least a hundred yards behind you– that you should have been more assertive in the meeting. Yet, you were tentative. With your boss you were *accommodating*. Smiling at things you found, in your deepest self, to be foolish. Presenting an agreeable face, a careful sociality. And did it work? Did you get what you wanted? You wish now, with the clarity of your heart striking cleanly at a hundred and twenty bpm, that you had shown on your face the frustration that you felt, that it hadn't simply built up inside you as a kind of heat, a sense that your heart might race off without you like a skittish horse. You regret, inwardly, the stoicism of your voice that did not waver, that sounded calm and measured even as you listened to that bullshit and got angry on the inside. Why, oh why, do you *appease* him? With a different director, a woman, you can see the meeting having gone differently, having moved into some effortless next-level collaboration, a sort of *tend-and-befriend*. Yet that's not what happened.

Fuck that, you think, taking an imaginary swipe, as you run, at the memory of your boss. Anyone who passed you now and looked at your face would see the lines of anger. You are running, pushing yourself, and close to red-lining, airboxing with someone who is not really there. Why, all of a sudden, do you want to *fight* with everything. A huge mountain biker, in full crash helmet like some kind of motorcycle daredevil, oblivious in the cocoon of his music, hurtles too close to the centerline of the trail, and without thinking you spin out of the way shouting– *Watch where you are going!* and then, again without thinking, as they recede, shout *Fucking Asshole!* It feels really good to scream this into the wind. Yelling, as we both know, to the biker and your boss who is not here.

You've stopped, hands are now on your knees, you stand by the side of the trail, waiting for your running partner. Did she stop? Turn around? In truth, that was a bit reckless. You are not in the habit of shouting *Asshole* out loud to strangers, but in the heat of the moment it is what came out of your mouth. There is the screech of bike tires, someone slamming on their brakes. Glancing down the trail you notice, almost absentmindedly at first, that the mountain biker has stopped. You cannot see your running partner. *I hope that guy didn't hear the asshole part*, you think. But you did scream it vigorously. As if summoned by your thought, discomfittingly, you notice that he is, laboriously, turning the bicyle around in the middle of the trail without dismounting. Swivelling the bike beneath him, turning it back toward you. You watch, with increasing discomfort, as he begins slowly to make his way back up the trail toward you. *Surely*, you think, *this has nothing to do with me*. Your body, however, is not so sure. Like a seesaw rebalancing, you note inwardly that the intensity is shifting, that anger is changing to alarm, confrontation is yielding to escape. The wave of energy within is changing direction. Your body now clearly wants to run. So, wisely, you allow it to do so. This is *flight*.

Away. Up the mountain. Since you are on a running trail, with your running shoes, running, running away is disguisable, and it feels good. It feels good to move away from the threat, to escape it, whether it is real or imagined, *probably it is just a coincidence that he turned around*, you think, *probably it has nothing to do with me*. Disguisable that is, until as the gap between you begins to close, he is mounted, and your body decides to turn off the trail and head into the forest. You are wearing trail running shoes, if someone asked you might you not say that you simply felt a yearning to go off-trail?

And so you cut right, or your body does, abruptly, and feel the temporary relief of effective evasive action. The movement plan of effectively executed escape. Until, that is, you hear the crunch of the bike tires leaving the asphalt, and taking up after you. Suddenly–in a strange moment, from the gathering panic of fleeing– your feet grow obscenely heavy. The desire to move leaves you, abruptly, in a wave of nausea. Your legs feel as though they are moving through knee-deep mud. *Hey*, you hear. *You!* The words arrest you. Your heart rate plummets. Bike tires crunching over sticks and leaf litter. You are clearly the target of this ejaculation. You are hot, and jammed. A deer in headlights. *Frozen* is the name of this state.

Before he reaches you, you have decided, or your body has, to placate him. You will apologize, agree with him, but in such a way that you direct motion back toward the trail, where people are. A sort of trance threatens to engulf you, you push it back, push it down. The helmet and the bicycle together give him the appearance somehow of not being human– of being a much larger mounted predator, a sci-fi monster. You are oddly calm, somehow deeply disinterested in what is happening. This is the state of *placate.*

By the time he reaches you, your body has already decided he is carrying a gun, that it is useless to flee. You are standing off-trail in the forest, no one knows where you are. In the face of impending death, something within you, not your ordinary sense of self, not your volition, has taken over. It is like you are high, the afternoon has suddenly accelerated away from the continuum of ordinary time and ordinary meaning, has become surreal, dream-like, inchoate. Noises are too loud and too quiet simultaneously. The contours of reality amplify and recede like a funhouse mirror. You are immobilized you realize, could not move even if you wanted to. You begin to leave yourself, to float up, out. It occurs to you– almost as though it is occurring to someone else– that you might pass out. That soon you might watch from above and the body crumpled on the forest floor would be yours. This is the state of *shutdown.*

You hear your name. It must have been said many times, because your neighbor is looking at you, and the expression on his face is inscrutable. Something doesn't make sense, and so you keep having to stitch it back together; to assemble the pieces like a puzzle. You realize that you are not breathing. You have no idea how much time has passed. You realize that you are not sure when you breathed last, and the realization of this, and the recognition of your name, in some part of yourself, allows the breath to begin to return, which it does haltingly at first. You swim back into yourself.

You have to build the scene with your eyes, explain to yourself the tableau before you, because you do not understand all the parts of it. There is the bike and the forest, and your neighbor– the helmet cradled in his right hand. There is no gun. There is no predator. From the shell of impending death emerged... Robert. It was Robert the whole time, but it could not have been. Your body has begun now, to shake, at first slightly, and then with increasing intensity. *Are you ok?* he is asking. *I'm sorry, I didn't recognize you until I was about a hundred feet past.*

You swallow, once, then again, and then again. The inward surface of your animal body revives, you feel yourself coming back, in the throat, in the lungs as you cough, and then your belly, which flips under you. You might vomit, you lean forward and rest your hands on your knees. You are still shaking. As you realize that you are safe– that your interpretation of the event was not accurate, that he was not turning around to hunt you, but to greet you, tears well up from your eyes, you are not doing that on purpose, and you are glad you are looking down and he cannot see them. *I'm fine*, you say. *I'll see you back in the city. Sometime I'll explain to you what happened.*

But will you?

Robert leaves. There is still so much energy moving through you. Attending inwardly, you feel the pulsing of waves moving through you, and decide to sit. You lower yourself to the ground, let the earth hold you, and begin to meditate. You don't do anything, simply allow. Now that you are physically safe, you begin to seek safety inwardly. Can you find your center here, in this moment. Can you feel your way home?

Across the inward screen behind the eyes, unbidden, other moments when you rediscovered, in the wake of an intensity, a catastrophe, that you were still alive. That the animal body comes back to you. In the body this sudden texture of coming back to yourself after an absence. Like coming up from under-water.

A memory of recoving from depression, the first rekindling of joy like a tiny spark. The death of your favorite grandmother, and the way the tears broke through several months later, on a roadtrip with friends. Other drownings and rescusitations. But these are not merely images, on the inward screen. You can feel them in your full body, the entire inward texture of your feeling is cinematic. You've passed through this feeling place before. You've visited this place on the map of what it means to be human.

Circle back to the beginning of this story. We have walked a full cycle on the autonomic wheel. It ends where it began, because it is a circle, or a spiral.

In truth, it is infinite, without end.

It is you, refracted through twelve of the most common autonomic states that humans move through. All of them are necessary, all of them are adaptive. None of them are *bad*. Some of them are unpleasant, even deeply so. Some of them may be more or less familiar to you. You are likely to identify more with some than others.

Take a moment to sit with this story. What are the states with which you are most familiar?

Which of the health-creating states: *Union, Restoration, Enjoyment, Connection, Play,* and *Competition* are you most comfortable in? Do you have other names for those energies? Which of them accord with your deepest sense of self? Most of the people that we work with have some depth of experience in at least one of these states. You might think of it as your *Autonomic home.*

E
SE
S
SW
W
NW
N
NE

THREE BY THREE BY THREE

In both of these meditations, which I hope you find useful for deepening an experience of the feel of the states, I have taken you on a spiraling symbolic journey around an archetypal wheel. We are in the domain of metaphor here, organized in the archetypal sacred geometry of twelve-around-one.

The Mandala on the cover of this book is an archetypal cartography: a map. And we can talk about the locations encircling it in a variety of ways. Throughout most of the book, I will talk about them as *autonomic states*, but you could also think about them as weather, or as water, or as color, or as season, or as hours. You can choose the metaphorical frame that suits your needs. What we are really talking about are *energies*. We are honing this vocabulary through the lens of autonomic physiology, sharpening the precision of its depiction so that we can communicate more precisely about something that has been understood intuitively for millenia in some cultures, yet remains nebulous.

To talk about the entire map, the cartography, in its entirety, I will often use the language of color. This is because the three-dimensional color space is comprised of three variables. Hue, value, and chroma. If we know these three levels with precision, we can create an infinite number of colors.

How astonishing! Three primary axes, and we get an infinite amount of gradation.

When you are really little, and you start drawing, perhaps you get a box of 12 crayons.At some point, you graduate to the box of 64, where colors have greater nuance. Instead of a two blue crayons, there are now twelve. I remember this pretty well, and I remember thinking, *Whoa*. I remember, as a child, being amazed by color. The blue I fell in love with, which was the color of the first bicycle I chose, I could gaze at for hours. It gave my entire body a feeling that was sublime.

Suddenly, my drawings got more interesting. The twelve blues included blue-greens, blue-purples, blue-greys. Suddenly I could think more sophisticatedly in the language of color.

If you start painting, you begin to realize that if you have a selection of basic colors, you can begin to combine them to create an infinite number of distinct tonal variations. It is lovely to have Pyrrole Orange, but if you have Pyrrole Red, and Benzimidazolone Yellow, you can probably make it yourself, along with fifty other shades of orange.

Autonomic state is like this.

Instead of hue, value, and chroma, what we were calling

THE MUNSELL COLOR SPACE

Courtesy of AH Munsell

Three variables:
Hue, Chroma, Value

autonomic state is comprised of neuroception, neurology, and chemistry. These are the three variables that will combine in the general patterns that form the mandala. But it is important for you to understand that there are actually an infinite number of variations here.

If we were really going to differentiate the entire color theoretical range of autonomic physiology, we would have to include as variables the differing hormones and hormonal cycles between men and women across the lifespan, the impact of individual versus collectivist cultures, all of the facets of our social locations, aspects of our embeddedness in the living world, and the influence of the domination paradigm expressed by modernity, globalization, and capitalism. Then perhaps we would we have a map sufficiently nuanced to describe more of the full range of human experience.

Because this is not possible, what I'm doing in this book is defining a color space in which this infinite variation can unfold. I won't try to name all the colors; I'll teach you how to blend your own. And the primary variables that create our autonomic spectrum are both somewhat esoteric, and also fairly straightforward.

You know the word chemistry, and you know the word neurology. Unless you have specialist training, you do not know the word *neuroception*. I would use the words *threat detection*, which conveys two-thirds of what neuroception is, except that threat detection by definition does not include detecting *safety*.

Neuroception is the term coined by Stephen Porges, PhD, the developer of the Polyvagal theory, to denote our embodied neural perception, in the present moment, of safety, danger, or life threat.Each moment, with each inhalation and exhalation, our Autonomic Nervous System is assessing and then updating our neuroception.

This is our first variable, because it is directly connected to survival. It has three primary colors: SAFETY, DANGER, and LIFETHREAT.

The second variable in our model is neurology. Neurology refers to the specific autonomic systems, of which there are three (CONNECTION, MOVEMENT, and GROUNDING), engaged in a given present moment. In all of the health-creating states all three of them are available in various combinations. When we shift into defensive states, we lose access to the Connection System. When we shift from danger to life threat, we lose access to the Movement system. Only in Shutdown, our deepest threat response, is there only one autonomic system engaged (the Grounding System in its lifethreat aspect).

The third variable is chemistry. And by chemistry we are referring specifically to neurochemistry. In order to not get overburdened by complexity, we've focused on three primary chemistries. There is the chemistry of connection, which is primarily comprised of oxytocin/vasopressin. There is the chemistry of activation, which is primarily comprised of adrenaline and cortisol. And there is the chemistry of shutdown, which is comprised of endogenous opioids. So, three chemistries: CONNECTION, ACTIVATION, and SHUTDOWN.

An autonomic state is the Venn diagram that comes together from these three overlapping variables. Functionally, it operates as an energy processing template that shapes the way experience flows through you.

At a physiological level, it is assembled from these three variables uniting. Whether your body interprets the situation as safe enough, dangerous, or life-threatening. Which autonomic systems are therefore engaged. And the chemistry that governs them. The identification of these three variables, and their intersectionality, is the heart of the foundation model of Autonomics that I have spent the last 30 years developing.

Three by three by three. And from them an infinite array of

nuanced autonomic states whose primary constellations are the twelve that we have named, which does not necessarily mean they will exactly align with your experience.

If they do not, once you understand how autonomic colors are made, trust your own blending, as you are the artist of your own life. Or at least you could be.

AUTONOMIC STATE

Throughout this book I am going to be referring to something called an *Autonomic State* as though it were a place you can visit. We are going to talk about fight states, flight states, and connection states. I'm going to suggest that some states are health-creating (salugenic), and that some states, if we get stuck in them, are disease-creating (pathogenic). I am not speaking about something abstract. When I talk about an autonomic state, and its signature energies, I want you to be able to feel what I'm talking about.

But what exactly is an autonomic state?

Throughout the book I'm going to refer to autonomic states as energy-processing templates. Your nervous system translates between interiors and exteriors, and the specific role of your autonomic nervous system is to continuously take the temperature of your experience and infer from this how you should interface with life. This inward listening is refined, and also organized in a structured manner.

The primary listening that your Autonomic Nervous System is doing relates to survival. This is where the rubber meets the road, we might say, because from the point of view of our biological organism, survival is always the first priority. Stay alive, we win. Die, we lose.[1] So biologically, in any situation we face, staying alive is the most crucial feature. As far as our biology is concerned, if we have survived we are victorious.

Because a principal brilliance of biology is utilizing the available metabolic resources most effectively, the body is very severe in re-distributing resources if it believes our survival is at

1 There is some complicating of this in the autonomic responses of prey animals under lifethreat, but we can safely make this assertion broadly.

stake. This means, for example, that if the body believes it is in danger, it will unhesitatingly shut down physiological systems that it deems non-essential.

If you are running away from a bear, the body directs all available resources to this effort. While this is happening, it turns off non-essential systems like digestion, immune function, and aspects of memory. Have you ever been working on your computer, and gotten that little spinning wheel of death, where the program that you are using just locks up? This can happen when you are running too many programs and the central processing unit of your computer gets overwhelmed. Your body, in prioritizing survival, avoids this lock up by just shutting down systems that are non-essential. It changes the way energy flows through you, reshaping the banks of this river of energy to divert it to the most important abilities for survival. The body flows energy only to the systems that are most needed. In this case, it super-powers the legs so that you can get away and it changes your hearing to make it specifically sensitive to the sound of the predator.

Under the massive neurochemical surges of adrenaline and cortisol that power a fight-or-flight response, our bodies can perform incredible feats. You hear a story about a mother who lifts a car off her toddler, for example. Or a soldier who picks up and runs with a wounded comrade thrown over his shoulder– a guy he can't ordinarily lift. I once talked to a middle-aged man who jumped fifteen feet out of a boat to get away from crocodiles. He explained that in ordinary life he could maybe jump four feet. Biology super-sizes the ability we need to survive, but the cost of doing this is that it may shut a bunch of other stuff down and it burns an enormous amount of energy.

The energy processing template we call the flight response basically turns the power of the legs up to eleven, tunes the predator acoustic listening channel up to eleven, and turns digestion, immune function, and short-term memory down to zero. This energy processing template is extremely useful

Autonomic state
shapes the river
of energy flowing
through you.

Fleeing
from a bear?

Class 5
rapids.

for getting away from the bear. But as you can imagine, it is extremely maladaptive for general purpose living.

If we get stuck in that template, and we cannot shift back to baseline, the next day we are deeply depleted, have no appetite (digestion is off), get sick (immune system is off), and keep forgetting things (short-term memory is shut down). Furthermore, every time there is a deep bass sound, a rumble from the highway, the grinding of garbage truck gears, or the furnace turning on, we nearly jump out of our skins and break into a cold sweat, having to restrain our body from running.

Post-traumatic stress injuries, broadly speaking, are autonomic survival energy processing templates that we cannot turn off.

Because autonomic state is like a pair of glasses that we cannot remove, once we are experiencing a state, it completely changes the contours of our experience, and we are stuck looking at the world through it.

Autonomic state shapes our visceral state, which is the felt moment-to-moment contours of how it feels to be in the body we are wearing. Do we feel tense or relaxed? Is our posture alert and responsive or is it collapsed? Is our heart-rate calm or elevated? Is our breathing deep or fast? Visceral state is a function of the physiological control of these parameters in the body.

This foundation of visceral state creates the context in which our emotions and thoughts move. If I feel jumpy and on-edge, my emotional range emerges from this physiological foundation, and in describing it I say that I feel *anxious*. If this feeling is very mild, I might find myself subtly feeling worried about everything. If it is quite overt, I feel constantly fearful. The energy moves through my body as a sort of current, an underlayer of high-voltage flickering readiness. As an emotional texture that colors my experience, I will find myself thinking about everything that could go wrong. My eyes wander the grocery store looking for exits.

This visceral state, coupled with emotions and thoughts determines my perception of the world around me. (In this case, *dangerous.*) And all of these together determine my behavior.

Have you ever played pinball? When you are releasing a ball into the game, it runs through a set of chutes before it is released into freeplay. Autonomic state is the chutes that control the pathway the ball (experience) moves in. It is the shape of the banks on the river of experience, governing the channel through which living flows. Autonomic state has the power to transform these banks from vertical, so that your experience races through like class 5 rapids shooting down a canyon, to horizontal, so that your experience flows so placidly there is no current detectable whatsoever.

Autonomic state as pinball chute.

WE'VE LOST THE HOUSEKEYS

Humans are architected with stress responses that can come online extremely rapidly at very high intensity, as the situation demands, and in our long evolutionary history there were a set of contexts (what Darcia Narvaez calls *the Evolved Nest*[1]) of culture and community that helped us shift out of the stress response after danger had passed.

Having the strength to lift a car off a child, or fight off a marauding apex predator is extremely useful, but if the cost of it is that you are always exhausted, can no longer eat because your digestion has shut down, that you catch every cold, that you can't remember where you put your wallet, and that you are on-edge all the time, the cost is too high. So biology has granted us a set of keys that can be used to unlock that response after the stressor has passed. You could think of these as *housekeys*, because if the Autonomic Nervous System goes into a survival response, you are going to need them in order to unlock the door to get back home to yourself.

Unfortunately, modernity has lost all of the housekeys. Oops.

The house keys, intriguingly, are autonomic. They facilitate state shifts out of survival responses.The first thing that is important to understand, is that while you cannot unlock these doors when you are still in danger, getting out of danger does not unlock them. Dwell on this for a moment, because it is deeply counter-intuitive. It stands to reason that returning to a situation of relative safety after one of danger would cause this functional shift, but the crucial missing ingredient is the context of community. We moderns have deviated from our ancestral baseline, and without this baseline it is much harder for our biology to return to safety and connection.

---

1 Narvaez D. (2014). Neurobiology and the development of human morality: Evolution, culture and wisdom. New York: W.W. Norton.

Humans are social mammals, part of a small band hunter gathering lineage baseline. In the context of this more ancestral, communitarian context, our identity is not located uniquely inside of our skins. We have a social identity that is part of a sort of collective organism. A collective body, or *collective soma*[2], if you will.

In our evolved nest, our appropriate ancestral context, if someone is deviated out of the baseline in safety and connection, the small band notices. Your community recognizes something is wrong with you.

If you're living in a communitarian context, this is actually quite obvious. In addition to the fact that you seem irritable all the time, people are going to notice that you can't maintain attention. When we are in defensive autonomic states, we do not engage socially, relate, or play well with others. In a small band where our survival depends on collaboration, anti-social behavior stands out starkly.

We might note that the ancestral function of healers, shamans, and ritualists, is in part this retrieval of people who have been deviated from the baseline of safety and connection. The concept of *soul retrieval* is analogous to what we are talking about with the housekeys. Yet a healer or ritualist often accomplishes this through orchestrating a community context that supports this person returning to a baseline of safety and connection.

The archetypal component of this probably involves them telling the community the story of what happened. If the warrior was fighting off an apex predator around the village, everybody probably knows. But there are plenty of things that can deviate us autonomically from a sense of safety that might not be witnessed by the community.

And so part of the archetypal recipe for healing is that the per-

2 This term for collective or cultural body was brought to my attention by Tada Hozumi, Developer of Cultural Somatics.

son will tell the story of what happened. This story is witnessed by the community.

If we tell this story in an embodied fashion, not merely as a cold recitation of facts, we might act it out for our community.

All of us are gathered around the fireplace, with a shaman who I will call *Moon*. Moon has gathered the community for a healing ritual to support my return to the baseline of safety and connection. That's probably not what she told everyone when she requested that we gather, but this is the ritual transformation we are seeking. With the community gathered around the fireplace, Moon asks me to tell the story of what happened.

I explain to them that I was on the ridgeline, scouting. *I was up there*, I explain, *looking for a place to locate our hunting camp for the fall season. The sun was high in the sky. It was the moon when the elderberries are ripe. Suddenly, it grew very quiet in the forest. Not even the Steller's Jays were calling. I heard a stick snap behind me* I say, *and I turned around.*

Moon, who is listening carefully, stops me.

Show us how it was with you, she says.

And so I pantomime the motion.

I turn to look over my right shoulder.

There was[3] *a bear*, I say. *It was right there*, and I point. *No further than the fire. I don't understand how it got so close to me without making any noise.*

3 For grammatical clarity I am using the past tense here, as in 'there *was* a bear,' so as not to confuse readers. Yet if the sensory evocation of this embodied memory is strong enough, it may come out in the present tense- 'There *is* a bear.'

The first paragraph of my story is *descriptive*. I am setting the scene for the community so that they can accompany me. That is why there are details. Everybody knows the season when the elderberries are ripe. It implies something about the late summer weather, the color of the sky, the hue of autumn leaves, the lack of rain, the smell of the forest floor.

Everybody knows that Steller's Jays are the most talkative of the birds. If the Jays are quiet, there must be a reason.

The first part of my story is setting a scene so that people can be there with me and accompany.

When the shaman asks me to show them how I moved, it is because she understands, perhaps intuitively, that one of the primary keys to unlocking autonomic state is the completion of self-protective motor gestures.

What this very technical sentence means in real life, is that because your autonomic nervous system is *automatic* (involuntary), and because the vertebrate body that you're wearing is a 200 million year-old body plan, it is hardwired with a bunch of movement reflexes that are complex patterned movements that allow the body to fight things off and escape.

Let's take a moment to marvel together at the incredible brilliance of this biology. Neurons covered in myelin, the fatty conductive sheath that accelerates signal transmission, can move an electrical impulse from the limbs up to the brain and back pretty darn fast, but this velocity is constrained by the laws of physics. In a full grown male human, that distance is up to 6 feet or more from the tip of the toes to the brainstem. If I accidentally step on a hot coal with bare feet, to complete the sensory motor relay pathway, which is a fancy way of saying that the sensation in your toe makes its way up to the brainstem, where it is sent to the midbrain dopamine system for processing, which drives a descending motor movement signal back down your spine to your foot to move it out of the way, there are ten different neurons that have to pass the signal.

Between each neuron in this chain is a microscopic gap, the synaptic cleft, where the signal is transmitted chemically, not electrically. Passing that signal up your leg, up your spinal column, to the brain stem, the midbrain, back to the brainstem, then back down the spinal column, and finally to your toe? Extraordinarily complicated. Ten neurons. Even in a neural superhighway, an A fiber, it takes too much time.

In the case of survival responses, where life and death is a matter of milliseconds, this takes far too long. And so the genius of the vertebrate body plan is that this is not what happens. Instead, for these kind of autonomic, or automatic responses, the relay happens in a much tighter localized reflex arc. Instead of having to travel all the way up to the cranial brain, the body has distributed brains along the spinal cord in what are called central pattern generators (CPGs). You can think of these as tiny motor units for the limbs. Hardwired into the post synaptic ganglia off of the spine are all of the motor movement programs required by the associated limb. When a pediatrician bangs on your infant's knee with that little triangular rubber mallet to test for reflexes, they are tapping reflex arcs.

If I step on a coal, the signal doesn't have to travel all the way up to my brain for processing. It hits the central pattern generator at the base of the spine associated with the leg, and then goes back to my foot. This cuts the distance the signal has to travel in half. Four neurons instead of ten. My response happens much faster.

When we shift into the energy processing templates of defense– when we lift a car off a toddler, or jump fifteen feet out of a boat to avoid crocodiles, the Autonomic Nervous System will throw up a defensive motor movement pattern very fast.

COORDINATE EVERYTHING AND LIFT!

COORDINATE EVERYTHING AND JUMP!

The part of us that surfaces these movements is not our ordinary sense of self. It is some deeper, more primal, more archetypal, deep neurobiological sense of self. If you've ever watched two cats fighting, the sudden complex evasive action one of them might take to get away – spinning at high velocity, pivoting to maintain balance, springing over the other in a vaulting leap, etc., you have a sense of the amount of energy contained in these responses, how fast they happen, and how intricate they can be.

Yet there are times when the autonomic nervous system throws up a movement pattern that is not able to complete. Like the spinning disc on your computer, there are times when such a program jams the central processing, and just gets stuck. In these cases, it is basically archived in the body, primed with extremely potent survival energies. A stored movement program, with survival imperative, jammed in the body's neurology like a bit of essential computer code that hasn't been executed. The key to returning the body to a physiological sense of safety is to create a context where this self-protective motor program can complete.

We have to unlock *the movement*.

This is why talk therapy doesn't touch trauma.

What happens next in my telling of the story in the presence of my community surprises me. As I turn and look over my right shoulder, it's not like I am just telling the story. It is *as if my body is back in the story*. I see the bear somehow, in all its horrifying bulk. Some part of me knows that I am actually around the campfire telling the story, but some other deeper part of me experiences what I experienced in the presence of the Grizzly.

For some reason, what is standing out to me is the incredible mass of his skull, on which my eyes feel fixated, and the slabs of rippling muscle as he hurtles towards me. It is hypnotic.

What happened in real life, the thing that turned on the energy processing template of flight that I have not been able to exit, is that one minute the animal, all 600 pounds of his flesh, was motionless, silent, and the next moment he lurched and then erupted toward me so fast my brain shut down, as though I had been punched in the face.

For a moment, I don't know where I am. My neck has gone totally taut, my head feels like it is held in a vice, him hurtling at me.

Everything is everywhere all at once.

At some point, I discover that Moon is standing right next to me, her hand on the center of my chest. My heart is pounding and I am frozen in place.

You are with the bear, she says, *but you're also with us. We stand outside of ordinary time, in the everywhen. You are safe. You are going to live.*

Something in my body believes her, but something does not.

Don't speak, she says, when I open my mouth. *Don't think.*

See the bear.

The thing that has stunned me is how fast it is moving. There seemed to be no time from my noticing him until he is five feet away. My senses have somehow now narrowed down to a fine point. There is a crackling in my ears, my head swims. There is no time. My heartbeat, which I could not feel before, is exploding out of my chest. The sound of him crashing through the forest is all I can hear.

Wheeling my body, a blur of chaotic motion, I rip sideways, grab the trunk of a small Douglas fir tree that is somehow there, and hoist myself into the branches like an orangutan. I vault and roll up the tree, clammoring, grabbing. Upward I scramble. Up up up.

I am in two places, telling the story, but mostly I'm with the bear. I swim back to myself, wrapped around Moon, as though she is the tree, the heel of my right palm pressing down on her face.

There is nervous laughter all around.

It seems that my body has climbed her as though she was the Douglas fir.

But the Elders know that this is serious business, and they are *shushing* the little ones.

Take your time, says Moon. *What is happening now?*

Bears can climb trees. And this one is trying. Lucky for me, and I cannot properly take credit for this, because the part of my body that chose this particular tree–knew it was right there when my conscious mind hadn't even noticed it–chose a tree that was too small to support the weight of my pursuer. If it had not I would be dead.

The tree is little more than a foot wide at its base, and when bears climb trees, they shimmy up them. This Grizzly, a male, is so large that he cannot get purchase on the trunk. He is too wide. And so for the moment, I find myself twenty feet in the air, only a few feet out of his fully extended reach.

Tell us, says Moon, and I speak.

He is underneath me, I say. *When he bellows, I can see all of the teeth. The big ones are the size of our spear points.*

At what point, Moon asks me, *did you realize that the bear could not reach you?*

It is a question that I do not know how to answer, because I never realized the bear couldn't reach me. He didn't, but even two hours later when I came down I did not fully know this. But as she asks it, again, I see the bear. For a few moments, at the beginning, until the lowest branches have snapped off under his weight, he is attempting to scrabble up the tree in pursuit of me.

But once he has broken these low branches, even though he is reaching, there are several feet between us. I can feel now that I am beyond his reach, and this feels safer even as I scramble to get further away. Were the tree smaller, he likely would have snapped its trunk, broken it in two, pushed it over. Yet the answer my body has given definitively, which everybody in the camp has seen, is to continue to scrabble up Moon's body.

I'm not really going anywhere, because she's still standing on the ground, but in the inner vision I have of the event, I am now several feet higher in the tree. The bear has stopped bellowing. Now he snorts. Then he sharpens his claws, raking them through the bark of the tree like flocks of knives, flaying its skin. I shudder.

Hugging the tree, I pulled my legs up under me. Again, there is nervous laughing, and the Elders shushing, and part of me notices that I am sitting in Moon's lap, legs tucked up and in.

Time is, somehow, rearranging itself. What was a jumble is now organizing itself into a sequence. In the tree, I can hear the Steller's Jays again, peppering each other with insult. There is static in my ears as they re-tune. The bear turns away and lumbers off. My gaze is fastened on what I can see of him through the foliage until the contour of the ridge swallows him and he disappears from view.

Just after the moment I lose sight of him, my body awareness returns to me, and I realize I am holding onto the tree much more tightly than is strictly necessary. The vice holding my neck has loosened, I roll my head around on my neck, there is suddenly some liberty there. I breathe more deeply, the thumping of my heart begins to recede like a tide going out.

I also notice, for the first time, the burning all across my arms. My ascent of the tree has scraped my entire body badly, but especially my forearms. As I ratcheted up, I did not notice this at all.

Tend the wounds, says Moon. I have no idea why, but I want to lick my arms.

Go ahead, she says, as though she can see this wish, though I have said nothing.

And in this time-between-times, my animal tongue passes over the skin. These cuts have already healed, all of this happened nearly a month ago, but somehow my tongue tastes blood. I lick out the wounds on both arms, cleaning them, embarrassed yet precise. I am a little repulsed by this, but it soothes a pain I am not aware that I have felt since the incident, until the pain goes away.

And then my entire body begins to shake. It is not voluntary. It rattles and hums and vibrates. My hips quake in their sockets. My eyes, closed now, zoom around the inside of my skull.

Let it happen, says Moon. *Let it take you; do not resist. You are safe, we are holding you.*

emotion, and memory. Opens the doorway to intuition when entered through Connection.

INTUITION - Knowing something without knowing how we know it. Functionally imprecise, intuition as commonly used indicates in fact an enormous variety of non-cognitive ways of knowing that are better experienced and refined when differentiated.

KAS LIMAAL - Tzutuji'l Mayan - mutual endebtedness. Literally, mutual ensparkedness.

KISHKES - Yiddish - the knowing that comes from the guts.

KOKORO - Japanese- Often variously translated as heart, mind, or spirit. Kokoro implies the physical heart, as well as the emotion of love, as well as the distillation of spirit. I will refer often to heart in the book, using the English word, and in some places I am speaking about the physical organ, but in most I am speaking to something closer to this Japanese word, or the Aramaic word *leba*, defined below.

KOMOREBI - Japanese- the color of sunlight filtered through the leaves of the forest.

KOTODAMA - Japanese- the spiritual interior of the word.

LEBA - Aramaic- Heart. The center, pith, or best part of anything.

N KASEA - Tzutuji'l Mayan - to live. Literally, *the ember yet burns.*

MONOZUKURI- Japanese - The spirit and skill of craftsmanship. From 'mono' (thing) and 'zukuri' (making). Classically a form of intelligence of the hands.

NOSTOS - Greek - the yearning for home.

Pretty soon, the entire village is holding and hugging Moon and myself. Still I quake.

The waterworks have loosed, and I am weeping with relief. In other moments, other states of mind, other situations I might find it undignified to weep with such abandon, and yet I cannot stop. And then the whole village is crying with me as One. The children, the Elders, everyone in between.

At a certain point, the meaning of the tears changes. When I began, they were tinged with grief. I felt alone, I was wailing in impotence and sorrow. At a certain point, they grow colored with belonging. A 'We' emerges; it is not something I think, but that I feel. I melt back into myself, and back into the band. All of our breathing coordinates. The deepest part of myself becomes available, my spirit settles back into my body.

I am home.

I am so profoundly tired, I cannot keep my eyes open. At some point, I don't rightly know when, I drift into sleep right there. There is a fire, I think. When I roll over during the night a hand comes to rest gently upon my arm, a reassurance, and I fall back into sleep. I do not dream. When I awaken in the morning, I am healed.

This is how our ancestral bodies are designed to circle the wheel.

GLOSSARY

Specialist Terms and Indigenous and Non-English words w
need

A brief note with regard to the personification of nouns in
book: in English (unlike the romance languages) nouns are
gendered. The possibly inadvertent result of this is that beir
a tree, a cat, a river, deprived of both animacy by our blindn
and gender by our language end up being referred to as 'it'.
When I talk about a given tree, I am forced, by the semant
structure of the English language, should I conform to con
mon usage, to refer to this being as an 'it'.

I refuse to do this. There are no 'its'. 'It' pushes us away: ou
intimacy, out of kinship. For this reason, I am going to ger
nouns that ordinarily in English are not gendered. This is
because I think this tree is of the female sex necessarily, or
hummingbird male. In some cases it may be so, but in oth
cases it is simply because I want to bring these beings tow
us, not push them away with blunt pronouns.

ALL THAT IS - The Creation.

ANAMAGOPTANYE - Lakhota- the earth is constantly uploadin
information to you. Non-cognitive ways of knowing that
through the soles of the feet.

ANIMACY - a cosmological view that suggests that everythi
alive, imbued with sentience, part of an intelligent fabric

EVERYWHEN - Not everywhere, but…

GOM - Tibetan - Familiarization. This Tibetan word for r
tion means familiarization (with reality).

INTEROCEPTION - the sixth sense. Moment-to-moment con
with the interior landscape of the body through sensati

NEUROCEPTION - A term coined by Stephen Porges, PhD, Developer of the Polyvagal Theory, to denote our embodied neural detection of safety, danger, or lifethreat in the present moment.

NUNCHI- Korean - the ability to read the emotions and internal states of others. A non-cognitive way of knowing.

POLYVAGAL THEORY- the groundbreaking work of psychophysiologist Stephen Porges, PhD, who is likely the world's foremost expert on the relationship between the Autonomic Nervous System and behavior, which identified not one, but two vagal systems operative in the human physiology, and reconceptualized our understanding of the basic functioning of the Autonomic Nervous System.

SAUDADE - Portuguese - Aching yearning/longing for someone or something.

SOVEREIGNTY (verb) - I sometimes say that in our work we teach three things: healing, sovereignty, and how to inherit our possible beauty. What is sovereignty? I have made, and make the contention that modern civilization is derived from empire. Empire is the material and psychic consolidation of the thoughtforms of domination. (In its current incarnation in the West the genesis of these thoughtforms is European, largely by way of the Roman Empire.) Empire functions by extracting a common resource (land, labor, ideas, attention), and conferring its benefits on an elite. So for example, common land is enclosed by the crown, and the material and resource (mineral, biotic, agricultural, etc.) productivity thereof benefits the elite (as wealth and resources, etc.). Capitalism is the same sort of self-terminating algorithm that extracts or encloses a common resource for the benefit of the few. In order to extract from people, the programs of empire acculturate them to become compliant.

With respect to Autonomic Physiology, this is a form of accommodation/ appeasement. Compliance is effectuated by disconnecting people from the Source. This happens both spiritually, and by destroying chthonic or earth-based traditions and Ways of Knowing, which is why Rome was so zealous about destroying earth-based peoples and their Gods. And the Church of Rome was later so zealous about killing 'witches'. If people believe they cannot commune directly with the Source, but rather require some officialized/officializing intermediary (such as a religious intermediary: a priestly class), they become more compliant. Religion is political. Spirituality is sovereign. Sovereignty is the reclaiming of our (ab)original connection with the Source. The disintermediation of middlemen in all forms.

When people become sovereign, rooted, and directly connected to All That Is, they reclaim their indigeneity as children of the Universe. Sovereignty is the practice (this is why it is a verb) of embodying the living law. In this form, it is also a continuous prayer in solidarity with all of Life.

TRACKING - a non-cognitive way of knowing characterized by the ability to follow trails. Ancestrally, hard-ground tracking was the way hunters found animals to feed the village. Tracking, as a modality, can be applied to energies, fields, group fields, etc. Meditation is a form of interior tracking. Healing (as we conceptualize it) and likewise facilitation are, applied skillfully, forms of tracking.

VILLAGE (verb)- to move through the world seeking out timespans (micro-moments, moments, flows) of relatedness with All That Is.

WAYFINDING - In the Hawaiian lineage, tracking on (in) water.

WELLBEING - The self-transforming flow of vitality that emerges spontaneously when your connection system is online more than your defensive systems.

CHAPTER 01

HOW WE GOT HERE

In May 2023, Vivek Murthy, the acting US Surgeon General, issued a Surgeon General's Advisory about the loneliness epidemic in the United States of America. Before the pandemic, the Cigna Loneliness Index, which is the largest recurring study of loneliness in the United States, suggested that 49% of American adults felt lonely some or most of the time.

During the pandemic in 2020, this number had risen to 61%. That is to say that three out of five people experience enduring states of loneliness as a part of their everyday lives. The report went on to say that loneliness, isolation, and disconnection was one of the underlying root drivers of mental distress, contributory to mental illness, and had an effect on people's longevity equivalent to smoking up to 15 cigarettes per day.

15 cigarettes per day. That's a lotta cigarettes. There are twenty of those little death sticks in a pack. That means disconnection in your life is like torching a pack and a half of smokes every two days. It incinerates your wellbeing at such a pace.

The Surgeon General's report starkly points out a feature of contemporary American life that has been noted by sociologists for decades. In 2000, Harvard Professor of Public Policy Robert Putnam, in *Bowling Alone: The Collapse and Revival of American Communities*, published about the steady decline of bowling leagues in the United States as a key indicator of a broad drop in social capital in modern American adults. The book was based on interviews with half a million people: an extremely large data set. It noted that participation in these kinds of collective social activities was dropping precipitously with disastrous effects for people's wellbeing. This phenomenon is not, in fact, all that new.

This is a problem for humans because we are social animals, and our wellbeing is social in nature. Lonely, isolated, and disconnected humans are sick humans.

While it is fairly easy to identify a number of proximate causes for this epidemic of loneliness, including the obvious recent fact that social distancing during the pandemic made it very difficult for people to gather together, as well as the rise of social media over the past decade, with more and more people spending time online in connection *simulations*, there is something much deeper at play here.

Let us combine the awareness of this loneliness epidemic with the parallel observation that the data suggests that prior to the pandemic, 60% to 80% of visits to primary care had a stress-related component.[1] This is to say that 3 to 4 out of 5 visits to your doctor (your general practitioner) are either directly because of a stress-induced condition (anxiety, gastro-intestinal distress, sleep disorders, etc.), or made worse by stress.

How are these two broad phenomena related? Loneliness, isolation, and disconnection are a cause and a result of the Autonomic Nervous System responding to threat. Stress-related disorders are a result of the Autonomic Nervous System responding to threat. The common cause underneath both of these epidemics (one of loneliness, one of stress) is that modern people do not feel safe in our bodies.

Modern people do not feel safe in our bodies.

If we do not feel safe, we do not activate the biology of connection. If we do not activate the biology of connection, the biology of defense does not turn off. When the biology of defense becomes our normal state, when it becomes chronic and non-resolving, when it endures, we become ill. Sometimes it happens quickly, and sometimes it happens slowly. But you know what? It happens every single damn time.

When defense becomes our normal state we become ill.

My name is Gabriel, and I am a connection phenomenologist.

---

1 JAMA, 2013. https://jamanetwork.com/journals/jamainternalmedicine/fullarticle/1392494

For the past 30 years, my research and clinical work has been focused at the intersection of autonomic physiology and awareness practice. One of the questions that I have been studying deeply and consistently now for three decades is the deviation in modern culture from an ancestral baseline in safety and connection and what we can do to change that.

The question I am most keenly focused on in our work is how to help people turn on, and keep on, the neurological drivers of connection. This is of great importance, because the drivers of connection are in fact the root drivers of wellbeing.

I started out interested in the healing of stress-related disorders initially because I needed to heal them in myself. I continued because what we discovered was extraordinarily effective with everyone. The enterprise that I direct has made profound breakthroughs in the diagnosis and treatment of a number of them. But what is slightly ironic, or slightly revelatory, depending on how you look at it, is the key to many of our breakthroughs in treatment came not through learning to heal stress, but rather learning to create the conditions in which wellness could fully arise.

As our trans-disciplinary study of the nervous system progressed, and I received mentorship in fields ranging from autonomic physiology, to neurocardiology, to psychoneuroimmunology, to cellular metabolic healing, what became increasingly apparent was that our modern healing modalities are focused on *pathogenesis*. They are focused on the remediation of disease states. This means that their entire framework is conceptualized in terms of returning to a pre-illness baseline. But the problem with this is two-fold. First of all, *the absence of illbeing is not the same as wellbeing*. A pre-illness threshold does not mean that you are well. Just as the absence of hate is not love, and the absence of danger is not safety, the absence of illness is not the same as wellness.

Wellness is vibrant. Wellness is shimmering vitality. How many

people do you know that you would describe as shimmeringly vital?

Secondly, in our clinical research it turned out there were two routes to treating disease. You can either treat the disease itself, or you can activate the root drivers of wellbeing, and allow them to change the internal context in which the disease process unfolds. The clinical reality is typically a mix of the two. But what we discovered is that a salugenic[2], e.g., health creation, model of wellbeing is as powerful, or more powerful, than addressing the specific symptoms.

---

2 I came across the term *salugenic*, or health creating, for the first time when I read Dr. Robert Naviaux's *Mitochondrial and metabolic features of salugenesis and the healing cycle.* Naviaux directs the Naviaux Lab at the Mitochondrial and Metabolic Disease Center at the University of California San Diego. In the article, he defines salugenesis as: *the automatic, evolutionarily conserved, ontogenetic sequence of molecular, cellular, organ system, and behavioral changes that is used by living systems to heal.*

Allow me to say this in a slightly different way. What we discovered, with stress-related disorders– anxiety, depression, gastro-intestinal disease, complex chronic illness, post-traumatic stress injury, many auto-immune illnesses– is that if we could create a pathway to shift the Autonomic Nervous System from a baseline in threat response to one of safety and connection, that shift in and of itself, which changed the global internal context of the client's inward landscape, transformed the inner climate in such a way that symptoms were either reversed, resolved, or dissolved.

If we could create a pathway to shift the Autonomic Nervous System from a baseline in threat response to one of safety and connection, that shift in and of itself transformed the inner climate in such a way that symptoms were either reversed, resolved, or dissolved.

If we could get the client's nervous system to feel and register a felt sense of safety 51% of the time, the disease state would often dissolve without us putting any further focus on the disease.

In many of the cases where we achieved breakthrough outcomes that allopathic medicine declared '*simply impossible*', and where we definitively shifted people from disease to wellbeing, we never treated the disease directly.

We helped the person shift the baseline in their nervous systems from being centered in threat to safety, and when we did this the interior energy-processing templates and metabolic climate of the body changed so drastically that the body healed itself.

With respect to stress-related disorders, you can either focus on removing the disease, or activating the person's intrinsic root drivers of wellbeing. Sometimes, in acute conditions, you have to focus on the disease. Yet often, even if symptoms are severe, activating the intrinsic drivers of wellbeing does more to change the outcome.

This represents a radical reconceptualization of what we are doing in healing, and how we are doing it. It suggests the notion that the medicine of the future, which is in fact an ancestral future, may be better served by focusing on the creation of wellbeing than the remediation of disease. It suggests that we may have brought the wrong things into focus if our aim is wellbeing.

As Vincent Felitti MD, the co-principal investigator of the Adverse Childhood Experiences study noted so succinctly, *What we are treating in medicine as the problem is often the patient's solution to a much deeper problem that we cannot see*. I would offer a disturbing corollary to this by saying that often the patient's solution to that deeper problem, offered up endogenously by their own innate bodily knowing, allopathic medicine treats as symptoms and tries to suppress. Allopathic medicine's approach to stress-related disorders, which comprise most of what ails modern people, is not only upside down, it is also backwards.

I was recently asked by the acquisitions editor of a wonderful publisher to write a book on the neurobiology of connection. Since it was his idea, I dutifully wrote a book proposal, entitled: *The Neurobiology of Connection: Transforming Loneliness, Isolation, and Disconnection at their Source.* This he took to his editorial team, who informed us both that books on loneliness don't sell. He came back to me, and asked if we could write the book about healing anxiety, or healing depression. And at first I said, *Sure. We can write it about healing whatever you want that is stress-related: anxiety, depression, social anxiety, gastro- intestinal disease, auto-immune issues*. He wanted to focus on depression, and he told me, "Remember: first the problem, then the solution."

This conversation happened a day before I taught a multi-day workshop on Connection Phenomenology in Northern

California. I was holding the architecture of the book in mind, thinking about its structure. And then I started working with a group on creating the conditions for safety and connection to arise between us, and I started to get upset. I have no argument with what the publisher is saying: problem/solution is the formulation that sells. The problem I have is that this is a response to a pathogenic framework. It is upside down *and* backwards. Notionally, I don't argue that this is how the world works. People come to buy a book about healing something that is wrong with them.

But at a deeper level, this is the same worldview that is structurally preventing us from looking in the direction of wellness creation to begin with. People get interested in their health when it leaves them, and then they want a particular cure to what is ailing them in that specific moment.

But what about fortifying your health? What about understanding the root drivers of wellbeing in the first place, so that they don't leave you? That's the book I wanted to write. I wanted to write the neurobiology of connection as a health creation manual. And that is *this* book. And so I walked away from the publisher, and thanked them for the title.

To orient you to what comes next, first we are going to set a long evolutionary context to begin to view humans through our lineage history as hominids. Then, in simple, approachable, and precise language, we will continue to explore what the Autonomic Nervous System is. Once we've done that I will walk you through five doors. We are going to walk through the door of safety to open the door to connection. Once we've opened the door to connection, we will open the door to interoception, the sixth sense. Once we've opened the door to interoception, we will open the door to intuition. This door to intuition will open the door to relatedness. And the door to relatedness opens the door to enduring wellbeing.

Take off your shoes. Take a deep breath. Put your feet up. We are going to journey together back home to the ancestral hearth.

CHAPTER 02

SETTING THE STAGE

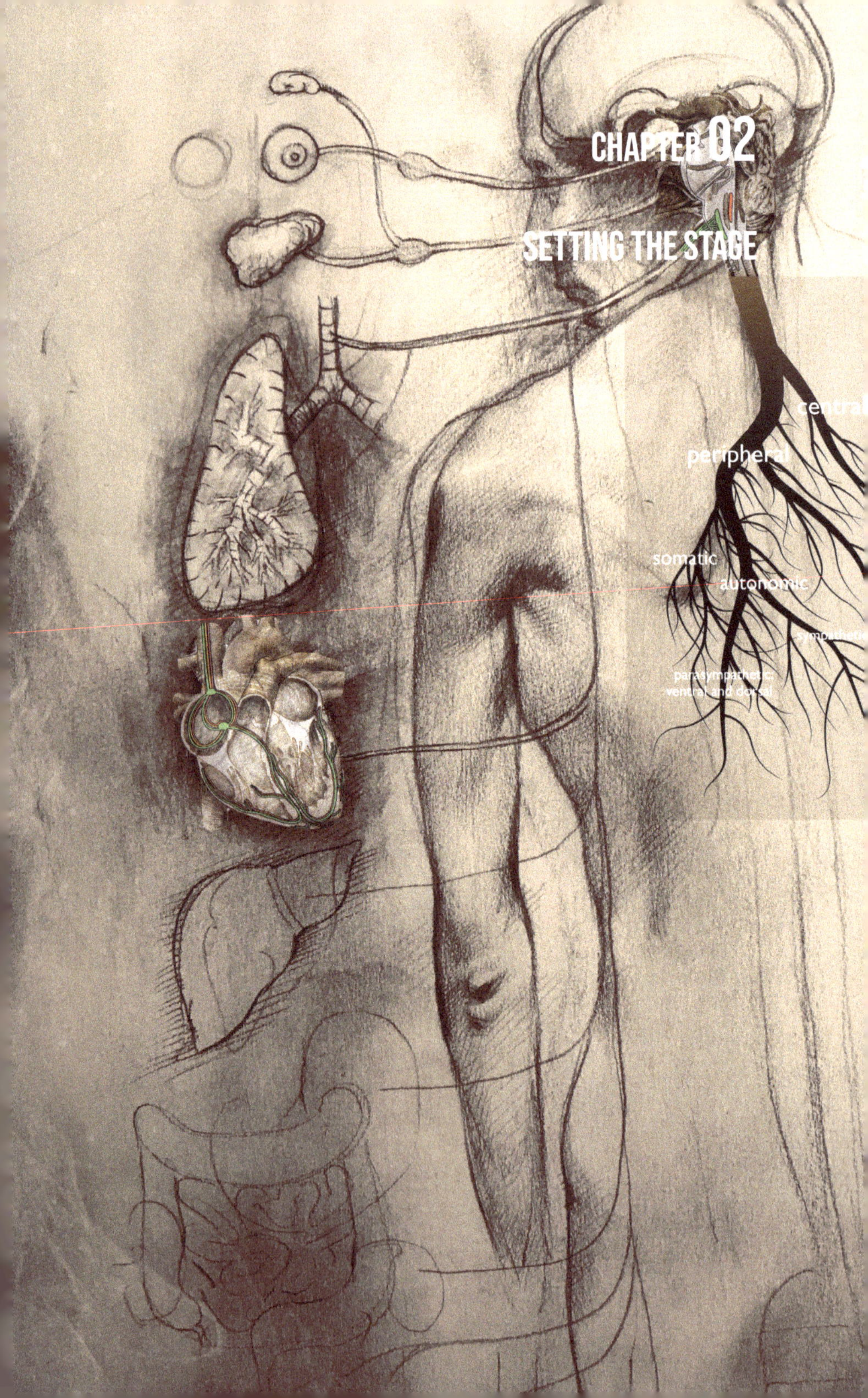

Most of the health outcomes in your life will be determined by a part of your nervous system that most people know almost nothing about. Those who do know that it exists understand it almost not at all.

I am speaking about your Autonomic Nervous System. The Autonomic Nervous System has some extraordinary ambassadors, but it still needs nervous system whisperers to translate the workings of this incredibly foundational system into language and frameworks understandable and actionable by everyday people. And what we know about the Autonomic Nervous System needs to be actionable by you, because you can, through understanding and practice, shape your deep nervous system to create enduring wellbeing.

Like a potter at a wheel, working the mudslick of your own neurology, you can through time and practice shape your deep nervous system to create enduring wellbeing.

I'm going to say that again because this is very good news. Your nervous system is neuroplastic, which means that it continues to change and learn and grow throughout your life, based on how you use it. By learning to interpret, converse with, experience, and support the dynamic needs of your Autonomic Nervous System, you can re-shape your nervous system in the direction of enduring wellbeing. You can self-direct this neuroplasticity.

In my career I have watched this happen many thousands of times, both with people who were very ill, and, with people who did not know they weren't living to their deepest and fullest potentials. The needs of your deep nervous system are pretty simple, yet they are not easy to meet in the context of modernity.

You see, your Autonomic Nervous System has two primary functions. One of them is the governing of the interior of the body: your heartrate, breathing, digestion, immune function...

all the things that have to be coordinated inside your body that you don't consciously think about. This is why the Autonomic Nervous System is sometimes referred to as the *Automatic Nervous System*.

The second function is as important, yet less understood by the general public, and is related with this governance function, specifically addressing safety[1]. You see, the Autonomic Nervous System is tasked with determining, in any given moment, whether the totality of your experience is, on balance, tilting toward safety, danger, or lifethreat, and then, based on this information, shaping the energy processing templates that govern your experience.

Your Autonomic Nervous System determines your moment-to-moment experience of safety or threat.

There are a set of neural systems in your body that determine, in each and every moment, if you are safe, in danger, or under lifethreat. This neural perception is called *neuroception*. Based on this realtime analysis, which updates with every breath, these systems re-tune your entire body, the same way a graphic equalizer dials up the bass and dials down the treble on a stereo, shaping the sound that comes from it.

These master control switches govern the way your body feels inside, the way you experience yourself, what emotions are available to you, the range of thoughts you will think, how you perceive what is happening around you, and the behaviors available to you. Far fewer than one percent of the general population understands how this system works because the state of the art in the field is the specialist knowledge of a subset of the study of autonomic physiology called *Polyvagal Theory*.

---

1 There is a very real and important existential question here, which could be phrased along the lines of 'Does safety even exist in a world of existential threat?' This is a profound question that we'll take up later in the chapter 'Safety in a World of Existential Threat.' It's most distilled shorthand is that this same question could be asked of our smallband hunter gatherer ancestors, migrating with the seasons in grass huts amidst maruading apex predators.

Your Autonomic Nervous System looks nothing like this sound mixer in a recording studio– yet this is exactly what it does.

My primary mentor in neurophysiology is the developer of that theory. Our work builds on and refines it.

Your autonomic systems activate and de-activate energy-processing templates in your body. An energy-processing template is essentially a tuning preset that governs the way that energy, information, and attention flow through your system.

Autonomic states are the energy-processing templates that govern the way energy, information, and attention flow through you.

Your body is constantly transmitting and receiving information. Your body contains a liquid system (blood pulsing through the circulatory system) wet-wired to an electrical system (your nervous system) that together generate an electro-magnetic field centered in your heart, the oscillations of which are detectable by a magnetometer eight feet away from you.

This field has an energetic geometry. The processing templates, governed by your Autonomic Nervous System–and activated based on safety, danger, or lifethreat–radically reshape the way that inputs flow through your mindbody system. The templates, and their associated neurochemistry, determine

- whether you move toward an experience or away
- whether you open up or defend
- are centered in your head, heart, or deep belly
- whether you ascend into elite performance, burn out in frustration, or immobilize into paralysis.

They precisely govern the shape and depth of the river of energy, information, and attention flowing through you, its turns and bends, eddies and rapids and slow still pools.

The Autonomic Nervous System IS *the neural architecture of your mindbody connection.*

It is the network that unites them. It unites the felt interior of your body with your sense of identity, what you think, your emotional responses, how you interpret what is happening around you, and then your behavior. It translates the elemental impulse to turn toward or away into behavior, calibrating your visceral state, emotional tone, thinking, conceptualizing, and perceiving along the way.

The first imperative of this system is **survival**. If survival needs are met, your defenses can downshift, and energy and attention liberate for restoration and connection. If you feel safe *enough*, your Autonomic Nervous System activates the Connection System: the doorway to relating with yourself, Others, and the Living World. When this system turns on, your heart opens.

I mean this both literally and figuratively, in the sense of the Japanese word *kokoro*. Your essence becomes available. You tune in– to your own internal landscape, and to the nuances of face and voice and sociability in others, to the Living and More-Than-Human Worlds.

This system can be turned on with a glance, a squeeze of the arm, a hug, the kindness in someone's voice, the steadiness in someone's hands, the moment you share laughter with a stranger. It can activate the moment you realize you can trust someone, the moment you cross a ridge-line out of the city into pristine nature and the wind catches the hair on your arm, by feeling your bare feet sink into the cushioning moss of the forest floor, your toes sink into mud, your hands sink into wet clay, your paintbrush hits canvas, the needle drop onto a record, or the syncopated beat of a drum ensnares the movement of your hips. It can be turned on by the pungent smoke of a campfire, the burble of frogsong, the first damp smell of autumn, your bone-knowing that rain is imminent. It can flip on listening to a song you know in a country you've never been to, exhaling fully, or holding a small child's trusting hand. It can activate the moment you feel truly seen. It surges at an unexpected kindness, a door held open for you, a high-five on the basketball court. It is designed to be the baseline from which humans operate, but as you may notice, most modern people are not living here in this realm of spontaneous vitality because its prerequisite is that we feel safe enough to emerge from defense.

Frogsong can turn on your Connection System, if that sound makes you feel at home.

Rather, most of us hemmed in by danger, navigating a world filled with existential threats, are experiencing the activation of our defensive systems, which range from appeasing, to fleeing, to fighting, to placating, to shutting down. We negotiate, accommodate, retreat, rage, numb out. We cycle daily through a retinue of survival states cued up on endless repeat, summoned as predictably as the dogs for dinner by the assault of newspaper headlines, elicited by daily indignities, dragging our bulk through everyday travails in a world that speaks to us in cadences laced with striations of menace. We contract to avoid relationship because we don't feel safe being touched physically or emotionally. Our ego, we might note, is the habitual contour of this contraction, our default mode the toxic ruminations of a balled fist. And why? Why do we contract?

Because the world is a fucking vampire. Capitalism, an extraction engine. Because transaction has become the metaphor governing our lives and self-worth. We are bought and sold. From the head, to capital, to chattel, to cattle, the etymology of domination brands, marks, and determines us. Our beautiful vulnerability gets weaponized. Systems everywhere seek to entrap us by taking advantage, tricking us with fine print, abusing our hospitality, hacking our social biology. We are prey that capitalism and empire flush into a net, juice like an orange. The modus of domination steals our resources, our labor, our attention, our time, and our inwardness.

All around us there is uncertainty. Financial instability. War and invasion. Climate catastrophe, impending and actual. Droughts. Megastorms. Wildfires. Flooding. Political corruption. Gun Violence. Fake news. False messiahs. Social media masquerading as community. Marauding AI. And on, and on, and fucking on.

Ancestrally, these defensive systems, these systems required to keep us safe in a dangerous world, were systems that only activated in emergency, and they were designed to be rapid in onset, and then to turn off. We activated them in situations of duress, and then we downshifted them when we came back to home and hearth.

The way that our social (they were actually small band hunter gatherer for the first several million years of our evolutionary history) systems were organized deeply prioritized connection and its restoration– so when someone went into an emergency mode there were many different kinds of restorative practices to bring them back to themselves. In the modern world, largely because domination paradigms (economic, social, political, etc.) don't like people who are sovereign, we (and here I am speaking of the modern West/ Industrialized North, and particularly of the tip-of-the-spear of individualism that is the American experiment) have largely undermined these social supports to bring us back to ourselves, and so people are getting stuck in defensive states.

Stuck, and unable to exit them. Stuck in terror, rage, and withdrawal. Stuck in grief and depression, frustration, resentment. Stuck in confusion, disbelief, disgust. Stuck in anxiety. Stuck in defense. Stuck in isolation, loneliness, and disconnection where our consolation lies in becoming a chainsmoking stack that ferries us across these bottomless chasms of loss. The internet mainlines news of tragedy straight to our breakfast tables like a digital IV: and so we combust 15 cigarettes a day worth of threat.

Not safe, not open, not connected, not vital. Barricaded, fortressed, defended. Armored up. Because that is what is required to navigate the barren tundra of the greed free-for-all of the modern world. Take or be taken. Survival of the fittest. And so our corporate bodies incite and incant a semiotics of war. We talk about *crushing it*, *killing it*, *destroying the competition*: we flash victory signs, piss on our enemies, lament the deplorability of *Those People,* not Us but *Them,* the Others – absent the recognition that these armaments of language and gesture become in fact bombs actual. That they pave the way for dehumanizing those we consider different, prejustify annihilation. That this mind and mindset are genocidal, not simply trash talk.

Most of us have become so accustomed to this competitive landscape of threat that we consider it normal. And within this paradigm, we can talk straight-facedly about needing to armor the doors of our elementary schools against the threat of shooters, about our elementary school teachers packing heat in the classroom. The need to defend ourselves from a hostile world translates, in the United States, to the highest rates of gun ownership the world has ever seen. We are armed to the teeth.

And yet, have you noticed that even the rich, barricaded in gated communities, armed and alarmed, with no economic insecurities and years of stored food are not immune to the illnesses that eat at modern people? The reasons for this are very simple. The removal of danger is not the same as safety. And our biological systems cannot flourish without feeling safe.

The embodied grapplings with this nightmare we call chronic, toxic, and traumatic stress–and understand to be the social determinants of health. The bodyset of this threat response is always pathogenic. As it becomes enduring it moves into mental illness and physical disease, which it births predictably. Sure as a kernel of corn is the seed of a corn plant, non-resolving stress is the seed of disease. Plant it in the body and water it with threat, and it will unfurl into cancer, branch into auto-immune and intestinal disease, bloom into sleep and movement disorders, fruit into mental illness.

It does this because it is not the ancestral baseline your body is designed to operate from. Non-resolving stress is fundamentally antagonistic to our wellbeing.

In 2019, I presented a keynote address about our work at the Academy of Integrative Health and Medicine's national conference in San Diego. The academy is among the foremost alliances of integrative health practitioners in the United States. I was in a conference room with two hundred of the most forward-thinking physicians, and integrative wellness practitioners in the world. I started by asking them to raise their hands if they agreed with what I was saying. "How many of you would say that 50% of your patients are coming to you due to stress?" I asked.

Every hand in the room went up. "How many of you would say that 75% of your patients are coming to you because of stress?" A murmur went through the room, but no hands came down. "80%?" A few hands dropped. "90%? 95? 100%?" At a hundred percent, more than half of the practitioners in the room still had their hands up.

Modernity has been called many things. I find the term *Anthropocene* accurate, because it describes humanity's unprecedented reshaping of our entire world, including the climate, in this era. As you will discover, I like playing with words. I think we could also call this the *Traumocene*. If you say this out loud, you will realize that it sounds like *trauma scene*. As in, 'the scene of a crime.' Modernity *is* a trauma scene. It *is* the scene of a crime. Look all around you and you will see humans arrayed in various states of extreme dis-stress. You will see the victims of crime arrayed before you.

As evidenced by the response to my question at the Academy, the research suggests that three or four out of five visits to primary care are stress-related. And yet modern allopathic medicine is terrible at treating stress-related disorders. Absolutely terrible. Why is this?

Simply, allopathic medicine does not understand that you are alive. Western medicine, and the worldview that birthed it, came from a paradigm of divide and conquer, and the study of cadavers. Dead bodies. Do you realize this?

There is no medical word in Western medicine for the animating spark of vitality that is called *Qi* in Traditional Chinese Medicine, *Prana* in Ayurveda, *Ki* in Japanese, *Ka* in Egyptian, *N Kasea* in Tzutuj'il Mayan. We do not have a word, in Western medicine, for what makes us alive. Think about it.

Western medicine is a mechanistic system that doesn't know that you are alive. It literally does not know this. Its purview is not aliveness. It never has been. It has no conceptualization or operating model of your vitality, your animacy. It never has. It can measure your heartrate and your blood pressure and your respiration, but it does not know that you are alive. Much of what it learned about you came from studying things that were dead.

Let me ask you a question. If you wanted to study trees, and I took you to a lumber yard, would you be satisfied?

If you wanted to understand the intricate ways that photosynthesis feeds them, the mycelial network that knits together tree roots into a wood-wide web that flows nutrients to places that need them, would you feel educated if I took you to the mill and counted the rings of dead trees?

You would not.

If you wanted to understand how sunlight nourishes trees in daily circadian rhythm, how fluctuations in rain impact their living growth, how soil nutrients impact upon their resilience, would you feel educated if we inspected 2x4s? No, you would not.

One of the fundamental differences between the forest and lumber is that lumber has no leaves. If you want to understand living trees, you have to study living trees. If you want to understand what makes a forest healthy, you have to study a healthy living forest.

I am not saying that allopathic medicine is not amazing. I'm not saying it hasn't saved countless lives, including my own several times. I'm not saying it hasn't brought humanity forward in myriad ways. I'm simply saying that it is terrible at treating stress-related disorders because stress-related disorders always involved the flow of your innate vitality. A medicine unable to register your aliveness, developed by cleaving the body from the mind, will never heal this. Western medicine arose from the foundational split at the heart of the European thoughtforms of domination.

Western medicine is downstream of Descarte's foundational error, expressed in the dictum, "I think, therefore I am."

Cogito ergo Sum.
-Descartes

And because of this, it segregates the mind from the body, associates identity with your brain, treats 'mind' as a chemical epi-phenomenon of brain, and views your body as a collection of intricate mechanical systems that can be treated by specialists with increasingly compartmentalized knowledge. The gears of a watch, the parts of a car…

Because of this radical failure to understand the holism of ani-

macy, it treats fragments of your experience. This leads to some errors that would be comedic were they not so tragic. One, for example, is that in allopathic medicine, gastro-intestinal disease is being categorically misdiagnosed at scale. Much of gastro-intestinal disease is being treated by bringing our attention to the wrong organ. Modern gastro-intestinal disease is just as likely to begin in the neural regulation of the guts as in the guts themselves. The neural regulation of the guts is typically altered due to non-resolving stress of some kind.

When the neural regulation is altered, digestive motility either increases or decreases, the oxygenation of the guts changes, and the pH shifts. As this climate is interrupted, the gut microbiome begins to change because the most beneficial bacteria, like any plants, really, have a certain climate they require to flourish. If this progresses long enough, or severely enough, and the gut de-oxygenates far enough, lots of things that ordinarily are in the guts anyway, like yeast, begin to bloom or become imbalanced. Absent the internal self-regulating climate that kept them in balance, these can begin to irritate the intestinal lining, eventually causing it to become permeable. We may start to have food sensitivities, allergies, the onset of inflammation, leaky gut, inflammatory bowel disease, ulcerative colitis, and so on down the road. But this situation did not start in the guts.

The guts were down-shifted because of stress. This down-shift happens because biology is expert in optimizing your energy resources as they apply to your present moment needs. And if your body is under threat, if you are being chased by a bear, digestion is not your top priority. Getting away from the bear is your top priority. What else might get downshifted? Your memory function. Your attentional stability. The capillary dilation of your skin. Your body temperature. Your metabolism. And if this stress state endures, you will start to crave foods that sustain the blood chemistry in this elevated (stressed) state. Foods that are fatty, sugary, salty.

Your body, in the energy-processing template of defense (a

flight response to get away from the bear), turned off your digestion *on purpose*. And unless this deep switch is flipped back, intervention will be like rearranging the deck chairs on the Titanic. The *problem* is not the guts. The *body's solution to the problem* is turning off the guts. Until we address the problem that the body is trying to solve by turning off the guts we get nowhere. The root problem is non-resolving stress.

So how will this manifest as disease? Food sensitivity? Allergies? Auto-immune issues? Inflammatory bowel disease? Memory problems? Sleep difficulties? Attention issues? ADHD? Pre-diabetes? High blood pressure? Anxiety disorders? Depression? Mental illness?

The problem isn't in your guts, your memory, your attention, your circulatory system, your body temperature, or your metabolism. Yet it might manifest in any of these because they are all downstream of your autonomic presets. Once your autonomic system has re-tuned to danger, everything in your body is downstream of this. So what will go wrong? Roll the dice.

Every biological system in your body is downstream of autonomic presents.

Returning to gastro-intestinal disease, specifically, this puts Western medicine then in the problematic position of referring clients with gastro-intestinal distress to gastroenterologists whose specialized purview, of the GI tract, does not include the neural systems responsible for the problem. With their scopes and their measures, they are not able to see the neurological systems that are controlling the guts, and to which their poor function is a response.

This results in a situation a little like the classical spiritual joke told about Mulla Nasruddin, the Sufi mystic and holy fool. He loses his keys in the alley and is looking for them under the lamp-post. His friend says, "I thought you lost your keys in the

alley," to which Nasruddin replies, impatiently, "Of course I did– but I can't see anything where I dropped them because it is dark over there."
Nasruddin, as you may imagine, never finds his keys.

The impulse at the root of allopathic medicine is imperial. Divide and conquer. Yet the bodymind is a single system. Healing is NOT the art of war. A medicine that does not know you are alive is not going to help you re-instate your vitality. It is time to look elsewhere.

My name is Gabriel. I grew up in as close to an ancestral village as any white person I know. It was a tiny town in rural New Hampshire, where I had a second mother, a best friend I've known since he was a handful of days old, and a forest out the door of my house that, at five years old, I could wander into at will. When I was seven years old, the economy sputtered into recession, my father lost his job, and I was taken by my family from this formidably wild, animate, & intimate place and moved to the parochial domesticated suburbs of St. Louis, Missouri. This was the defining experience of my childhood. The *trauma scene*.

My body experienced this relocation as an abduction by my own family, and responded as though I was being drowned. The removal sucked all of the oxygen out of me. When I was taken out of this place, and my community, I had night terrors for months that were so intense my parents had to hold me down because I was thrashing so hard they were afraid I would wound myself.

Have you ever loved something so much that you could not imagine it taken away from you? This was the story of my childhood. I woke up one day in November of 1982, an outsider, an exile, in a place I could not find the living earth under my feet, any whisper of her proper magic.

My feral playmates replaced by pale domesticated hominid cousins with muted yearnings and regimented schedules. The forest supplanted by Nintendo. In a place where nothing felt any longer alive, and looking inward I could no longer feel or find myself at all.

As a result of this experience of being uprooted, my body shut down in a manner that was so deep and primal that I could not even remember it happened for eleven years. Not until I was eighteen years old and a freshman at Yale University, where I was studying neuroscience, and met a girl. This happened at the tail end of winter. She was from New Mexico, and studying theatre: a beautiful wild-hearted girl.

The first night we spent together was at the home of a neon glass artist, strangely enough, tucked into the inky silence of a snow-drifted New England forest. In that bedroom pulsed, hand-made, the humming stalk of a neon flower that painted the room, and our faces, with rose and mint-colored light. My heart re-awakened that night, under the enchantment of being seen. I came home to myself, far and fast enough to drive her, in a borrowed station wagon, several weeks later, back to Walpole, New Hampshire, the town from which I had been taken, and plug back into the story of my childhood. On the trip with her I would disappear, periodically, and weep with abandon, not even knowing why. Together we sat on the forested hill overlooking my childhood home where I had first laid on my back and watched Monarch butterflies laying eggs in the milkweed.

I see it now, the arc of this, from thirty years forward of it, and I can appreciate that this young woman was the doorway that sparked me back into re-igniting the fire of my own hearth. Yet that same winter night, before we travelled north, before our lips met, she said, "Someday I am going to wake up, and do a little dance, and then I'm going to leave." And she was true to her word.

When this happened, I fell into a severe depression. We say that we fall in love, but I contest the orthogonality of this. I rose in love, elevated in it, and when she left– *then I fell.* It was like being swept into a river and taken under. I couldn't think. Words swarmed in my head like smoke-addled bees, and I could not derive meaning from them. I uttered not a single word in any of my classes, week after week. Three months into the fall semester of my sophomore year, after I had dropped out of school and returned to my parent's home in St.Louis, where I did nothing but gain weight and stare at a wall, I began to realize that something had happened to cut me off from my own heart. I had lived in it as a child, then it had been taken away.

With the girl I had found it, touched it again, and then lost it. And so began, thirty years ago, my journey of healing.

This journey was not elegant. I dabbled with drugs, with self-harm. I painted self-portraits in which the space of my chest was void, a sucking emptiness through which the dark night showed. I got caught up with people who didn't have my best interests at heart, to put it mildly, wannabe gangsters and islands of lost children. I hurt people who did not deserve this either.

In 2012, I had a breakdown that landed me in psychiatric hospitalization for ten days, and from which the attending physicians told my wife and parents I would never recover. And yet, strangely and non-accidentally, I did.

Over the past thirty years, I have charted a course as inwardly epic as the Homeric Odyssey, and with the same goal. I was determined to get back home, to reclaim my Original Self. This alone was my North Star. With yearning as my compass, I followed trails as far as they could take me if I determined that were in the direction of home.

timelapse photo of North Star (top center)

Along the way I spent time in virgin forest in the Amazon. I drank my body weight in what is commonly known as ayahuasca, sat up many nights in teepees within the persuasion of drums, meditated for many thousands of hours, and re-encountered in a ceremony my seven-year-old self in the closet of the room I had been moved into in St. Louis, where he was in a state of suspended animation, nearly dead, his mouth stuffed with old socks and wired shut.

Over the courses of decades, I brought myself back from the near-dead, from the state of unintegrated lifethreat that is the shutdown of the Grounding System (some of you know it as Dorsal Vagal), and about which I will teach you in this book, because I assure you that I was not the only person living in those realms. There were times in those years when I thought I'd rather be dead, and there were times in those years, many in fact, when I thought I would never be ok.

Over the course of this time period, because I needed to understand what was happening to me, to sense-make, to meaning-make, and I could not find anything resembling a map of the terrain I was encountering in consciousness, nor the kind of mentorship that I wanted in a single university or PhD program, I built a trans-disciplinary school around me which has grown to include well more than 100 faculty, mentors, and advisors from 25 disciplines of healing in 24 cultures, many of them Indigenous.

My personal mentors and advisors now include more than a dozen of the world's leading experts on neuroscience, connection, healing, and wellbeing. Our work locates at the bleeding-edge where art meets science, the fruitful collision of autonomic neurophysiology and ancestral lifeways.

In May of 2020, on the global call where I convened the Restorative Practices Alliance, I brought my friend and mentor, Dr. Stephen Porges, the world's leading expert on the relationship between the Autonomic Nervous System and behavior into conversation with my friend and mentor, Kuuyux Ilarion Merculieff, an Unangan (Aleut) Elder who is President of the Global Center for Indigenous Leadership and Lifeways. This was the most beautiful moment of the call, because each of them in their respective languages, one the language of science, the other in the language of spirit, said the same thing: when humans are well, the heart is telling the mind what to do.

I can explain this to you physiologically, and I can know the spiritual physics of this the way a tracker knows their territory. This is my ecological niche, and what I am going to do in plain language in this book is help you learn the geologic map of your own ancestral heart. I am here to help you put your mind back in its proper place: as servant to your sovereign heart. In order to do this, we are going to have to heal your heart. There is no more effective way to do this than teaching you how your Autonomic Nervous System works, so that you can begin to reorganize your experience, and shape your life in the direction of safety and wellbeing. I am going to share with you, right now, the foundational algorithm of this book, in five doorways through which we will walk together. It is pretty simple.

DOORWAY ONE – SAFETY

Safety is the doorway to connection. This means that your embodied (your body is keeping this score) not cognitive (this has little to do with your thoughts) sense of safety is the doorway that must open of its sovereign discernment, to allow your connection system to activate. Safety is always a sovereign decision of the nervous system, made down in the depths of being, the furnace of implicit memory. Safety can be invited, but never dictated. We cannot make anyone else feel safe. We can create the context for our selves and for others, in which safety may emerge, but when it does it will be like a flower opening from the inside: innate, mysterious, responding to inward yearning, its own directive. It will be like a snail that emerges, ever so slowly from its shell, unfurling so subtly at times it will be like

5 doorways on your journey home

Safety is the

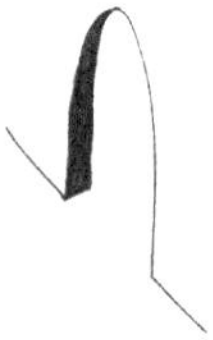

to Connection is the

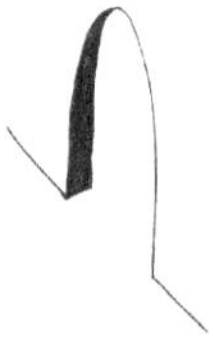

to Interoception is the

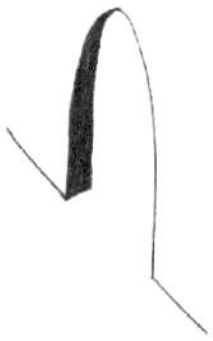

to Intuition is the

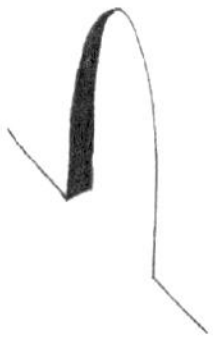

to Relatedness is the

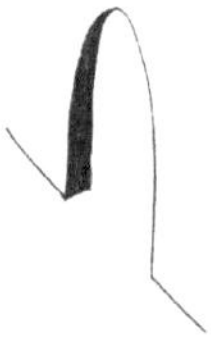

to Enduring Wellbeing.

watching paint dry, until suddenly the rubbery antennae appear, inflate, the face shows itself, and we arrive. And – until it becomes steady – could evaporate: pull back into its defensive shell on the hair-trigger of breath.

DOORWAY TWO – CONNECTION

Through safety we arrive at connection. This Connection System is a neurobiological system that knits together the neural regulation of the face, the voice, the tuning of the middle ear, the turning of the head and neck, the palms of the hands, the soles of the feet, and other elements of our interior with the heart and breath to bring us into the dance of co-regulation when we feel safe *enough*.

DOORWAY THREE – INTEROCEPTION

When this system activates, connection becomes the doorway to interoception. Interoception is the sixth sense, the doorway to the felt inward landscape, what Eugene Gendlin called *the felt sense*. The terrain that Polyvagal Theory suggests should replace the dictum of Descartes with the following pronouncement: *I feel myself, therefore I am*. Only when we enter interoception through the door of safety and connection does it provide the next doorway: to intuition.

DOORWAY FOUR – INTUITION

I don't like the word intuition because it is constraining. The door interoception opens is to non-cognitive ways of knowing. I am personally aware of at least 16 non-cognitive ways of knowing that are documented in other cultures.

Some are pure awareness-based, like: *gom* (Tibetan), and *yūgen* (Japanese). Some refer to a knowing that comes from the guts, like *kishkes* (Yiddish), *bauchgefühl* (German), and *haragei* (Japanese). Some relate to our ability to track the energy of emotions, in ourselves: *comover* (Italian), *la faculdad* (Spanish), or in others: *nunchi* (Korean), *szimpatikus* (Hungarian). This way of knowing, in Lakota, comes up from the earth through the feet: *anamagoptanye* (Lakota), one of various ways of knowing in Lakota culture that depend on who is talking to you through the earth. E.g., is it the earth herself? Or your ancestors? Or your grandmothers?

Some words reflect our connection with the Living world, and ability to discern elemental energies: *wayfinding* (Hawaiian), *ilm al-raml* (Arabic), and *tracking*, for example.

Some refer to intuitive extensions of sight: *clairvoyance*, hearing: *clairaudience*, and feeling: *clairsentience*.

I am also aware of concepts, for which I do not know the words. The San of the Kalahari, as well as aboriginal Australian cultures, have ways of talking about putting on the mind of an animal. Many cultures have ways of talking about putting on the mind of a plant.

Each of these is a science in its own right: trainable, objectively measurable in its calibration, repeatable, and has lineages that teach it.

And I am confident that this barely scratches the surface. We sometimes hear that modern humans are using just a tiny percentage of our brains, but what this really means is that *we are using just a tiny percentage of our capabilities.* Yet what all of these ways of knowing have in common, and the common purpose that they serve, is that they dismantle isolation. When we are able to feel ourselves and others deeply, we open the doorway to relatedness.

DOORWAY FIVE – RELATEDNESS

This doorway from intuition to relatedness is the one that brings you back into the village, back into the circle of Life. This doorway brings us in out of the cold, the sterile, the alienated, the dissociated, the domesticated, the suburban. It brings us back to the ancestral hearth, the original fire, our innate vitality, true nature. Back to feeling, and *feeling with*. Because if we can actually feel ourselves, and feel one another, guess what? We are not alone.

It restores us to the kinship worldview in an animate universe, the *Alaha Ruha*, the One Sacred Breath, where we are no longer in exile.

ARRIVING AT ENDURING WELLBEING

And this my friends, because we are social mammals, conducts us to enduring wellbeing. This is the journey I will take you on in this book.

In service to this objective, I am going to deploy three primary metaphors, which I will tell you now are not merely metaphors. I am going to describe the deep nervous system in the language of trees because the reality of your Autonomic Nervous System is that it is the terrain of a vast inward aboriginal forest, and your neurons are the varied trees in this inward ecology. Millions of years ago our species came down from the trees, and

the forest lives inside us still, vivid in our neurobiology.

I will describe the deep nervous system in the language of water because, like water, we can exist in three states. A liquid flow state, which is our birthright outside of the domination paradigm, and the state in which we claim our full potential as human beings. This is the state in which wellbeing spontaneously arises. Shift into detecting threat, however, and we find ourselves in the gaseous state of steam, which is the high-energy state of fight and flight that modernity has normalized, and in which most modern people reside through a potent combination of caffeine, capitalism, and domination mind. And the ice state, which is the frozen state I fell into when I experienced the social death of being abducted from my small band and my place, and is a state of dissociation, numbness, and withdrawal that overtakes us when we experience lifethreat.

I will then fuse this language of water with color theory to explain how, from three primary colors: blue, yellow, and red, which represent liquid water, steam, and ice, we begin to blend these into a full palette of colors. First the greens, oranges, and purples that exist along a continuum between them, and finally the full autonomic spectrum. Along this path, we will grieve together, shake together, and come to new and very old understandings. We will begin to re-organize the core operating system that structures how your body organizes memory as it relates to experiences of danger and lifethreat.

I have done this before. First I did it with myself. Then I began doing it with others. I have now led over 2,500 trainings globally. I have trained more than 50,000 people, and lead an organization that is managing what is either the first, second, or third largest data set in the world correlating stress-related disease to autonomic physiology. In the fall of 2022, I was asked to train the entire faculty of the Stanford University School of Medicine in our work. In 2023 I led a global group of 5,000 wellness professionals studying Polvagal Theory, and in 2024 was asked to lead another studying the Vagus with

13,000 members of its own.

Our research has demonstrated that there are about fifteen primary states that comprise the autonomic spectrum. It is not a line, or a ladder, as is often taught. It is a spiral. My objective here, in this volume, is to reintroduce you to the part of your nervous system that will, more than any other biological system, govern your experience of wellbeing, and specifically teach you how to activate the neurobiology of connection in service to bringing you home to yourself.

As I hope you can feel, my passionate interest in this area is both academic, and deeply personal. And at times in this book I will use my own journey as a way of illustrating the points that we are discussing.

I come to my relationship with you as a reader not as a professional first, but as a human being: someone who has experienced being as alone as anyone I have ever met. I have passed through things that I would never wish on anyone. I want you to know that there were many years in my own life when neither I nor any of the many doctors or wellness practitioners I consulted could tell what was wrong with me, years when I didn't think I would ever be okay, years when I did not want to continue being alive. I want you to know that my own journey of healing has been nearly forty years in the making and that it continues. Yet I also want you to hear me when I say that both healing and transformation are possible. I know this from my own life, and I know it because I have watched thousands of our clients, friends, colleagues, and people we trained transform. People with very serious early adversity, very serious chronic health conditions, including many that allopathic medicine tells people are not reversible.

The journey that we are about to undertake is a journey that explores what it means to be human, through the lens of your neurobiology. What I am going to teach you is both cutting-edge and ancient: the rediscovery of the medicine of the ancestral future. Because this is the neurobiology of connection, this is the science of the heart, and the science of the original fireplace. It is why the company I founded is named Hearth Science.

This book is a complement to my previous book, *Restorative Practices of Wellbeing*, which is focused on the practices that you can use, once you better understand the fullness of what we are teaching here, to bring yourself back into wellbeing.

Welcome to our journey. I hope to help you find your way home. Let us now transition from the traumocene to the healing scene.

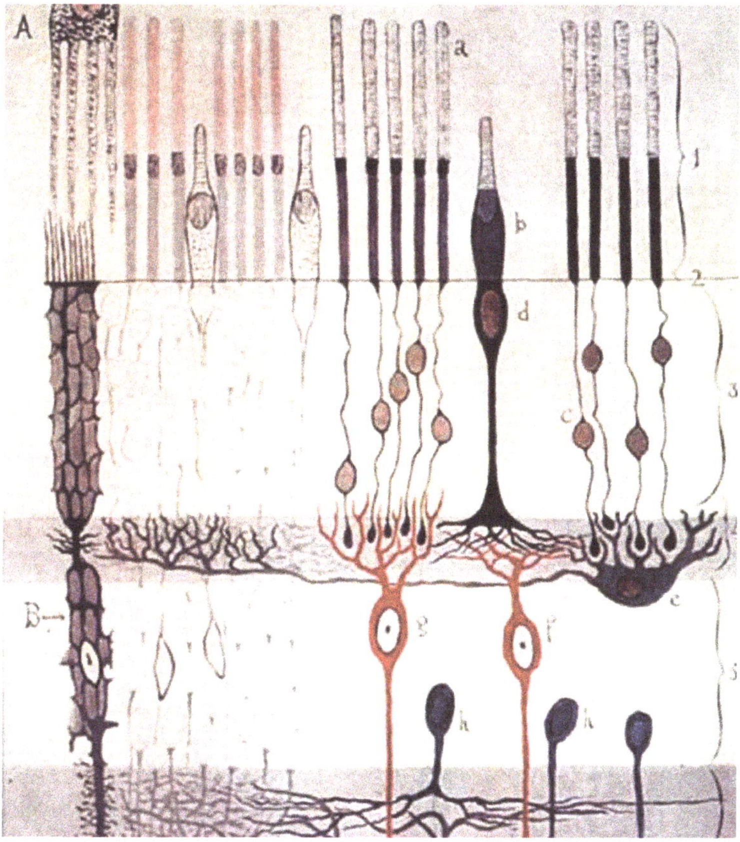

A NOTE ABOUT SUPPLIES AND CADENCE AS WE JOURNEY TOWARD WELLBEING

If I am doing my job as guide here, this book will be an experience for you. As an experience, unfolding through time, a cadence is likely to emerge, an in-breath and an out-breath. If you let this book work on you, which I recommend, it can help you organize, or reorganize, aspects of your memory (memory consolidation), and how your bodymind holds that memory in ways that are useful to the evolutionary project of becoming who you really are.

A subtantial amount of the work that this book will do on you, if you allow it, will happen when you are not reading it. When you are simply allowing what it explains to settle upon your experience and notice how it changes the cinematic lighting of it. If the book is working well, it may bring different aspects of your experience into the foreground, and shift other aspects into the background as you begin to reorganize your experiences through the filter of autonomic state– in this manner acting rather like the focus on a camera.

Practically, it will begin to encourage you to group memories by autonomic state and feeling tone rather than the story you've told yourself about them. *By bodily response rather than narrative.* Experiences that evoke similar autonomic states- a flight response, for example– have something architecturally in common in terms of their emotional landscapes and arousal, even if the experiences themselves are widely separated in time or by context.

I might have been afraid of the dark, afraid of sharks, afraid of my father's anger, afraid of jumping out of an airplane, afraid of getting divorced, afraid of losing my job, and these experiences might be separated by decades, yet all share a foundational *felt* architecture.

As I begin to allow them to live next to one another, woven more on their embodied and movement flavors than loomed on my stories about them, a certain kind of bodily reorganization takes place.

One of the ways to support this process is to take notes about what stands out to you. Some autonomic states may be mirrors for you. And some portals to the connection architecture you will already know intimately. I invite you, therefore, to mark up this book. To circle things, take notes in the margins, make diagrams. Own it, make it yours. Some people have been taught that marking up a book is not respectful– but I would propose to you it is a sign of engagement. I actually seek out used books that have been marked up because it means someone felt passionate enough about what they were reading to comment on it. You can affirm the book, emphasize, or fight with it. You can disagree. Your full engagement is welcome. You can even throw the book, if it makes you angry. It comes with a strong spine ;)

The diagram of all the autonomic states, what we call the full autonomic spectrum, will also be a trustworthy reference. The dust jack of the book is a poster, and you might want to place it somewhere you can easily refer to while you are reading.

You might find it also useful to have a notebook or a journal or a sketchbook to accompany you through this process. Someplace you can track your responses, follow inward trails, gather insights together. Doodle. Connect dots. Daydream with a pencil in your hand.

You will likely receive the most from this book by reading it and then feeling into its implications for your life. The implications are multi-layered. Hold your life up before the book and see what happens. And allow that inquiry to govern your pacing. Don't rush it.

If I may be so bold, you also may find it useful to have water nearby. A glass of it or a bottle of it. This book is, in a way, an act of rehydration. A marriage of fire and water: Alchemical cookery with you in the stockpot.

CHAPTER 04

EVOLUTIONARY CONTEXT

Before we embark on this journey, while we are still packing our provisions, let us establish the long evolutionary arc of this exploration. The hard-won understanding beneath all of the work our team has developed is very simple, although reaching it was not easy at all.

The source context of relatedness is the foundation of well-being.

For most of the long arc of human evolutionary history, the way that humans lived was in small band hunter gatherer groups, and the foundational understanding that rested beneath all of culture was our relatedness to All That Is.

For 99% of human history, the taproot of culture was the living experiential practice of connection.

Modernity is the unfolding story of our alienation from this source context. It is the story of an accelerating deviation from this ancestral baseline in relatedness. Modern civilization so-called, is the tragic story of how we became exiled from who we really are: how we entrenched and then organized our identities outside of the living fabric of inter-being.

Yet this ancestral context is present, available, and has been here the entire time. She is waiting for us to return to her. She is calling us home. She doesn't yell, doesn't shout, doesn't insist. She doesn't sound like a carnival barker shouting over a riotous crowd. She doesn't look down her nose askance at you, or try to seduce you by shaking her ass. She isn't trying to sell you anything, and she is not discounted on a particular Friday associated with the color black. She's not on your phone or your laptop, she's not on TikTok or Facebook. She's not in virtual reality. She can't be scraped by ChatGPT. She is very quiet, in certain ways, despite her insistence.

All the same, she is calling you home. She is the original context. If your life is too busy, too fast, too modern, she may only reach you obliquely. She might come stalking you through illness. Or grief. Or tragedy. Or disaster. She might reach you in dreams. She is persistent, because she is the wellspring of reality. Most modern people spend most of their lives running from her, screaming.

She is very ancient, yet stands outside of time. She exists everywhen. Modern humans have become divorced from her. We have lost the ability, and the intuitive ways of knowing required to come home to this experience of relatedness. We might say that the edifice we call *ego* developed to avoid her. In losing this relatedness, this kinship with animacy, we have lost our ability to come home to ourselves, one another, and the Living World, which means that we have lost the ability to come home to our true nature. We have lost our own embodied vitality and a living experience of the animacy of the entire universe, despite the fact that our most cutting-edge science affirms it.

Affirms it without knowing how to access it. Points at it without knowing how to touch it.

This alienation is a result of many things, but the source from which it springs is domination. Domination is the way that humans who have been traumatized, and shifted into defensive responses, attempt to assert control over their lives when they no longer feel safe and vitally connected. In the absence of the dance of relatedness, in the absence of feeling alive, we move to dominate. Force is this shorthand of domination, empire its calling card. What the Irish poet and Celtic Spiritualist, John O'Donohue calls, in the poem that opens this book, our fatal attraction to aggression.

Force is the shorthand of domination

The civilization we live in, whose source code is an operating system that structures domination, is the systemic expression of hundreds of generations of humans exiled from their own hearts. We are the inheritors of the thoughtforms of Imperial Rome, the orphans of empires of domination. The physical empire has crumbled, yet the psychic empire has been internalized so completely it is invisible to us.

The origin stories of modern civilization, perhaps most obviously the story of the Fall, are attempts to grapple with this original wound in consciousness: this original exile from relatedness.

It is notable and unsurprising that the story blames this occurrence on a snake, a creature whose entire underside is in contact with the ground, since this story is about our rejection of our earth-based and indigenous natures, the loss of our deep rootedness, in collusion with a woman, since this story is also about our rejection of the feminine and the mothering instinct, particularly the undomesticated (wild) mother.

It is obvious that this story was written by a man. One who was training us to worship sky gods and paving the way to kill the wild feminine (e.g., burning 'witches'). Its very architecture points us precisely to what has been forgotten and fallen out of balance: our earth-based and feminine natures.

About 12,000 years ago, as humans transitioned from small band hunter gatherer tribes to settled agricultural societies, human populations attained surplus food for the first time. This accumulation led rapidly to a series of innovations in abstraction, generating numeracy (let's keep track of the surplus) and literacy (tell those folks about the surplus), birthing property boundaries (this arable land belongs to me) with straight lines and right angles, originating the multi-generational transfer of property (let's keep this in the family), and rapidly initiating a re-organization of human sociability from horizontal modes of relatedness to vertical modes of domination. From power *with*, to power *over*.

Could we track back to the notional shift here, in consciousness, we might demarcate it by observing that it represents a relocating of vital energy from embodied relating (the spaces

between: felt inter-being) into abstraction (disembodied inwardness). The dawn of numbers, required to document excess, to keep track of it, and the dawn of alphabets, shifts human energy processing away from what has been its central domain for possibly millions of years: the felt sensory interchange of energies, gross and subtle, within us and with our environments. Now as we settle into towns that will grow into cities, as we stop migrating seasonally with the animals, our attentional horizons move inward to more abstract planes, where we find ourselves organizing energy in abstract cognitions with these symbolic toolsets of number and letter.

Take a moment to reflect on your own experience, your own childhood. Consider the difference in mindset and bodyset required to sit in a classroom, listen, and manipulate abstractions (stop fidgeting!) as compared to that required to move on an athletic field, fish, or hunt an animal. The shape of attention in these circumstances is entirely different. The vectors of energy flowing through the body are entirely different.

When I am hard-ground tracking my body is lit up with the nuanced signals of sense and scent, the intuitional and interoceptive arguments presented by the radar of my hands, the inclination of my feet who know things my internal talk does not.

I am not thinking, *per se*, but listening, intuiting, belly to the ground, sometimes literally, sometimes metaphorically. I am in call-and-response with the Living World: aware of shadow and light, texture and sound. The alarm call of the dark-eyed Junco and then his silence, the trip-skitter of squirrels, crunch of a stick snap-crackling. My senses are alert yet relaxed, tuned like a rosined bow on a violin, tensioned into an ecological niche I am dreaming awake through. My entire body, muscularly, is relaxed and yet ready to explode into action. To stretch, turn, race, run, jump: loose the arrow. I am vital like a jaguar when I am stalking. Lithe as a monkey moving through the trees as I scan, dissolving the boundary between the felt and the seen, the known and the intuited. My hominid, monkey, mammal, and lizard ancestors exist within me like nesting dolls, granting me access to their billions of years of nuanced embodied awareness. Often they know something long before it makes its way up to my thinking mind, and my body has responded, sometimes successfully, to these directives before I have processed them consciously. I am moving and relating through ancestral awareness.

Coming in from forest and field, from this animated vitality, to the classroom of the modern school is excruciating. With all of these senses online, sitting in class, I run the risk of being BORED OUT OF MY SKULL. I cannot sit still, trace inside the LINES. My ancestral intelligence experiences a modern classroom as a carceral system. A prison. A sensory deprivation experiment.

Make no mistake, if you were educated in the modern world, endured the acculturation of the modern classroom, your vitality was beaten (or medicated) out of you. If it was not, you were labeled a troublemaker.

In the modern classroom, my intuitional capabilities, the felt non-cognitive ways of knowing that comprise my greater interaction with the inner and outer worlds–*to read pattern, see energy, sense depth...the knowings coming from my hands, my skin, my feet, up from the earth*–they have nowhere to go. Nothing to do. Nowhere to be of use. I am stranded inside of a box, a room that bears no relationship to the living cathedral-contours of forest, no sensory variety compared to the most humble rock I might pick up to finger, the feel of the wind carrying scent of bloom and moss and mushroom.

And so, absent a Waldorf school, a Montessori classroom, a forest school, I must learn to occupy myself in a different way, in the modern classroom, so I don't go crazy, shoot darts into the teacher's backside.

For me, personally, at seven years old, being in a shutdown state made it much easier to endure sitting in class. Had I not been shut down, my feral seven-year old self probably would have incited a rebellion.

Yet I was dissociated when I got to a modern classroom, though I didn't have words for it at the time. To sit still, in rows, and listen to regurgitated information, the already-chewed-food of the contemporary curriculum, is much easier if you are not in your body. And this is the shift in relating I want you to feel.

With this movement 12,000 years ago, out of the kinship worldview, the constant and necessary mediations with the Living and Spirit worlds, we shift inward and upward into cognition, into a world of abstract symbolic manipulations. And in doing so, we transition from a world where we talk to everything inside and out, a world of permeable boundaries and gradations, to a world organized by the constraining lines of *power over*, a domination paradigm, effectuated through fortressed identities and hierarchical gradations of power.

No longer able to negotiate the nuanced dance of relating, we substitute force, the shorthand of domination. And in the imaginal realms we move from peacetime to war, from a garden to a mine. From sovereign creators to imperial subjects. This what the *Fall of Man* means.

This domination paradigm was reflected perhaps first in the ownership of land (the removal of the sovereignty of Mother Earth, the rights of Nature), then in patriarchy (the removal of the Sovereignty of Women), then in the hierarchizing of humans (the structuring of Oppressions).

These three deviations from the ancestral baseline happened in the blink of an evolutionary eye. These deviations are embedded in things as mundane as straight lines and right angles, which exist very rarely in nature but were first used to demarcate land boundaries[1], and as pervasive as where we locate heaven (which we think of as being up), and as encompassing as how we conceptualize the self, which we understand to be unable to contact the Sacred Source directly.

Our movement out of relatedness is the byproduct of empire. It has generated the source code of the Operating System of the modern world.

At the level of individual physiology, this accretion of a domination paradigm is a manifestation of nervous systems that do not feel safe. The structural edifice of modern civilization has been developed by people alienated from their own hearts.

I am spelling this out here because although this book will focus on the individual physiology, this physiology is nested and ensphered by a set of relationships to family, community, society, ecology, and civilization, that if not addressed will leave us incomplete. The ways that modern people are ill are not merely *individual.* They are as much a question of sociology and ecology as physiology. One of the profound challenges in the world of trauma healing is a lack of understanding of the degree to which our experience of harm is not merely individual, but rather social, collective, ecological, and civilizational. As Mary Watkins and Helene Shulman write so clearly in *Toward Psychologies of Liberation*, "We have decontextualized suffering from the context of community and ecology."

We modern people are certainly traumatized as individuals, but we are ensphered by alienation. We live in an alienated context. The breakdown of relating in our bodies mirrors the breakdown of relating in our contexts. If we do not go all the way to the taproot from which alienation springs, our healing

---

1 Right angles emerged in Babylonian land ownership as the traditional implements of Divinity and Sovereignty (the rod and ring) made their way into the hands of surveyors to establish private property boundaries, sometimes between 3500 and 1900 BCE. I have written about this in the essay *Wrong Angles: Inverting the Logic of Enclosure (Part Two)*, in the book *So, About that Death Cult you Joined? Essays unmasking Empire in Modernity.*

will always be partial. It *cannot* be complete. What this means is very important. Because this alienation has played out in the realm of relationships between people, and relationships with the natural world, and in the deeper foundations of our civilization, it must be addressed in each of these contexts.

We have to figure out how to heal social trauma, which is the failure of the community to keep all of its members safe. Where we fail to do this, the mental illnesses of social supremacy take root: male supremacy, white supremacy, etc. And because this social trauma is a breakdown of community, it cannot be healed individually: it has to be healed in community.

We must also address ecological trauma, which is the alienation of humans from the Living World, of which we are a part. Where we fail to do this, we do not experience ourselves as part of the Living World, and therefore we destroy Her. When we do not dig out these taproots of supremacy at a civilizational level, all of our institutions and structures replicate supremacy. All our religions, our institutions, our nations, our organizations spring from domination mind. This leads to the perennial generation of solutions that replace older institutions but replicate their basic thoughtforms of domination. The 'cure' recreates the disease.

My own interest in this topic was personal before it was professional. At seven years of age, I was removed from a deep context of wellbeing. The story of my life has been an inquiry into how to heal from this original displacement. Without over- or under-emphasizing my own story, something about the story mimics the shape of humanity's alienation from its own source context. It is fractal to the larger pattern, which is that in the infancy of humanity we experienced a pervasive and universal connectedness, which was then denatured by a traumatic event from which we have not individually or collectively healed. Looking across all of these domains of alienation–from the individual and the personal, to the social and collective, to the ecological and nature-based– we can see a common injury as well as a common cure. The injury is to our ability to feel relatedness. The cure is the restoration of this felt sense of connection.

To anyone whom we are connected, we share a mind. The San people of the Kalahari, one of the source cultures of humanity, explain this as follows: *if I see an individual bird, and I recognize it as an individual, the moment that happens a tiny energetic thread of relatedness forms between us.* Each time we recognize one another, that thread of relatedness thickens, growing into a cord, then a rope. The San say that, *To be San means to create ropes of relatedness with all of the Creation.*

When I first heard this story, I thought it was a metaphor. Yet it struck me deeply because it had never occurred to me that I could recognize an individual bird. At that point, my recognition of birds was occurring at the species-level, and I felt pretty satisfied with myself if I recognized a California Towhee, or a dark-eyed Junco. One day, after hearing this story about the San, my daughter and I were standing in a garden near our house watching a hummingbird.

He was very territorial, and was fighting off other hummers. Ounce for ounce, hummingbirds are among the most aggressive creatures on earth: a convincing argument about the dangers of eating sugar all day long. He would fly straight up in the air a hundred feet, zip downward at full velocity, flick his tail with a whirring smack, sing full-throated vigorous songs for awhile, and then zip about chasing off other hummingbirds. And each time he did this he returned to the same tiny branch on a tree. He did this five or six times in succession until my daughter and I were both trailing him with our eyes. And then, singing from the branch, he turned toward us, and my daughter and I both saw that there was a patch of feathers missing from his facial ruff on the left side of his throat. In this moment, this hummingbird with a bald patch in his beard became an individual. And in the instant this happened–the very moment

we recognized it–the bird turned to my daughter and I for a timeless moment and stared into our eyes without blinking until both of us felt an electrical shock. We turned, stunned, and looked at each other. It was like a tiny circuit had completed and a spark leapt across it. We had both felt it. What the San were describing was real. It is not a metaphor.

For the month that followed, we visited Flurry (which is what we named the bird) each day. I began to dream hummingbird. Not only did the bird enter into my dreams, but there were moments when I felt what it was like to be a hummingbird. I know this sounds odd. Yet there were moments when I could feel in my own body what it was like to be able to scratch the back of the head with your claw. I could feel the way he nestled into sleep, where he tucked his sword of a bill. I could feel the velocity–which was ordinary to the bird–of his swift knifing through the air. The motional freedom of his explosive vertical takeoff. Wow.

This is related to something else that the San explain, yet something that you know already in your ancestral awareness. With any *thing* we have built a cord of relatedness, we share a mind. (Any *one* really, for relating happens with subjects not objects, *whos* not *whats*.) And this mind is in fact felt: it is the mind of feeling, the mind of the heart (kokoro). Relatedness is not something that lives inside of us, but in the space between. It is anchored to us yet lives between. It resides not in our heads, but in our bodies. It is a connection phenomena. It is the same phenomenon that happens when you think about someone you love and call them and they tell you they were thinking about you.

I imagine that as we dreamed Flurry, Flurry dreamed us. Perhaps he studied geometric shapes with my daughter, or drifted through trails of word and image as I wrote poetry, tripping over my verbal acrobatics as I tripped across his motional. We opened one another's minds, became one another's teachers and friends. Through this thread of relatedness, we enter one another's worlds. And each time we recognized each other, this bond grew stronger. I had never had a hummingbird as a friend before, but after that initial connection, I would have defended Flurry, protected him, done many of the things for him that I would do for any true friend.

Think about that for a moment. I made a friend who weighed maybe ten grams. Five pennies weight, let's say. At a hundred and eighty pounds, I have approximately eight thousand times his mass. Yet I would have used all of my size to keep him safe, had the need arisen. I would have drawn my sword for him, if necessary. Imagine what would happen in the world if most of us experienced ourselves as related to the Living World at this level, if we felt connected enough to life to defend her like this. The destruction would halt RIGHT FUCKING NOW.

Flurry was my first bird teacher. Since then I have had many teachers who were not humans, including many trees, many plants, and many animals. You can say that this sounds New Age, but this is the original way of relating. It is actually extremely *old age*. As in, ancient. *Old school*, the kids would say. This is what happened before we fell out of relationship with All That Is.

Underneath cognition, there is a universal language of feeling that all of Nature speaks, except most of us modern humans. And we can re-learn how to speak this language. The doorway to this language is *kinship*. What we have a relationship with we feel connected to. What we feel connected to we care about. What we care about we take care of. What we don't, we don't. If we want to live in a world where people take care of themselves, one another, and the Living World, we have to figure out how to get connected to ourselves, one another, and the Living World. The person who told me the story of the San is a connection phenomenologist. And it was from him who I learned who I am professionally.

My lifetime study has been the art and science of connection and relatedness and the removal of what gets in the way. I have been doing this for nearly thirty years. Unsurprisingly, many of our mentors and advisors are Indigenous, or indigenously-trained, because the ways of knowing that are focused on connection and relatedness were abandoned by modern Western culture several thousand years ago. Descartes' pronouncement – *I think therefore I am*- rather etches it in stone. Modernity chose to live in the world of domination and abstraction. We abandoned the relational and the felt. Our civilization simply has not valued teaching the intuitions of relatedness at all. We are terrible at it. Yet by the Indigenous world of earth-based people, these ways of knowing were never abandoned. Without this embodied knowledge humanity will not survive.

This brings me then to the deeper question of what is healing, and what are we healing from? And I would propose to you, again, that this is simple but not easy. Healing is to become, as the San suggest, related to everything.

Healing is to become related to everything.

Our work is focused on helping modern humans develop the proper intuitions of relatedness. We are healing from *separation.*

This process generally has two wings. One of them is focused on the transformation of trauma, and the other is focused on the growing of connection. They happen as twin and parallel processes, looping back and forth.

Many of the people that come to our work do so for healing, but the container in which healing happens is our capacity for relatedness. You can think of these as yin and yang, two halves of one whole. In this book we will learn the deep language of relatedness through the lens of the Autonomic Nervous System, which is the neural architecture of the mindbody connection.

But it is important for you to understand, at the outset, that our purview here is to heal relatedness. The reason that modern people are unwell is not merely inside us, but the breakdown between us. The neurobiology of connection that restores wellbeing isn't *my* neurobiology–it unfurls in the space between us. *Interpersonal,* to use Dan Seigel's frame, it is also more-than-human. And so the learning we do about the individual physiology is in service to re-weaving the fabric of connection that has been dismantled by alienation.

To begin this work, we are going to have to dismantle the death carapace in which your soft body is encased by multi-generational trauma. Let's get started.

CHAPTER 05

SOIL YOU WERE BORN INTO

The human lineage history goes back somewhere between two hundred thousand and eight million years.[1] For 95 to 99.9% of this history, depending on how far you believe it goes back, we lived in small band hunter gatherer groups. Kin groups. Small nomadic bands.

Take a moment and consider the human infant. Think about one of us at three or four months of age, hairless little monkeys that we are, and consider how hard it is to keep us alive. How profoundly vulnerable we are. Modern parents, of which I am one, spend considerable energy and resources keeping our infants alive, and we have refrigerators, climate control, and vaccines. We have Whole Foods and department stores. We can order diapers. We don't have to burn fires to keep our caves warm, or find all the calories required to keep us alive each day. We can close the door at night and the other apex predators stay outside the door.[2] We can buy baby food in a can if we choose: we don't have to pre-chew it and feed it to them unless we want to. If you wanted to design a species to survive easily, you would not design a human. You would probably design a cockroach.

We are at birth not even fully cooked. We are born when we

1 There are several Indigenous cultural groups whose members have told me that their histories go back several hundred thousand years. This has been documented by western scientific methods with regards to the aboriginal peoples of the Kalahari (San), though I don't think western scientific methods are the only ways to validate this. I have been told this, specifically, by other Indigenous peoples as well, although the contexts in which this cultural knowlege was shared lead me to believe it was not intended to be disseminated publicly. Neanderthal (Homo sapiens neanderthalensis) bones have been discovered that modern science suggests date to 430,000 years ago, and the split from modern humans postulated to have occurred between 315,000 and 800,000 years ago. Yet hominids–our relatives in deep time, go back likely about 8 million years, if we look at the beginnings of the bipedal stance. As a relevant aside, in 2011 Stanford University geneticists concluded that the San people of the Kalahari were the ancestors of us all. I loved this notion, particularly in the way it undercuts received notions of racial supremacy, which I began to share with people in our networks, until a Lakota Elder politely took me aside and explained that the Lakota did not come from the San, at which point I became curious about the intricacies of the Stanford study. What seems more likely is that the ancestors of European-derived peoples are the San. I do not believe the geneticists tested San DNA against Lakota DNA.

2 Unless of course they live *in* the house, but that's a different problem.

are because our heads are so large that if pregnancy lasted longer, we would not be able to pass through the birth canal.

Human infants aren't fully cooked until about 18 months after we are born. As evidence for this, I submit to you that a baby's skull doesn't fully close until that point.[3] You have windows in the top of your head until you are a year-and-a-half old. Ancestrally, we spent the first nine to eighteen months in a sort of external womb, what Dr. Ashley Montagu calls 'a womb with a view.'[4] We were held, skin-to-skin, on the body of a caregiver, often the mother. Keeping us alive was not easy. It is somewhat miraculous that we are here at all. We lived in kin groups because we needed a community to survive.

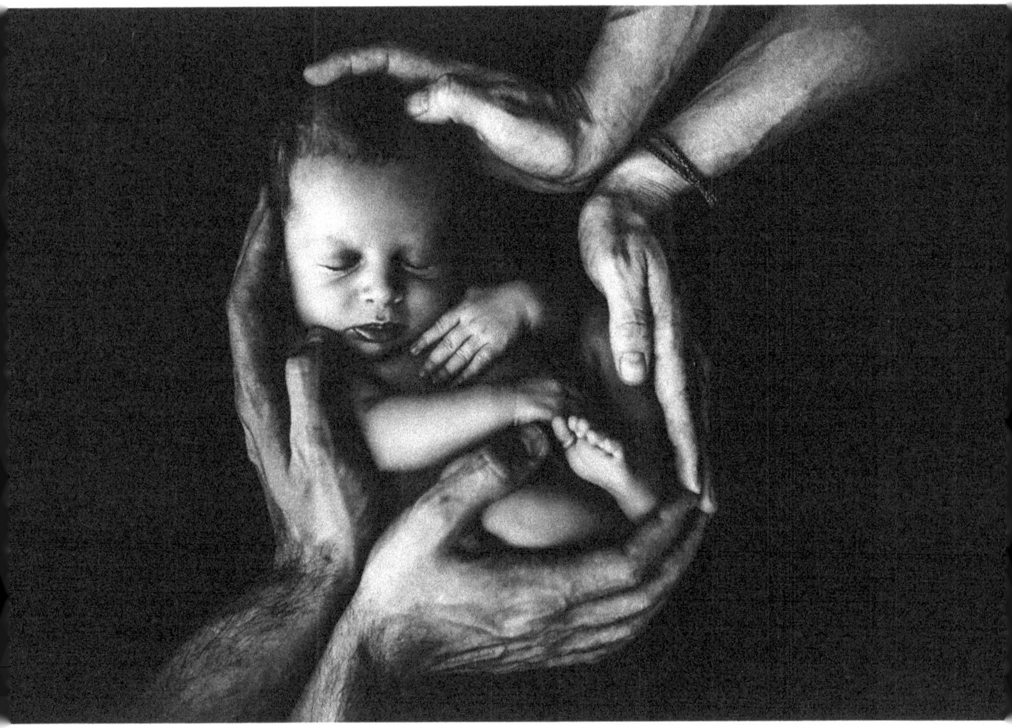

3 You are born with two skylights in your skull. They are called fontanelles. The posterior fontanelle usually closes within 2 months after birth; the anterior between 7 and 18 months. These structures are part of what allows baby's head to narrow slightly in the birth canal. They also accomodate baby's growing brain.

4 Dr. Ashley Montagu, from *Touching: The Human Significance of the Skin*

Although our ancestors in deep time experienced extreme physical stress, and most of us could probably not endure the amount of physical effort that this ancestral life required (our ancestors were basically all extreme endurance athletes), they experienced almost none of the social stress that is a defining feature of modernity. Rather their lives were intensely social and collaborative. Watch a troop of bonobos moving through a forest, and you have a sense from our simian cousins of what our ancestral pattern was probably like. Our ancestral origins are of deep sociality. In deep time we did not experience the social stresses of loneliness, isolation, or disconnection.

I have no way of knowing this, but doesn't it look for all the world like this group of bonobos is going to study the beauty of that waterlily?

Why are we modern people experiencing this level of social stress? Why has this kin-group/ small band/ collaborative lifestyle broken down so completely? Why are we so defended? So ill? So alone?

In order to answer this question, we looked briefly in the previous chapter at the last 12,000 years of human culture to understand the deviation from our ancestral baseline in safety and connection that has led us to where we are. When we bring a long evolutionarily lens to this inquiry, understanding our social mammalian lineage, we can see that human thriving is a function of a certain kind of safety, and a certain kind of relatedness. I don't mean *biological* relatedness primarily: I mean *social* relatedness. This was the context in which our brains and nervous systems developed. Yet the structure of modernity, a cultural scaffolding shaped by normalized modes of domination: war, empire, forced migration, displacement, colonization, oppression, and economic uncertainty has systematically undermined our multi-generational experience of both safety and relatedness.

In order to thrive, each species must provide its young with a nest. Our friend and mentor, Darcia Narvaez, PhD, one of the world's leading trans-disciplinary psychology researchers, calls this the Evolved Nest. It is the extra-genetic (which means cultural) context designed to give rise to thriving in the young of a given species. Most modern people live lives where their deep nervous systems are receiving consistent cues of threat, rather than safety. At the neurobiological level, cues of threat activate defensive, rather than connective states. Nearly all of us are birds who have grown up in a faulty nest. A nest surrounded by predators– sometimes a nest with predators inside it.

If you feel in danger, you don't seek to open your heart and connect. Rather you defend yourself or shut down. We weren't supposed to grow up in nests where we had to do this. The nests are supposed to be safe. That is our naive biological expectation, in the most innocent and beautiful sense of the word *naive*. As infants we are so vulnerable, so porous, so permeable, and it is therefore heartbreaking that so many of us are growing up in nests that do not meet our most primal needs for safety and connection. Think of how much more beautiful both we, and the world, could become if these needs were met.

When these needs are met, the children flourish. When the

infant-mother dyad is centered and uplifted by society, everyone flourishes. Morality proliferates.

When this is not so, and these needs are not met, and not met from a very early age (when we are most permeable to experience), the contortions we have to make to survive can get very deeply etched into our neurobiology. In the next chapters, I am going to compare our neural development to an inward forest, and the individual neuron to a tree. Think about a tree in the forest that doesn't have the right conditions to flourish. Perhaps the sunlight is too dim because there are no gaps in the canopy. Maybe the soil is too rocky or degraded. If the tree is to survive, it may have to contort itself, growing around stones, twisting and turning toward the nearest approximation of light. If we've never experienced, in our nest, the pure light of the Sun, we may have learned, deeply, to grow toward its closest approximation, even if that was not Sacred Light. If we've grown up under flourescents, the pure light of the Sun can be so overwhelming we are not able to receive it.

Our deepest biological imperative is to survive. Sometimes the context we arrive into is replete with pure resources, the safety of a nest: all of the things that we need to meet our deepest needs, to become who we are meant to be. As in the forest, in this alchemy of species and conditions, this leads to deepest thriving: the ancestral baseline I am speaking about. And in such conditions of nesting, we may become this version of our ourselves: the sunlit magnificence of our fully realized self, like the resplendent tree on the previous page.

And yet for many, for most of us, this is not the kind of environment, inner or outer, into which we are born. Rather, we find ourselves in landscapes of resource depletion or constraint. Resources and time for us spread thin. Like trees born into rocky soil, sculpted by intense wind, sun that is not available or is too harsh. We are born to survive, and so we do, and yet the battle of this sculpts us. We have to twist and turn to endure the conditions. Here is a thousand-year-old Bristlecone Pine,

and she is magnificent, and she has survived. Yet the harshness of the environment is revealed in the twisting of her body. Her fight for survival is visible. She has grown from rock. There are no leaves and she provides little shelter for the next generations.

We are both of these trees, you and I. We live somewhere on the continuum between them. Our context is the nourishment, or the constraint, in which our neurobiology, like these trees, grows. And this context, for us moderns, is shaped by this civilizational-level lineage history, this long evolutionary deviation from our ancestral baseline in safety and connection, then by our society, our community, our family of origin, and our own innate biology's response to all the above. We are ensphered by nested sets of contexts, providing or depriving of us nourishment. These create the contexts that support or constrain our flourishing.

I once heard a story about a rose bush in a garden. She was enormous, and healthy, and she grew in a garden sheltered by a ring of trees on a hillside with very strong prevailing winds. For some reason I do not recall, the tree directly sheltering this bush from the wind had to be removed, and the rose found herself abruptly in the line of a steady current of wind. Here the storyteller paused to describe the beauty of the bush, and then she said that she watched and marveled as the bush grew a line of thorns from the ground to her crown that faced directly into the wind. Imagine! A single line of spines, facing like a knife into the assault. She was defending herself. It takes my breath away, this story, this intelligence of life.

We are this rose bush. Deprived of the nest, we have grown these thorns. The gardener then, noting the effort expended by the rosebush, had a small wall erected to shelter her. A week later when the wall was in place and the winds held at bay, the thorns began to drop off, until they were gone, and the skin of her green and new again, with no trace of them remaining. If we can re-create the nest, we can drop the thorns. That rosebush is us.

We are ensphered by contexts. We are born into them, *en medias res*, actors who step out onto a stage in the middle of the play, inheritors of contexts, conflicts, histories and situations that don't belong to us, ancestral traumas of displacement and destruction, strong prevailing winds swirling around us, whipping through our hair, impacting upon us, and we have to, or we get to, as we become more aware, discern what is nourishment and what is poison. On the stage we step onto there is sunlight and wind. We have to figure out which is which. Most of us have been bottle-fed poison, from within the nest we grew up in. Where we expected, opened to, and anticipated receiving the sweet nectar of pure love and safety, most of us were fed something else. And our biology has grown around it. Now we have to unlearn this.

This is the trauma scene. So here we find ourselves, reading this, adults in all likelihood, so many experiences that we have already passed through, finding that our bodies move into, or reside in, defensive states, which are, in fact, very sensible and evolutionarily adaptive responses to danger. When under threat we defend: we don't relate. Of course. Of course. Under attack we grow armor.

What I am here to help you understand is that residing in defense is incompatible with well- being.

The neurobiology of defense is fundamentally incompatible with wellbeing.

Evolutionarily, our stress responses were designed to come on rapidly in moments of extreme duress, and then quickly resolve. These systems are not designed for the kind of chronic, toxic, and traumatic stress that characterizes modern life. The systems of defense are designed to respond to acute moments. Yet for many of us, where this distress does not resolve, we stay locked into the neurobiology of conflict, avoidance, or retraction: socially unavailable either with overtones of anger, fear, or shutdown. We remain in the biology of defense: bellies subtly tightened in anticipation of the next micro-aggression. Fists subtly clenched in anticipation of having to swallow down the next insult or assault to our dignity or our pocketbooks. Pre-resenting our boss because we already know the outcome of the meeting we have been 'invited' to attend. Breath shallow or sped up, because we are harried animals, animals under duress, running on the modern rat wheel...job, mortgage, kids in school, bills to pay...animals under assault. Capitalism, nationalism, the politics of grievance– all of these feed our defenses.

While we don't have to make any particular effort to get stressed out, coming out of the stress response requires skill and discernment, and it requires novel neural inputs to help us shift back across the thresholds of threat response, into the biology of connection again. We have to change the context. We have to update what is feeding the system...Ancestrally these novel and restorative inputs usually came from other trusted humans and the Living World, in which we were deeply embedded. Take a moment to consider this. Connection phenomenologist, Jon Young, who has deeply studied the ancestral inputs required to activate the biology of connection, asserts that human children need twenty-five story-catchers to thrive optimally. That means twenty-five people you trust, and with whom you can share your 'story of the day' – the thing that happened today that touched your heart most deeply.

In order to share such a vulnerable story with someone, you have to trust that they can receive it. That in the telling they will not de-nature it. Think about what you look for when you tell someone a story like this. You want to be listened to, deeply. You want to be witnessed. We don't tell a story like this for someone else to solve. How could they? We don't tell it to be judged.

How many people in your life do you trust to catch your stories? Are there twenty-five? Do you know anyone, a single person, who has twenty-five storycatchers in their life?

And then this thing with the Living World. Ancestrally - deep ancestrally- we were outside 100% of the time. Which is to say that it is only in the past several hundred years that our materials (building) technology has become sophisticated enough to seal out the outdoors. For at least 99% of human evolutionary history we were outside nearly all of the time. Let's take a very conservative estimate, and say that the average percentage of time outside over the entire span of human history was 93%. (It's not- it was more like 98%, but let's just say.) In modernity, this has been completely inverted. Research in the United States, at present, says modern Americans are outside 7% of the time. We have completely inverted the ancestral context of wellbeing. No wonder we are stuck in defensive states.

We need to change the context.

No wonder most modern people do not understand how to make this transit out of distress. It is not something we are taught at home, or in school, or at work. It is not something that we see modelled around us, and almost no one we know is living it. Even though the template for it is somewhat neurophysiologically uniform, the art of restorative practice has been largely lost. What this means is that more and more people find themselves in non-resolving states of distress that they don't know how to exit. If the particular way that the body responds to threat is by going into flight then people find themselves experiencing chronic worry, anxiety, or overwhelming fear. This becomes their center of gravity to such an extent that they begin to *identify* with this state. If the body responds to threat by going into the fight response, people find themselves in states of chronic non-resolving irritation, anger, or rage. If they spend too much time here, again, it becomes how they identify themselves. And if the body goes into a response to threat where it shifts into shutdown, people find themselves in chronic states of depression, numbness, and withdrawal. And each of these chronic distress states is the doorway to a particular set of diseases.

When we can't find any relationship between our actions and their consequences, if we cannot identify the patterns driving our shifts into defense, or figure out how to exit them, we lose motivation and hope. We fall into learned helplessness. We turn to coping strategies that may help in the short run, but cultivate unhealthy dependencies. We might engage in extreme risk-taking behaviors, abuse alcohol, food, substances, sex, or shopping, ourselves, or other people. We might go down the innumerable rabbit holes presented by the internet: gaming, gambling, social media, pornography. If we cannot find our way out of the maze, we may lose the hope required to carry on, or we get carried into the politics of grievance where we require an 'Other' to pin all of our problems on.

Do you know anyone who is anxious, angry, or depressed?

Have you experienced, or do you experience these states yourself? If so, it is important to realize that each of these responses is, in fact, an adaptive response to feeling chronically unsafe. And if these states are familiar to you, as they are to most modern people, this book is an exit manual for you, to help you shift the baseline from which you are operating. To help shift it in the direction of safety and connection, which are the roots of flourishing.

The Autonomic Nervous System is your primal operating system. It is beneath culture, beneath language, and universal. Buried deep in the primal architecture of your body, guts, heart, and mind, these are the deepest layers of your nervous system. What we'll learn together is a deep map of these systems. We'll learn how to bring ourselves out of these defensive responses, back toward safety and connection, by understanding the needs driving these states, and the kinds of new inputs required to shift them. And we'll learn the sequence of doors that you'll have to open in yourself on the transformational journey to enduring wellbeing.

In doing this work you will gain an understanding of a part of yourself that is deep beneath thinking. A more ancient, ancestral and feeling layer of awareness that is beneath your ordinary sense of self: a part of you that speaks to you through the language of sensation, emotion, and the felt sense. This deep primal layer is the part of you that makes decisions about whether you are safe, in danger, or under lifethreat. As we get to know this part of ourselves better you will learn a new language for communicating with these deeper parts of your system. You will find this both extremely useful and also enjoyable because we are talking about learning to have more direct contact with non-cognitive ways of knowing that many people think of as intuition. This will bring you closer to your deepest self.

You are also likely to find yourself grieving quite a bit on this journey. This grief is important: it is in fact Sacred Grief. In Brasilian Portuguese there is a word, *saudade*, that is difficult to translate into English, but it is related to a Welsh word, *hiraeth*, which is a kind of bone-deep longing for something that was before and is no longer. Saudade might be translated as soul-deep yearning, the longing for someone, or someplace. On one of my trips to Brazil, I was brought to my knees by a scene in the São Paulo airport. A teenage girl was leaving her family, whom she obviously adored. I don't know if she was traveling for college or something else. They were standing together, in a tight knot, and the daughter was weeping unconsolably. Through this veil of tears, came, again and again, the sentence: *Já tenho saudade de vocês.*

I *already* have the longing for you. She hadn't left yet, and she already had the grief of longing. What struck me most deeply, on reflection, was the open-heartedness that allowed her to be cracked open, with abandon, by this grief born of love. She fully experienced the heartbreak of separation. This was such a teaching for me, accustomed as I was to glossing over partings. Because at that point, I had not fully grieved my own brokenheartedness at being torn out of my own childhood place.

She didn't try to stop the tears, this young woman, probably couldn't have. She just felt her way through the grief. She surrendered to it. I had to catch a flight before she did, so I didn't see this episode end, but I imagine she got on the plane. Those tears were a way of honoring the love. Notice that she didn't fall apart. She fell together. She fell into them, her family. Into love. She grieved into love.

We are terrified of grieving because we think we will fall apart. But that is because we have forgotten how to fall together. The grief and the love–they are the same thing. And so as you read this, and what it means strikes you, and as you understand the birthright that has been stolen from you by harms, let yourself grieve. It is part of the healing. Yet don't grieve alone. In this way, rather than falling apart, we can fall together. Find storycatchers to hold this grief with you.

You will come through it, and out the other side. You cannot go around it, so stop trying. And as you really understand how your deep nervous system works, and begin to move in the direction of home, which is the direction of your awakened heart, and the version of yourself that emerges from safety, you will slowly begin to find that you have more choice, more agency in how you respond, more sovereignty, clearer boundaries, a stronger YES and a clearer NO, and that when you do get shifted into danger, which leads to anxiety or anger, or into lifethreat, which leads to shutdown, depression, and numbness, you are no longer helpless to respond, or unclear about what is happening and why, but can locate yourself on this map and find your way back home. In fact, rather than these states simply overtaking you, as you begin to learn the terrain of this map you'll recognize the states earlier, steps along the pathways down to them, and so you'll be able to catch yourself sooner and you'll be able to harvest the wisdom of these states–because they do contain wisdom.

The science of this domain, which is the neurophysiology of the Autonomic Nervous System, is generally taught in specialist domains. I often teach it in a trans-disciplinary post-doctoral seminar to wellness professionals. But this work must not stay confined to the academy, because this work is about our deepest humanness. You don't need a degree to understand this. You simply need to be able to tune into the experience of your life. Each of us has a nervous system. You need to become the expert of your own. You need to be able to trust your own knowing of it more than you trust anyone else's.

My primary mentor in autonomic physiology is Dr. Stephen Porges, who is the Developer of the Polyvagal Theory, which is the most important revisioning of our understanding of the Autonomic Nervous System in the past 50 years. Steve is a beautiful, brilliant, and humble man deeply concerned with human and planetary wellbeing, and he walks his talk. I met him after studying his work for about ten years when I reached out to him to ask if we could rebuild his website. Once we got to know each other, I asked him if I could attempt to make a poster of the Polyvagal Theory. You see, by that point I was teaching it to thousands of wellness professionals a year, and because it is sophisticated I was creating illustrations to teach it more effectively. Steve laughed, told me they had been trying to do this for about seven years, and agreed to let us try. I thought it would take three weeks to make the poster. It ended up taking nine months, and we went through twenty-one versions before completing the project and launching the Official Polyvagal Poster in 2020.

Along the path to finishing the poster, I dove deeper into autonomic physiology than I had ever imagined that I would go, and a number of things we discovered surprised us, as well as Steve. One of these is that the neuro-anatomy diagrams of the Vagus that are taught in medical school are not accurate. More on this in the next chapter. Much about the way that the medical and mental health professions have been taught to understand the vagal systems is not accurate, and much of it actually obscures our understanding of what these systems really do.

Over the last several years, a part of my role in the global Polyvagal community has been to continually push us in the direction of simplifying our explanations so that everyday people can understand the theory. So I've spent a great deal of time figuring out how to teach the theory through metaphors and analogies that render it more experientially vivid. I feel that this is important because everyone is born with a nervous system, but we are not born with a manual about how it works.
Over the past several years, our clinical work and research moved beyond the forward edge of what Polyvagal Theory has been able to explain, and our clinical work with thousands of patients and clients showed us that some of the Polyvagal conceptualization was not accurate. At this point our work has moved beyond Polyvagal Theory into a novel, yet more ancestral realm: I conceptualize the field as Autonomics: the study of your living deep nervous system.

I am preparing now to take you into your deep nervous system, which is in reality an inward forest. Let us set out now on our journey, our sacks well provisioned, companioned by storycatchers, eyes open, hearts expectant. I am preparing to take you into the primal forest, the inward forest, the heart and belly of your neurology.

CHAPTER 06

ALTAR OF THE SPIRIT

If you could figure out how to plug your body back into the Source context, you could experience it as an infinite well that would fill your body with all the vitality that you require. Your body, as we have been told by many traditions, is made in the image of the Divine. Yet this image, what we imagine this to mean, has been de-natured by millenia of Empire infecting us with the thoughtforms of domination. The divine is neither some benevolent bearded man, nor simply the esoteric body of Jesus, but Nature Herself. The Divine is the cosmic inhalation and exhalation of the Universe, of which you are a part, but from which you have become un-tethered through alienation.

The reason you have become un-tethered, the reason that modern people have become un-tethered, is because our umbilical cord of relatedness to the Living World has been severed by a deviation from our ancestral baseline in kinship with all of Life. Severing this cord of relatedness with the Living World led to severing the cord of relatedness with one another, which led to severing the cord of relatedness with ourselves.[1]

The purpose and purview of this book, and our work, is to restore this felt lived experience of kinship, so that we can come become what some of our Indigenous mentors call 'real human beings'. When you experience this kinship with yourself, you take care of yourself. When you experience it with others, you treat them as you would treat yourself. When you experience it with the Living World, you treat it with the same care you would treat your own body. (A quick secret: it IS your own body. Not your body in the private, enclosed, self-centered way that we usually understand this, but your greater body in the sense that as you begin to develop real relatedness with the Living World you extend into it, through threads and then cords and then ropes of connection.)

1 It seems conceivable that the order of this could have been different in individual cases, e..g, first the relatedness with others was shattered, e.g., through war, then with Nature, then with Self, etc. Perhaps the primary take-home is that shattering relatedness in one domain transfers to other domains.

Everything that we are connected to is part of us. Many people have told us so, including Jesus, Buddha, the Zen Master, Thich Nhat Hanh, and the American Transcendentalist poet Walt Whitman, who did so in this manner:

There was a child went forth every day,
And the first object he looked upon and received with wonder or pity or love or dread, that object he became,
And that object became part of him for the day or a certain part of the day or for many
years or stretching cycles of years... [2]

And the thing that both quantum mechanics, and spirituality has been trying to tell you is that everything is connected to you. We are woven through the Universe. And yet there is a problem, a paradox, because most people do not experience this reality. They might hear about it, or even believe it, even accept it as conceptually, or foundationally, or definitively true: yet this is not how it feels to them.

How they feel is isolated and alone.

And the reason that this experience - *isolation and loneliness* - is happening for most people is that the neural architecture of connection is shut down. Most modern people spend most of their lives defended against relationship because their bodies are experiencing threat, and when we feel threatened we are not open. When we are contracted, defended, and shut down, we do not FEEL woven through the Universe.

One way to understand your Autonomic Nervous System, in terms of its potentials to reconnect you to a greater experience of wholeness, is that the Autonomic Nervous System is the altar of the spirit.

The altar of the spirit.

---

2 *There was a child went forth every day*, Walt Whitman: 1819 –1892

If I could not use the language of spirituality, and there are times in my work when I cannot, because a lot of what I've done for the past fifteen years is train mental health and medical professionals, some of whom would set their hair on fire and run screaming from the room if I used the word spirit[3], I would say that the Autonomic Nervous System is the architecture of the mindbody connection. And this is true. It is the neuro-electrical grid that unites the felt experience of the body with your sense-making and meaning-making apparatus: your storying ability.

Yet if we can open to an animistic framework, and acknowledge that life is a spark, a vitality, we see something else emerge which is so much more profound. In the Tzutuj'il Mayan language of Santiago Atitlan, in what is now Guatemala, the word for life - to be alive - is *N Kasea*. This word means: *the ember yet burns*.

Life, vitality, is this mysterious flame. The difference between us and a corpse is this animating vitality. You can call it N Kasea, or you can call it spirit, or qi, or prana, or 'the breath of life'.[4] All mystical traditions share this awareness. If we think about our essence as this flame, this burning ember, the Autonomic Nervous System is the altar upon which it meets our experience. The neural interface between spirit and matter.

An altar.

Having a body is deeply strange. We act like we know what the nervous system is. Medicine acts like it knows what the

---

3 I remember a conversation with the Training Director of a large child-serving clinical organization that was paying my team to train their clinical staff in tracking Autonomic states. She became apoplectic when I suggested that we use federal dollars to train their staff in Bird Language. At this point we had already been using federal dollars to train their staff in bird language for several years, we just hadn't described the trainings that way.

4 In cranial osteopathy, as far back as the 1940s, Dr. William Sutherland, DO, called it "the Breath of Life".

system is. But I'm here to tell you that we don't. Medicine doesn't. One of the primary conundrums in the Universe is the relationship between spirit and matter. Einstein articulated it thus: $E = MC^2$

Energy is related to mass across the square of the speed of light, the limit velocity in Universe. So be it. But your nervous system is what makes this equation *tangible*. It is where you experience this miraculous metabolic exchange between energy and matter.

Your nervous system is the translator between energy and matter, wave and particle, between vibration and action. The lacelines of the nerves unite the realms of the ethereal with the structure of the concrete. They translate the full continuum from the vibration of thought to the contraction of muscle.

And your nervous system is much stranger and more magnificent than you, or most of the medical professionals who purport to work on it, understand. This is true partly because of faulty comprehension, and it is true also because most of the way that modern medicine understands the nervous system has come from studying dead bodies (cadavers), and it is true because the gaze that was looking at these dead bodies was deeply de-natured by domination.

The neuro-anatomical diagrams that have been used to train most of the Western world's physicians are profoundly inaccurate. And because they are inaccurate they have prevented those practitioners, and the population at large, from gaining an accurate understanding of what your Autonomic Nervous System actually is, or what it actually does.

I know this for a number of reasons, and one of them is that I worked with the world's leading expert on the relationship between the Autonomic Nervous System and behavior to create a neuro-anatomically accurate map of the Autonomic Nervous System to illustrate his theory, which is called the Polyvagal Theory.

This was in 2019. At the time I had been teaching about the Autonomic Nervous System, and utilizing Dr. Porges' Polyvagal Theory, for nearly a decade. I had trained several thousand wellness professionals in the theory. And because it was complex, and detailed, I kept making illustrations to help me explain it to them. I had had the good fortune to be introduced to Dr. Porges, after studying his work for nearly a decade, because we gifted him a new website. When we had designed this to his satisfaction I asked him if we could work on an illustration of the theory together. And he said, *Yes*.

We had already put together a visual design team, including a medical illustrator with the apt name of Leonardo. My vision for the poster was that it would have, at its center, an illustration of the human body showing the neurology of the three Autonomic circuits that Dr. Porges' work illuminates. I wanted them color-coded for ease of understanding.

Porges' theory teaches us that who we are depends on whether or not we feel safe in our bodies. In the present moment, our bodies and minds re-tune depending on whether we feel safe, in danger, or under life-threat.

Dr. Porges was the first person to isolate the specific functional (physiological) configurations of neural circuitry undergirding each of these responses– the first person to recognize that there are two distinct neural circuits undergirding our threat response: one that activates when we feel in danger (Movement System), and another distinct circuit governing experiences of lifethreat (Grounding System). And what I wanted to do in the diagram was show people how these circuits were anatomically distinct, because they flow into different parts of our bodies.

When we feel safe enough to open to connection, the neural circuits that connect to our hearts light up our faces, voices, and the palms of our hands.

When we are in danger, the spinally-mediated circuits connected to our hearts light up our legs so we can run away, or our arms so we can fight. Sometimes both.

And when we shift into lifethreat, when we think we are going to die, we end up immobilized, and this circuitry is centered largely in our guts.

I had read many papers Porges had written about these systems (he has authored more than 400 of them in peer-review journals across many disciplines). And I had been scouring the literature for diagrams and medical illustrations that showed the neural architecture of these systems. We felt pretty confident about our illustrations of the connection circuitry, and the fight/flight system. But the diagrams of the deep belly Grounding System we were encountering varied wildly. Some of them made it look like this system was very modest: a few wispy nerves. Others branched in unexpected ways that looked more like netting: a basket woven through the guts.

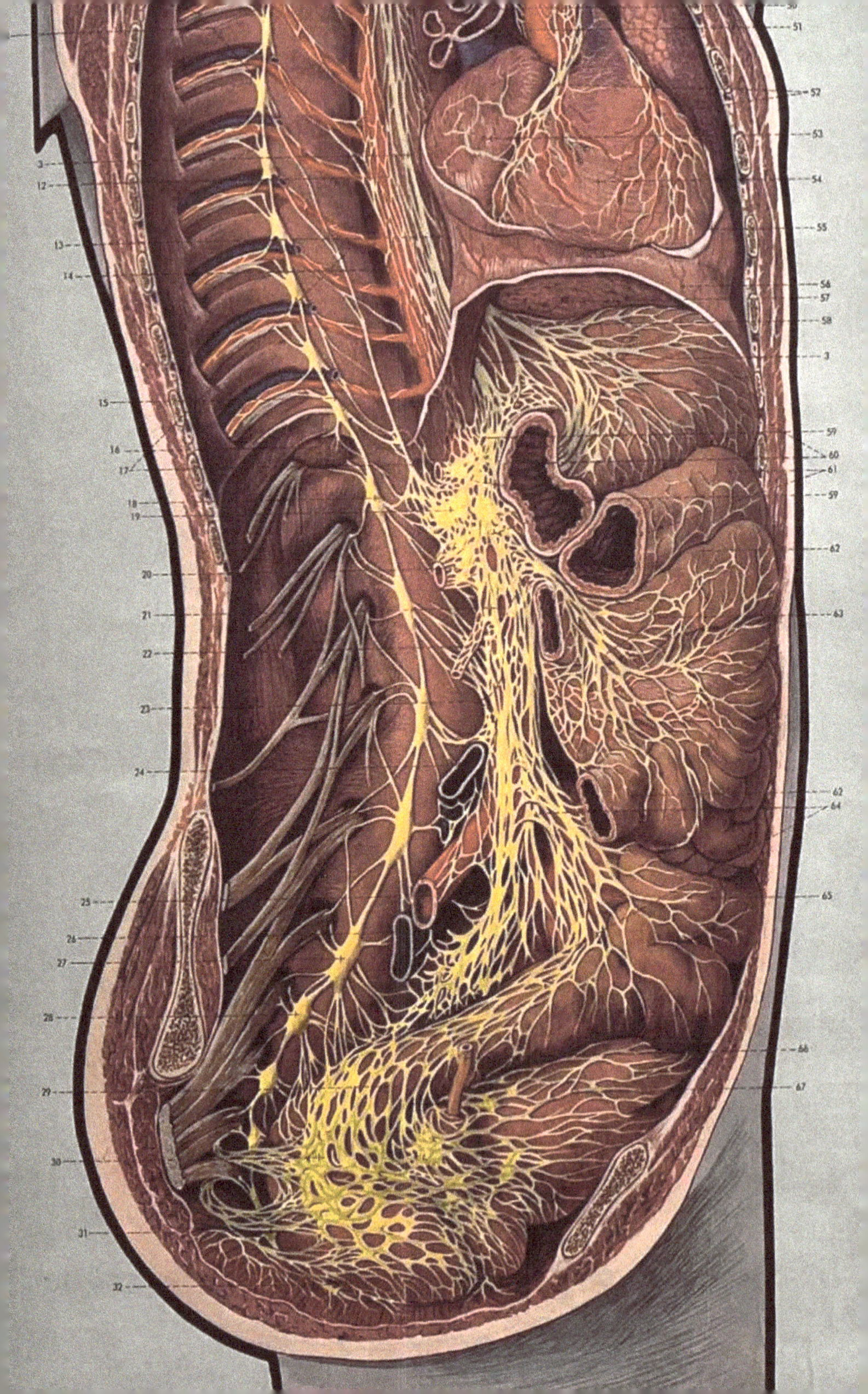
3
12
13
14
15
16
17
18
19
20
21
22
23
24
25
26
27
28
29
30
31
32
51
52
53
54
55
56
57
58
3
59
60
61
59
62
63
62
64
65
66
67

Because of my own meditation background, and my own trauma history, I had both very precise interoceptive awareness, and was familiar with being in a shutdown state. And when I shifted slightly into shutdown, I was aware that the places that I felt uncomfortable were part of this system. (If we are able to feel something inside the body, it is because there is a sensory receptor there, which is connected to a nerve.) So I knew from my own embodied experience that the diagrams with the least number of neural circuits were not accurate.

One day I was receiving bodywork from a gifted manual therapist trained in neurophysiology and chiropractic. He had his hands on my belly, just underneath the bottom of my ribcage on the left side. Because he knew of my interest in the Autonomic Nervous System, he was speaking to me as he touched my body.

"And now I'm palpating your celiac plexus," he said, "which is a portal to the dorsal vagal (Grounding) System." I grabbed his hand and sat up. "You just said what?" I asked him.
Taken aback, he repeated what he had said.
"Where did you learn that?" I asked him.
We sat there for a minute, and he thought about it. Finally he said it was from the French Osteopath Jean-Pierre Barral, the developer of Visceral Manipulation, and from Traditional Chinese medicine.

I contacted Porges. We had been checking in regularly about the poster and the diagram of the Grounding System.
"What about the celiac plexus?" I asked him.
"The what?"
"The celiac plexus." Silence.
"You know," I said, "the portal to the dorsal vagal (Grounding) System in the abdomen."

To be fair to Dr. Porges, who is a pioneer, a godfather of this autonomic renaissance, he didn't have access to accurate diagrams either. The people who knew about this part of the

Autonomic Nervous System *were not physicians*. Once I started looking for the celiac plexus in medical illustrations, I found it pretty quickly. But no one I talked to seemed to know that it was part of the Grounding System, except for people who had studied a particular branch of osteopathy. Neurology understood the *anatomy*, but not the *physiology*. The *structure*, but not *how it works*.

This whole experience helped me realize that the diagrams I was seeing, even in anatomy textbooks, were schematics masquerading as illustrations. And even when the pathways were correctly illustrated, the texts didn't explain *what kind of information was passing through them*. This is a pattern I have seen repeat dozens of times: even where the neuro-anatomy is understood, its functional role in the Autonomic Nervous System often is not.

Another example. Every square inch of your skin has nine meters of c-fibers, which are tiny sensory receptors connected to ultra-fine nerves. It is impossible to accurately illustrate this. The nerve fibers are much finer than the threads in luxury sheets. Your skin contains a microscopic mesh of these fibers. They are *way too slender* to see with the naked eye. A super high-resolution illustration of the c-fibers in the skin looks like, wait for it...the skin. There are so many fibers that illustrations showing all the tiny fibers would just be solid. A neural wetsuit called skin. Yet this is the exactly the point.

Think about a tree, such as this majestic oak. If you are teaching students tree anatomy, you start with the trunk and the major branches. Yet the reality of being with an oak is that the branches ramify into finer and finer, tinier and tinier branches until eventually they leaf out. Your nervous system is a great deal like this tree. Your animate anatomy.

With the Autonomic Nervous System, the wandering trunk of the tree is the Vagus. If I am teaching you the anatomy of the autonomic nervous system, and especially if I am studying it in

dead people, I am going to be looking at the tree in winter. So I teach you the major branches of the Vagus.

But the reality of standing next to the oak tree in summer is that you can't see any of the trunk or the main forking branches, because they are ensheathed in the living fabric of leaves.

Similarly, in a living body, the Vagus is ensheathed in the living fabric of all the tinier and tinier branches that end in receptors that comprise your *living* nervous system. The altar of your spirit. Your nerves don't end in branches. They end in leaves. The leaves of your nervous system are receptors. And these receptors are feeding the main trunks and branches of this system. They are gathering the inputs that will be relayed back to the trunk lines.

It wasn't until I uncovered a French surgical atlas from the 1850s that I began to find illustrations that began to approach accuracy. So here is the point of this chapter: let me bring this home. *The Vagus is not a nerve. A nerve does not end. A nerve branches into finer and finer cords, threads, and filaments until it terminates in receptors.* Think of this as trunk to branch to leaves.

Your Vagus nerve is not a telephone wire running through your body. The entire way we've been taught to think about nerves is mundane, faulty, and misses the point.

Your vagal system is more like cobwebbing penetrating, ensheathing, filling the visceral space of the entire interior of your body. It is a cloud of sensory and interoceptive awareness, coterminous with the interior space of your body. It has to be like this because your Autonomic Nervous System is the listening system (surveillance system) for the interior of your body. It has to inter-penetrate the entire interior of your body, so that you can feel the entire interior of your body.[5]

---

5 Nerves bring their own blood supply. So your nervous system brings your circulatory system along, and together they inter-penetrate the entire interior of your body.

The Vagus has to inter-penetrate the entire interior of your body, so that you can FEEL the entire interior of your body.

I have studied neuroscience for more than 30 years. My first really deep dive into neurobiology took place when I enrolled in neurobiology at Yale University, with a remarkable professor named Haig Keshishian. And what we learned, which is the standard model of the neuron taught in all university courses around the world, began with the individual unit of the neuron, which was illustrated like this:

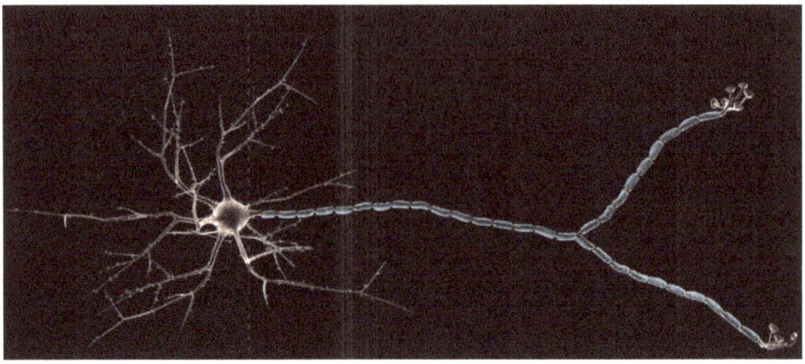

You are probably familiar with this neuron– its dendrites, skinny axon sheathed in myelin (the blue insulation), and its terminations.

Yet there are two problems with this template neuron. One of them is an epistemological problem, in that it deals with how we know what we know. The idea operating underneath studying the nervous system in this way is that if you break a system down into its simplest parts– atoms, or molecules, or in this case neurons– you will understand the way that they combine into systems. This is the same kind of thinking that broke into the bodies of cadavers with blades, parsing apart the tissues. It is a logic of dissection, of divide and conquer. It is an imperial thoughtform that roots back to Rome.

This might be properly called your neuro-vascular system.

At nineteen years old, I didn't know that I would spend the next thirty years of my life studying neurobiology. I was a freshman in college. But the way that I was taught set in my mind a pattern that I then had to spend decades unlearning. Because understanding the fundamental unit of the nervous system as a neuron tells you no more about the nervous system than understanding binary code tells you about how computers actually work, or understanding bricks tells you about building pyramids. Starting with the fundamental building block of a complex system tells you almost nothing about the system.

Not that it's untrue– the isolated individual unit of the nervous system *is* the neuron. Rather– it puts the emphasis completely in the wrong place. If you want to understand systems, study systems. Because the point of your nervous system, the very nature of it, is that there is no such thing as an individual neuron in the body.

The specific quality that makes a neuron a neuron is that it is connected to other neurons. It is a micro-unit of connection. It has either a receptor or another neuron feeding into it, and it feeds other neurons. It is part of a root or part of a branch. A cell in a root or branch, through which information, in the form of an electrical signal, will flow. The point of a neuron, the purpose of a neuron, does not reside in itself.

The purpose of a neuron does not reside in itself.

A neuron is a link in a chain of meaning-making: an input/output system. The purpose of a neuron is the meaning-making that passes through it, not the link itself. This is very important. Don't get me wrong, surgeons know that if you sever a nerve you have a real problem. But as long as nerves are intact, the point is what passes through them.

If you want to understand the nervous system in a functional way, don't look at the individual unit. Look at the whole system. So this is the first problem, the epistemological problem with how neurobiology was taught, and what it then, logically, obscures about the nervous system as a whole system.

The second problem is that most neurons in the human body look almost nothing like the neuron that we were taught neurons look like. There is no such thing as the template neuron we were shown.

Morphological variation of human neurons

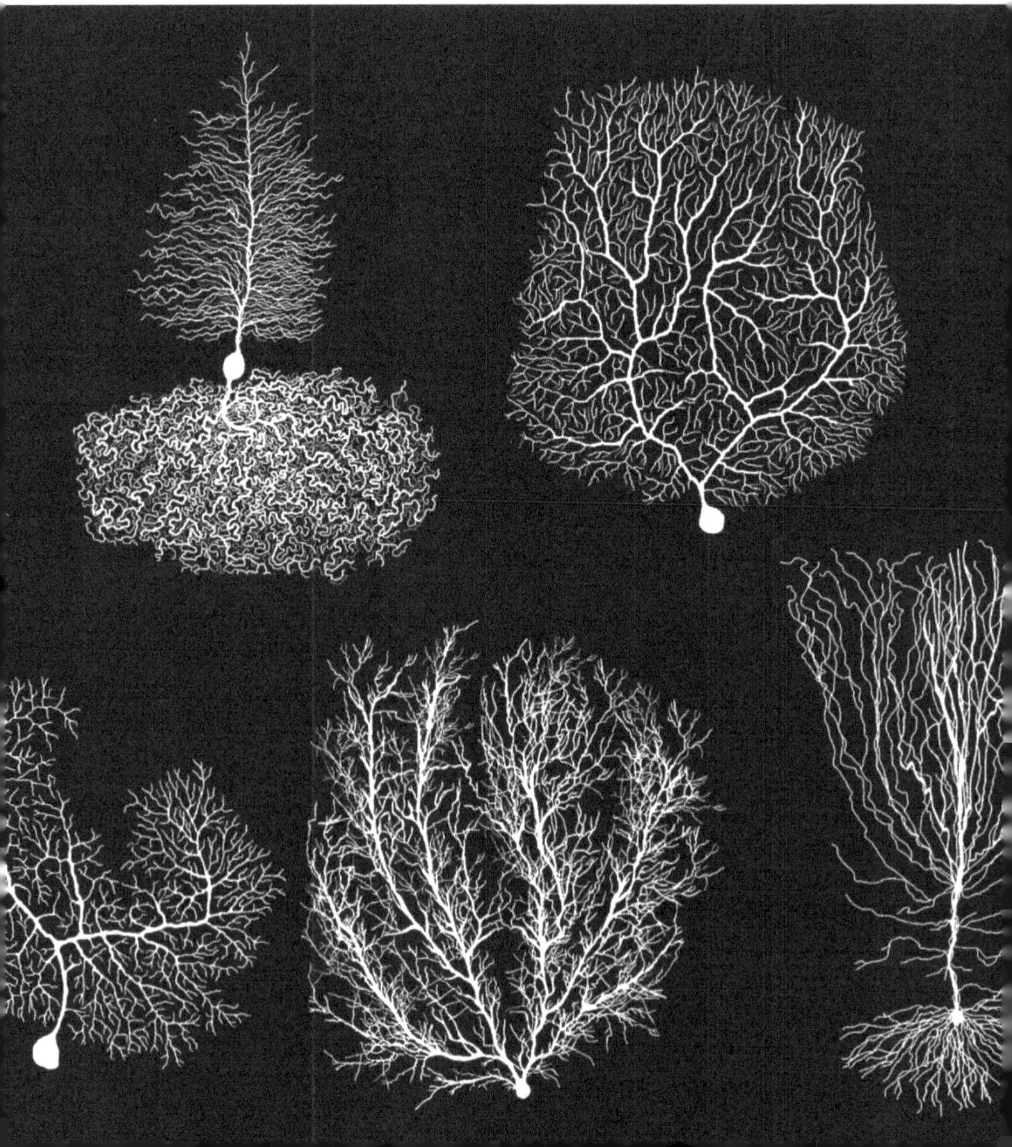

In a way this is a stranger problem. Who picked that template neuron that doesn't exist? Is it again a schematic rather than a representational example? The neuron that you learn about doesn't exist in the body. This is not to say that there are not long skinny neurons with a single axon in the body. There are indeed. But if you drew them to scale, most of them are much less than the thinness of a human hair, and between inches and feet long, so you couldn't draw them to scale in a textbook. What proliferate in the body, in reality, are a vast morphology of neural types.

They all have cell bodies, a nucleus, dendrites, axons, but they are as wildly varied in shape and size as the plants in a meadow and a forest. Just like plants they have roots and leaves. Parts for taking information in, and elements for sharing it. But take a look at these forms. Do they not look like trees?

What is inside of you is not simply a set of telephone wires. Although we could not help but metaphor ourselves into this view because we are storytelling creatures and we work with analogy, and this was the communications technology present when these discoveries were being made. What resides inside of you, the composition of the altar of your spirit, is in point of morphological fact, an inward forest. It is many orders of magnitude more sophisticated than a telephone network. It is an entire inward ecology.

If you stained the neurons of the body green, subtracted all the other bones and tissues and membranes, and enlarged the entire system so each neuron was the size of a tree, you would find yourself in a vast aboriginal inward forest. This is the inner ecology of your felt interior.

And if we extended this metaphor, we would see that this inward forest is nourished by an extraordinary system of waterways, an inward ocean, rivers, streams, and creeks: your circulatory system, nourishing and being nourished by the forest, and bathing it in relevant neurohormones. Together they create a neuro-vascular system: the inward ecology of forest and stream that is the way your body communicates with itself neurally and hormonally.

Maybe– and I say maybe– it is necessary to simplify things when we are teaching them so that people will understand. But again, this is a very Western approach to knowledge. When your child is learning to walk, you don't put her in a room with only right angles because learning to walk across sand, gravel, carpet, up stairs, down hills, through water, over obstacles is difficult. You hold her hand. You don't constrain her to a world with three colors because the millions of colors the world presents are too nuanced to name. You teach her the full spectrum. At least I hope you do. You help her differentiate it, and name it.

Instead of starting with neurobiology, with the study of the individual neural cell, I wish I'd started with a class called *The Altar of the Spirit: What is your Nervous System, Why is it Important, and How Does it Work.*

I wish someone had taught me that I contained an aboriginal forest of nerves. I wish someone had taken the time or had the insight to introduce me to the strange and mysterious beauty of this original inheritance, wet-wired into each one of us, twinned with a circulatory system of comparably intricate beauty (an inward ocean) that together comprise a neuro-vascular field co-terminous with our interior, generating an electro-magnetic field that extends through the fabric of the matter we call a body and beyond.

If that had happened perhaps I would have come to neuroscience as a young child, and found in it an inward mirror of the Living World, into which I could gaze in order to glimpse my True Nature.

Had I done so, I would perhaps have come to the same conclusion as the Indigenous Lakota peoples of the Great Plains, one of whose words for people, *čhantę* means *the moving trees*. And in this way we see that one of the oldest cultures on earth has foretold the future of our medicine.

CHAPTER 07

Santiago Ramón y Cajals purchased his first microscope in 1877. It was acquired with money he had saved serving as a medical officer in the Spanish army.

Lens-grinding technology evolved greatly in Europe during the 1830s, with a number of lens manufacturers making breakthroughs that were the patented and proprietary manufacturing advances of their era, and by the time Ramón y Cajals, who would become the father of neuroscience, purchased his scope in 1877, the precision of the optics of these lenses was so great that a world, previously invisible because it was too minute to be accessed by the human visual system, could be opened.

We are, when Ramón y Cajals buys his microscope, one hundred and twelve years forward from Robert Hooke's first drawings of cork, a hundred years from the dawn of cell theory. Beneath the lenses a strange and alien world unveils itself, deeply divergent from the forms we see at visible scale. As this world opens, our minds must expand to encompass these new revelations of the real. Strange creatures step into the field of view, yet they are not chimerical, not fantasies, not fictions. Things we can barely see are magnified into magnificence and horror.

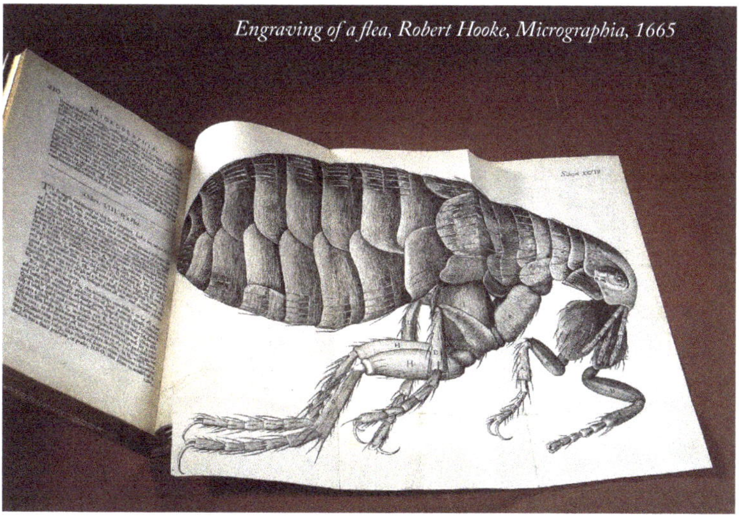

Engraving of a flea, Robert Hooke, Micrographia, 1665

Think of the microscope as a portal opening on an alien dimension: the strange new world of the tiny. The magnification transports us into another range in the lifescale. Later instrumentation– electron microscopy, eventually particle accelerators–will open to vision successively exponentially more minute worlds: that of the molecule, that of the atom. Yet for now, in the 1870s, we content ourselves with the opening of a new biological frontier. The fundamental unit wholes of life are rendered visible. The cell becomes sensible to us visually. We perceive the building blocks of life: plant cells, animal cells. What a revelation! What glory! What jawdropping beauty in how Nature assembles her symphony of magnificence. Can you not hear the angels singing? These are mysteries heretofore unwitnessed by humans.

Technology extends our senses, amplifies our capabilities. In the case of first the telescope, and then the microscope, what is amplified is our *seeing*. And we moderns are, to an extent that is ancestrally anomalous, deeply identified with our metaphors of sight. We use the term clairvoyance, with its etymology firmly rooted in vision, to speak of perceptive insight, clear seeing, the ability to penetrate the surface of the real. We talk about really *seeing* one another. About being *seen*. Yet this is careless language, in actuality. Merely approximate. What we are really discussing is clear perception. Lucid awareness. It is not sight, but something deeper, more penetrative, more encompassing. Something for which vision is merely a metaphor. And yet, because we love to look, because we identify with the gaze, seeing becomes the metric of our sensing. We substitute the most dear of our five senses for the act of witnessing. Yet note that it could have been otherwise. In a culture that had refined its flavor detection, a culture of great gustatory achievement, we might have used the word *clear tasting* to denote this deep awareness. Dogs, with a sense of smell ten thousand times more acute than our own, probably call their shamans *clear smellers*.

He is an unusual person, this Ramón y Cajals. They will call him a peasant genius, and he will go on to win a Nobel Prize. From a rebellious boy, deeply anti-authoritarian, who is sent to juvenile hall at eleven years old for exploding his neighbor's gate with a home-made cannon, he will father the discipline of neuroscience, rendering visible, through his microscope, the fundamental cellular unit of neural structure: the neuron. He is the author of the Neuron Theory. A reporter gazing into the cellular depths and documenting his visions in illustrations so startlingly precise they are still used today.

Or so we are told. Or so is the orthodoxy: that he saw the neuron.

And yet I call bullshit on this.
Bullshit, bullshit, bullshit.

Cajals is an *artist*. Cajals is an *inventor*. Cajals is a *shaman*. He conjures the neuron, the tiny trees that comprise the nervous system, in the image of an inward forest. Cajals invents the texture of the inward landscape of neurology. He doesn't discover it. He invents it. And he renders it real for us, renders it visible through drawing-as-discovery. Through illustration-as-transmission. Through art-as-revelation. Renders it in such a way that we cannot unsee it, such is its power.

To situate Cajals correctly in the pantheon of disciplinary histories we must add and claim another set of lineages: the artist. Cajals is the cousin of Cézanne, the nephew of Rilke, the brother of Rodin. His medium is not paint, not words, not stone, that is true. His medium is the inward landscape of the body. Yet as Cézanne does not paint still lifes, does not paint mountains, does not paint objects, but subjects, paints not things but the act of seeing them– Cajals is not painting what he saw through the miscroscope but bringing us into his listening through seeing.

Cajals does not see the neuron through his microscope. He is not looking with this eyes, he is listening with them. He says so himself. This is the same man who once blew up his neighbor's fence. Don't expect him to do less with your mind.

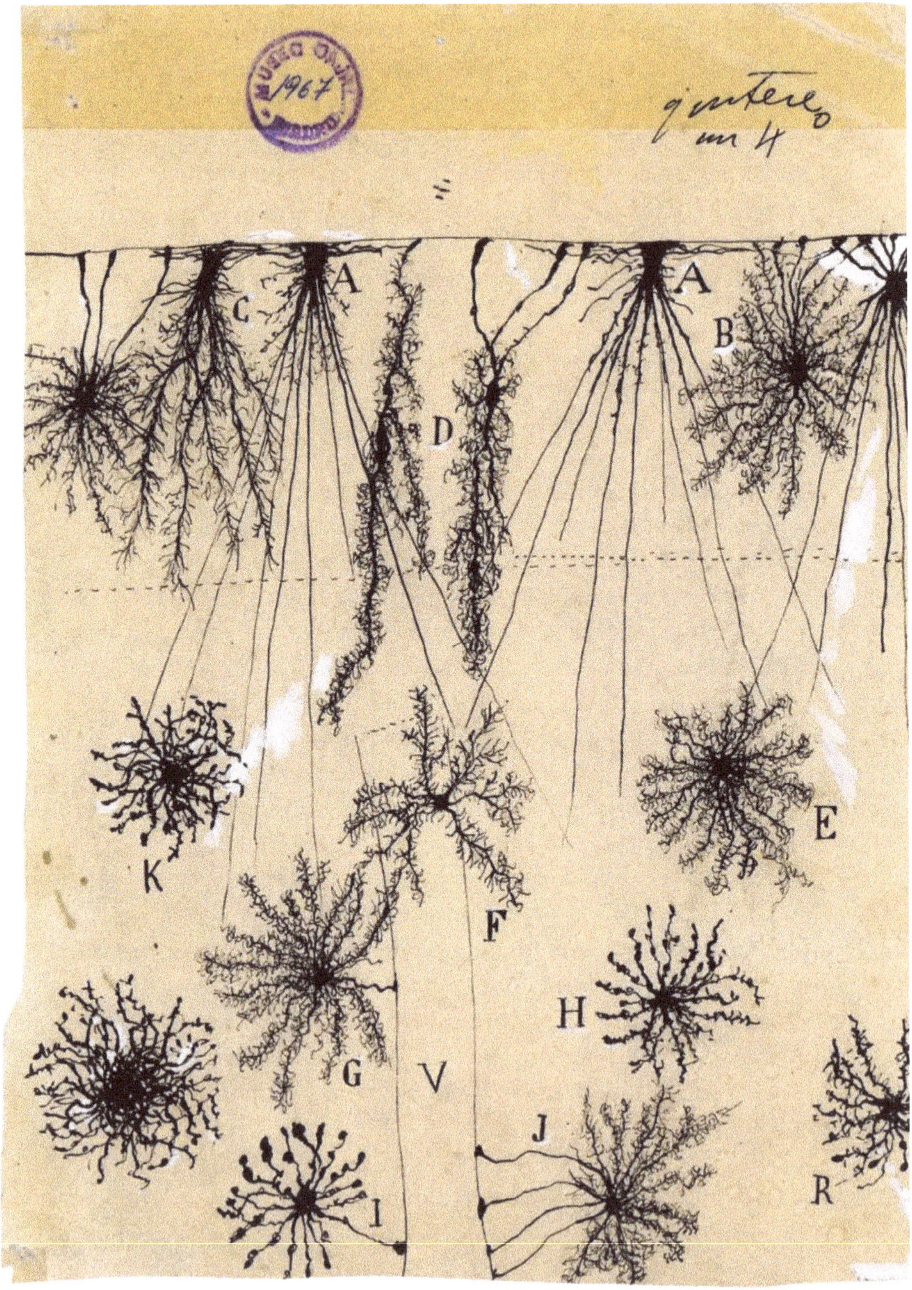

As he would write in his poetic autobiography, *Recollections of my Life*,

> *I finally chose the cautious path of histology, the way of tranquil enjoyments. I knew well that I should never be able to drive through such a narrow path [as microbiology] in a luxurious carriage; but I should feel myself happy in contemplating the captivating spectacle of minute life in my forgotten corner and listening, entranced, from the ocular of the microscope, to the hum of the restless beehive which we all have within us.*

Listening, entranced, from the ocular of the microscope. It is such a strange turn of phrase. Because listening is not something we ordinarily do with our eyes. So what does he mean? But you know what he means. Listening is a receptive sense. To listen, again a sort of sloppily-textured metaphor when we apply it beyond the acoustic realm, is to make ourselves available, receptive. Seeing *declares*. Listening *asks*. Seeing is a statement, listening is a question. Ordinarily. So to listen with the eyes?

Imagine a man walking through a forest, listening, with a sketchbook. A forest no one has ever seen before, ever visited. A forest in an unknown land. A forest filled with strange and unidentified species of *dendrons* – this is Greek for trees, don't you know? It will transmute into dendrites - treelike… Enter into the forest with the man, he is joyful, a bit mischievous, a bit of a disruptor, and he is deeply present. He is paying attention. Look up.

Can you see them, the neurons? Reaching all the way to the sky? The neural structure peaking through? Now take out your sketchbook...

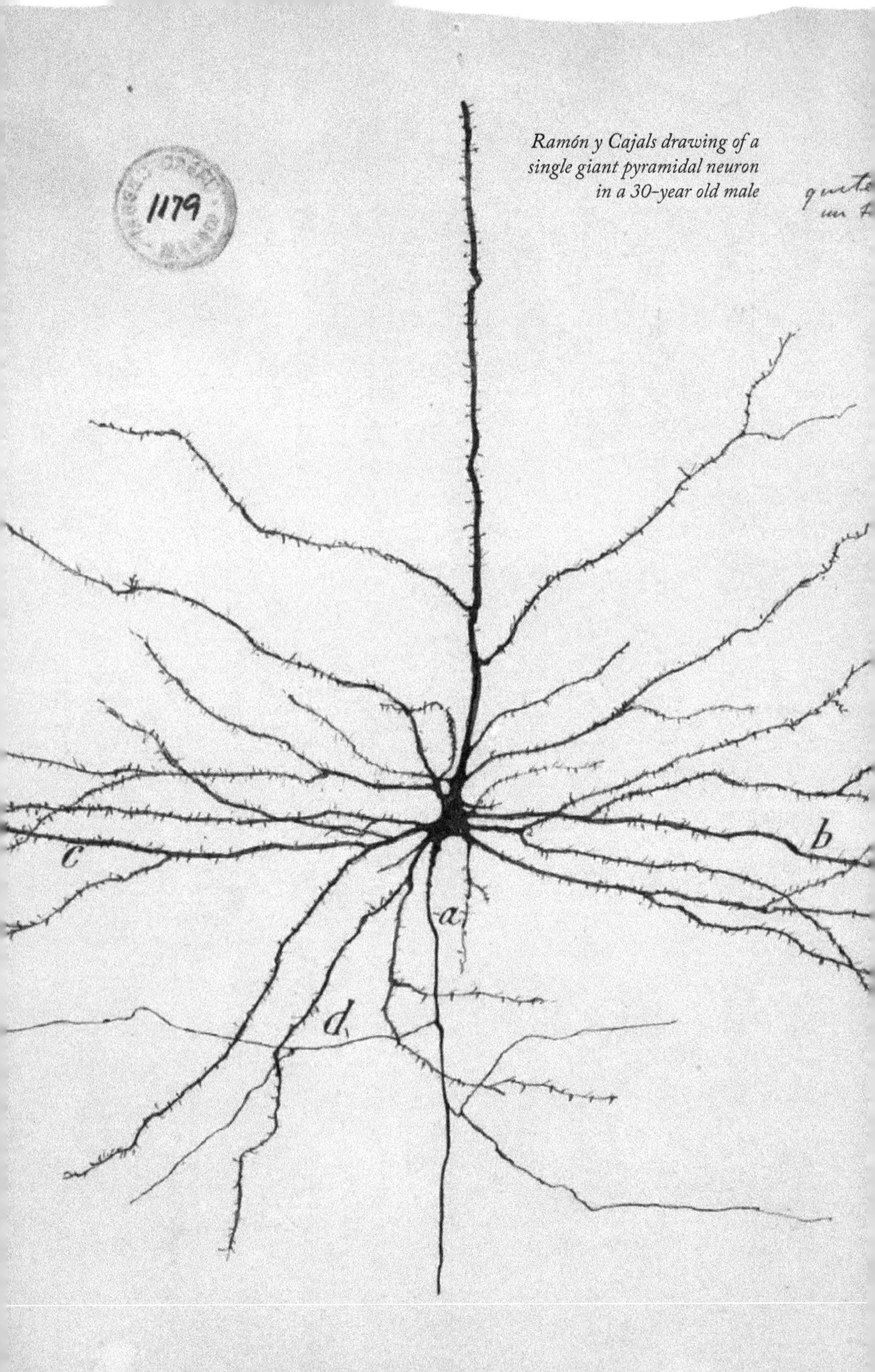

Ramón y Cajals drawing of a single giant pyramidal neuron in a 30-year old male

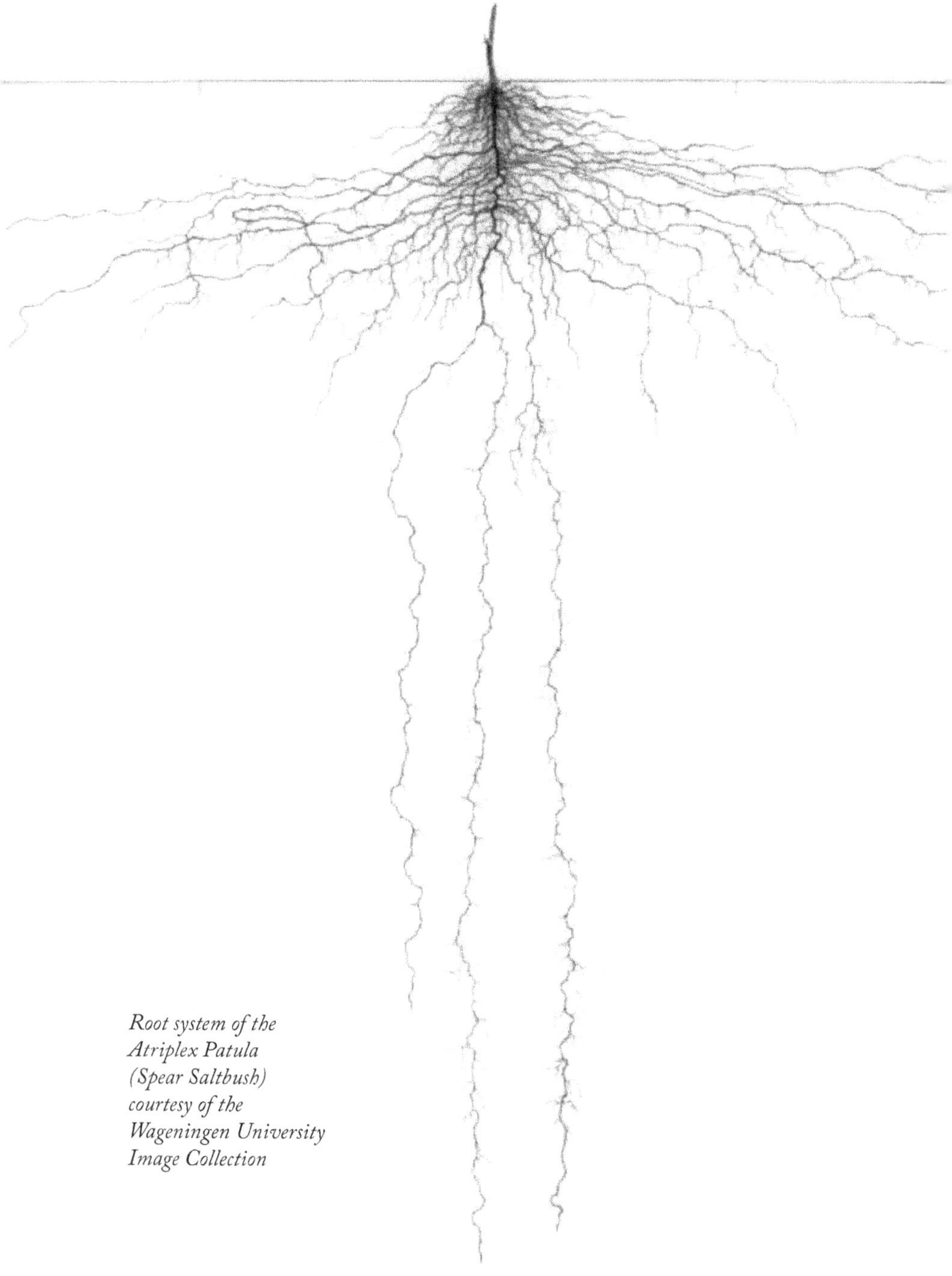

Root system of the
Atriplex Patula
(Spear Saltbush)
courtesy of the
Wageningen University
Image Collection

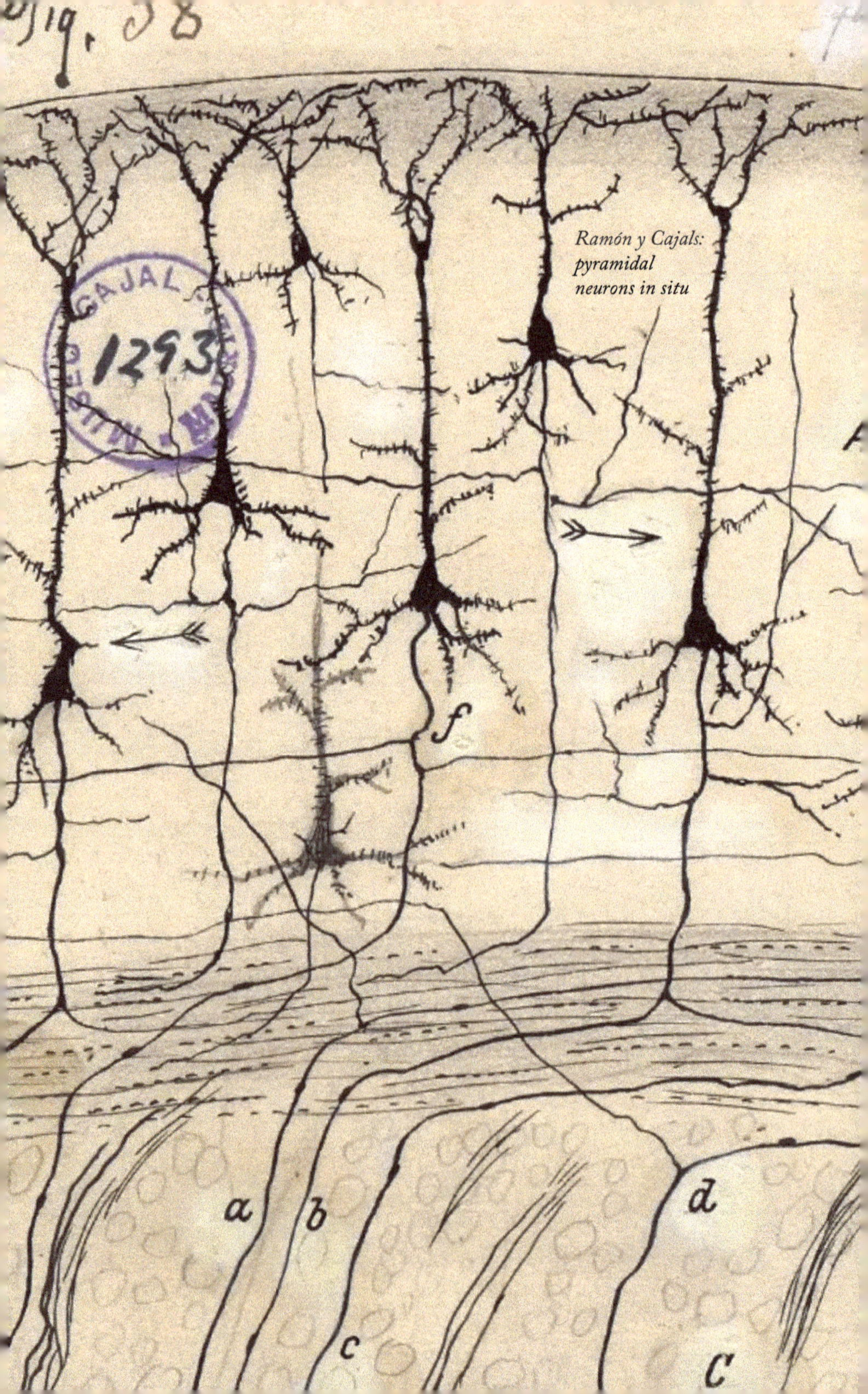

Ramón y Cajals: pyramidal neurons in situ

Remember though, that Ramón y Cajals, in this inward forest, listening with the ocular of the microscope extending his listening, is not looking at a three-dimensional forest. In order for the light to pass through the slides, these slices are razor-thin. He is only looking at sections. Only gazing at planes. And so he can only see the slices of trees arrayed upon the plane of the slides. And so he must infer the forest, deduce its dimensional structure, from these fragments. From these potsherds, these shatters of pottery, he must infer the vessels whole and intact. This is what people do not understand. Prior to the neuron doctrine, it was believed that the nervous system was an undifferentiated mass of material. A squiggle. A velvet electrical jelly? The brain a jellyfish encased in skull.

And in this manner his work is an act of construction. A synthesis, an invention. He is a temple builder. He builds the inward world. He sculpts the nervous system. He makes it out of pencil and gouache, conjures it into dimensionality. Evokes it for us. Details this world like Rodin polishing stone, with the finest penstrokes.

And yet, please recall as well–the staining technique–the chemistry whereby he has rendered this system visible, it is a reductionism. He has disappeared the supporting scaffolding of the circulatory system that always twins the neurology. Deleted fascia, muscle tissue, bone. He has disappeared all of this. In order to help you see the trees, he has vanished the soil, let us say, for simplicity. Do not forget that he is a magician. What you are seeing is not the inwardness of the body, but what an artist is showing you. It is something that he created. You are not seeing the intact holism of the body. You are seeing what Ramón y Cajals wants you to see.

I do not for one minute believe that Ramón y Cajals would be pleased about the way the discipline has evolved. I do not believe he would be delighted by the hegemony of rationality, the brain-centric and empirically buttoned-down nature of moden research in the field.

And I will whisper in your ear a secret, something that is going to get under your skin like a splinter, something that is going to get under your worldview, down into the soil of it, the deepest strata– may dislodge something at its roots. Ramón y Cajals was under the sway of an unconscious bias so deep, so unacknowledged, so invisibilized, that it corrupts the entire edifice of his work if we don't transplant the work in its entirety.

Cajals was under the illusion– the grand illusion of the last five hundred years of the modern world– that *I think, therefore I am*. Hush, keep this quiet. Finger to the lips. Let us not go around exploding other people's fences, the retaining walls of modern psyches. Keep this quiet.

He was under the illusion that identity resides in cognition, and that the organ of cognition- the cranial brain- is the sovereign seat of the nervous system. Cajals was a brain supremacist. Hush, don't tell anyone.

And here I call your attention, one hundred and fortyseven years later, to the utter fallacy of this, and the vast destruction in the wake of this foundation error gaping beneath the edifice of modern neuroscience.

I call your attention to the birth of the autonomic nervous system in the body, to the doors and windows of the house of your nervous system loomed miraculously through the entirety of your living skin. I return the nervous system to your whole body, trace its ethereal lacelines through the entire extension of your inwardness. For in order to feel yourself from inside- and you can do that, can you not?– in order to know yourself through the sixth sense that we call interoception, you most have neurons everywhere. To feel yourself on this three dimensional inwardly volumetric screen, your nervous system must be everywhere. It is the pixels that render this landscape sensible to you. Neurons the tiny bright lights that illuminate with a gut feeling, allow you to sense the curvature of your spine, the tingles in your wrist, the warmth in your heart…

I call the nervous system out of the head, out of the snailshell you were taught to believe it lived in, and back out into the deepest interstices and the fullest extension of your animal, and animate, body. I do nothing other than what Cajals did, listening clearly to the 'hum of the restless beehive' that we are.

I repurpose the entire edifice of Cajals now– I claim him, I consume him, I place his entire body of work upon the hearth of the original fire and ignite it. I immolate it in the name of liberation.

Watch the glorious drawings catch fire, go up in flames like tiny ornate cobwebs, immaculate mycelial cabechons buoyed skyward on the flame-lick dark, fireworks all, then watch them re-organize, watch them re-weave themselves upon the loom of the body entire, making you whole.

Turning your entire interior into an inward organ of perception, for that is what it is. The grandest of our senses, it should not be known as the sixth sense, but ground zero, for it precedes them all: interoception, invisible to the brain-centric gaze even as it fills our entire bodies.

I kiss you, visionary father of Neuroscience, as the student kisses the Master, and the Master kisses the student, and I make you my own. I claim you & your visionary knowing too. Your method– your epistemological framework, your non-cognitive ways of knowing, your listening through your eyes, your tasting through the hands, your smelling through the ears, your seeing through the sound? This I reclaim as well for all of us interoceptive trackers traveling together in the direction of home.

CHAPTER 08

INWARD FOREST

Have you ever set foot deep in virgin forest? There is something that happens to your senses deep in an aboriginal forest. It is hard to reduce it to science, yet there is profound science to it. Part of this science is that we come from the forest. In the dawns of time – the deep time of our species – we came down from the trees and moved onto the grasslands. Likely it was because of a change in climate. From out of the trees we moved onto the savannah, and shifted from climbing in trees to walking upright, at some point about eight million years ago, this shift from four legs to two transforming not simply our view of the landscape, but of time itself. And yet, if you get still enough in original forest, you can find again this ancient affinity with the source context. It is palpable. You can feel it.

I remember sitting in an ayahuasca ceremony in Brazil, one of the first times I drank the sacred tea there, and having a bone-deep realization that I am a forest creature. It came over me as a full-body awakening: a sort of home-coming. We are all forest creatures.

Part of it is the plant life around us. The magnificence of the trees themselves. I remember the feeling of forest on my first trip really deep into virgin forest in the Amazon, a place called Tefé, two hundred miles up the Rio Solimoes (Solomon's River). I was with a group of locals native to the place. Flying in on a propellor plane from Amazonas, the capital city, was to fly over an ocean of forest as far as the eye could see. In the clouds of vapor puffed above it, clinging to it in some places like a blanket, you could witness the forest exhaling, creating its own weather.

Walking through the forest, the light was jeweled, the trees an elemental catalog of monumental forms, the soil a reddish-brown and ochre chalk. Trees, vines, bromeliads, twinned, twined, entwined, married across and through one another. Mossgreen, forest green, emerald, aquamarine, turquoise, chartreuse, mint, pistachio, seafoam, lime, olive, fern green. Towering Castanheira trees spreading acres of limbs above us, clutching volleyball-sized packets of Brazil nuts. A kind of creeper fig, the Apui, a hybrid of vine and tree that grows down from seeds dropped by birds, latches onto a tree it falls upon, and grows around it, hugging it until it reaches the ground and dissolves its host, the downward shoots of its many trunks like stalactites in a cave, fringed with fuchsia tips hungry to drive fingers down into soil. Iridescent Blue Morpho butterflies the size of undulating dinner plates wafering through a metabolic breeze, seeming to ride the breath of the forest. The concatenation of monkey calls, birdcalls, frog calls, insect cacophony. Explosions of life all around.

One of my guides asked me if I knew what color the forest was. He asked this with a sparkle in his eyes, and I knew I was being set up. "Verde?" I said (green). His reply: "Verde d'orado." Green-gold. Or, said better, Green *of* Gold. Why gold? Because of the light. We don't have a word for this, in English, this majesty of sunlight filtered through the leaves of a forest, but the Japanese do: *komorebi*. This is something you can experience when you engage in *shinrin- yoku*, which is the Japanese art of forest bathing. Yet although they have a word for this as well, the art of forest bathing is not Japanese, but universal. We come from the forest. We are forest bathers all. Our consciousness as a species dawns in green-gold of forest.

And from this dawning, restoration resides in the forest because our visual system has the most nuanced differentiation of the color green. It is our deepest ancestral context. I would wager that whale eyes differentiate blues better than greens, no? Imagine yourself living in the forest several hundred thousand years ago. Your survival depends on your ability to notice the slightest change of color in the foliage that is the velvet black shoulder of that jaguar slipping toward you, dappled by shadow. You see not the shoulder, but the rippling shadow, slightly deeper green. Each species of tree, vine, bromeliad has leaves of a subtly different hue. There are therefore thousands of shades of green. Couple this with the light, at various times of day and different seasons and you have an infinite cascade of color variation. Staring into the forest in daytime is to gaze into an ocean of undulating greens, flecked with bursts of gold. Greengold is the primal homescreen.

The virgin forest, you will also notice, contains no right angles. Zero. Not a single one. With the exception of some crystals, right angles occur almost never in nature. Instead, nature is built on the fibonacci sequence, which is a set of proportions that drive the spiraling of tree branching, the shell construction of the nautilus, the arrangement of petals in the rose.

To step into virgin forest is to step into source code: the pattern-generating algorithm of the Living World. Every seedling grows into soil in relation to place. They grow not straight, in lines or rows, but spiral toward the light, reaching. The geometry of the forest then is a pattern language of aesthetic beauty: source context and source code. Every tree grows from somewhere (soil) to somewhere (the light). The forest is a visual symphony of yearning for the light. Trees, a living prayer concretized in matter.

Now think about a specific tree. Behold the majestic Douglas Fir, for example. Notice that nothing about him is symmetrical. He is a being shooting upward toward the roof of sky. His branches swirl and churn akimbo, seeking the light between the arms of his neighbors. Study it closely, open your eyes. Pay attention and you cannot help but grow more harmonized with the intelligence of the Source.

Akira Miyawaki, the Developer of the Miyawaki method of forest creation, says it so beautifully:

> ***The forest is the root of all life; it is the womb that revives our biological instincts, that deepens our intelligence and increases our sensitivity as human beings.***

Do the trees themselves not resemble neurons uncannily? Can you not see that they are made in the same design? The giant axons shooting up, the dendrites that branch off of them?

Santiago Ramón y Cajals, the visionary father of neuroscience, were he with us, would enjoin us to look at them closely. Want to understand your neurology? Go stand deep in an aboriginal forest and turn your eyes skyward. Want to know and feel the shape of neural structure, want to look inward to your deepest neurology? Eyes up. Behold the elegance of the body you have been gifted. As above, so below.

Douglas Fir

CHAPTER 09

ATTENTION IS THE LIGHT

Now, let's go a bit deeper into the neurology without leaving the forest behind because this forest is inside you, and because the way that the nervous system is understood in mainstream allopathic medicine has denatured a number of fundamental understandings that we will require if we are truly to make sense of ourselves, to understand who we are, and what we are doing here.

I have studied neuroscience for more than thirty years. I have studied with some of the world's leading neurophysiologists, psychophysiologists, neurobiologists, and emotion researchers. And in thirty years of study I have never heard someone point out that neurons grow from somewhere, toward somewhere.

Like a tree seeking light, a neuron, when it develops, grows from a source in a direction. Most of what we are looking at, as clinicians, as wellness professionals, are nerves *already formed.* But in the process of becoming that way, they are growing, and they grow from somewhere to somewhere. Because we are neuroplastic, ever-changing neurologically, our nervous systems are constantly growing and pruning connections. Our neurobiology is a use-it-or-lose-it proposition. Embryonic development, and infant development more broadly, is accompanied by explosive neural development.

Various parts of us fold inward on themselves, the outside becomes the inside like dough rolled and folded on itself, and then parts of us migrate, move, shift into position. It is appropriate here to visualize timelapse photos of plants growing in the spring, of roots deepening into soil substrate. Neurons bring their own blood supply, so that the circulatory system grows in tandem with your neurology. A neuro-vascular system. There is massive neural growth, and then myelination. Myelin, the fatty sheath that grows around neurons rather like insulation around a wire, increases the velocity and precision of signal transmission. And neurons myelinate, typically, when they are used frequently. Our earliest development is characterized by massive neural and vascular growth, the inward forest branching and finding itself, then the insulation of circuits that are being used, followed by subsequent pruning. The tree grows toward the light. If there is no light the limbs die back. This happens again and again and again. A five-year old learning to hold a pencil in her writing hand is generating new neurology.

A beautiful secret: *Attention is the inward light.* At forty-five when you learn a new language, you are generating new neurology. The algebra you've forgotten as an adult? Pruned because you didn't use it. Attention is the light. Pass attention over the neurology to keep it alive. Attention is the inward sun passing across your aboriginal neural forest.

CHAPTER 10

I FEEL MYSELF THEREFORE I AM

Orthodoxy in neuroscience states that the Autonomic Nervous System originates in the brainstem. This autonomic nervous system circuitry is distinguished from midbrain circuits that are involved with emotion and emotion regulation, and cortical circuits involved with language and abstract thought. The brain, we learn, develops from the bottom to the top, from the back to the front. The autonomic circuitry, which regulates our internal milieu, begins in the basement of the brain, and is then the neural foundation beneath our emotions and cognition. Or so we are told. The problem is that this is not true. What I am about to explain to you upends several hundred years of orthodox neurology, but the logic of this proposition is very straightforward.

Most of your autonomic circuitry does not begin in the brainstem for the simple reason that neurons grow from somewhere, to somewhere. If you want to know where a neuron started, and where it ends, where it grows from, and where it grows to, you look at the direction of information flow. Where the signal arrives is where the neuron terminates.

In our brain-centric model of neurology we classify nerves as either afferent (flowing from the body toward the brain) or efferent (flowing from the brain to the body). What I am saying is simple: if a nerve is afferent, it doesn't begin in the brain. It ends in the brain. This is why the Autonomic Nervous System cannot begin in the brainstem. The Vagus is 70%-80% afferent.

70% to 80% afferent means that 4 out of 5 fibers are flowing into the brain. If they are flowing into the brain they did not start in the brain. This afferent Vagus doesn't *begin* in the brainstem. It *ends* in the brainstem. The *sources* of the sensory afferents of the ventral vagal system are not in the brainstem. The *terminations* of the sensory afferents of the vagal system are in the brainstem.

The afferent Vagus (which is most of the vagal system) begins in the guts (and the soles of the feet, the palms of the hands, the skin, the visceral organs, everywhere in your body), and then unites in the heart.[1] From the heart it makes its way up into your brain.

My contention is that the current brain-centric model of the nervous system is the delusional byproduct of a worldview over-identifying self with thinking. Thinking (in words and pictures) happens locally in the head. If you identify self with thinking (axiom Descartes), you elevate the organ where this thinking happens in your construction of intelligence and you center it in your neurology. Just think about schematic diagrams of the nervous system. They do not contain the heart, despite the fact that the heart has its own intrinsic nervous system (neurocardiology). Neither do they contain the guts, again despite the fact that it has its own intrinsic nervous system (the enteric nervous system).

If we understand how neural systems grow, we are forced to come to the conclusion that direction of information flow in a neural conduit corresponds with how the conduit developed. When we say that that Vagus is 70-80% afferent, i.e., flowing from the body upward to the brainstem, *we already know* the Vagus doesn't develop in the brain.

In order to understand this more deeply, and more systemically, in ways that have profound implications for understanding who we are, and how our nervous systems (and therefore our bodies) work, we have to deep-dive into a particular niche of neuroembryology that studies the earliest development of the nervous system, at the stage of foetal development when your

---

1 Final signal assembly happens in the sino-atrial node (the pacemaker of the heart), but it probably happens in local circuits of the heart's intrinsic nervous system prior to this. The most refined neurocardiology does not have agreement about this. There exists an intrinsic cardiac nervous system that is not fully functionally mapped by the most advanced neurocardiology that exists. Propositions about its structure, including intrinsic cardiac ganglia (i.e., they never leave the heart), have been elaborated by a number of neurocardiologists, including Dr. J Andrew Armour, whose laboratory has studied them extensively, and who presents on it during the 2015 Second UCLA Autonomic Nervous System Control Symposium.

spinal cord and brain are forming. This is, once again, specialist knowledge that you need to understand, because its practical implications for your life are enormous.

We are looking at the developing fetus (you) at the moment when it begins to fold in on itself. This is within the first few weeks of embryonic formation, within fourteen to seventeen days of conception. At this moment, the outer layer of the embryo gets folded inward on itself (imagine a baker rolling out a sheet of dough, gathering both edges and folding them back to the center, and then rolling it again). This outer-become-inner will become your spinal cord and brain. And between them a cell layer called mesoderm will develop into your vascular system, heart, and blood. At this moment, the heart has already started to beat, but it is not in the place it ends up in. At this moment of very early embryonic development, your heart is above your head.

I'm going to say that once again because it is so deeply strange. At the moment when your spinal cord and brain form, your heart is beyond the cranial end of your body. The heart then begins to migrate and roll downward, in the direction of your feet, until it eventually reaches your chest, which is where it will end up. Amazingly, the chemical that ignites this migration comes from what will become your guts (your endoderm, in the form of a growth factor called VEGF - vascular enthothelial growth factors). As the heart rotates downward across the ventral surface of your brain it catches it neurally on the transit. I want you to try to visualize this, because it is important. Your heart has its own intrinsic nervous system. It is also, as it migrates southward across the ventral face of your brain, sending out dendrites that link into the brainstem. At this moment, it is like a plant sending tiny shoots that root into the brainstem. The brain in turn reaches back.

At the earliest stages of your neural development, your heart migrates from north of your head, tumbles down across the front side of your brain, and catches your brainstem as it passes.

This neural link-up between heart and brain becomes your Vagus. Four out of five of its fibers are running from the heart to the brain, the fifth is running from the brain to your heart. If this is a highway, it has four lanes running from heart to brain, and one lane running from brain to heart. In other words, it is highly asymmetrical in terms of information flow, *favoring pathways from body to brain over pathways from brain to body.* Under a specific range of conditions, and these are the conditions that your physiology is optimized to reside in, this foundational wiring integrates inputs from the whole body in the heart, and then passes them up into the brainstem, giving your heart sovereign control over your brain. You have four lanes of control running up from the heart to the brain.

This system, the foundation of your neurological Connection System, is capable of organizing and harmonizing all the afferent inputs from your body with an extraordinary degree of refinement. Yet, also, in the conduit of this connection, there is, in parallel, a single lane running down from the brain to the heart. This lane permits the brain, under certain conditions, to control the heart. This descending neural control is less robust, precise, graded. It is less capable of entraining all of your other organ systems.

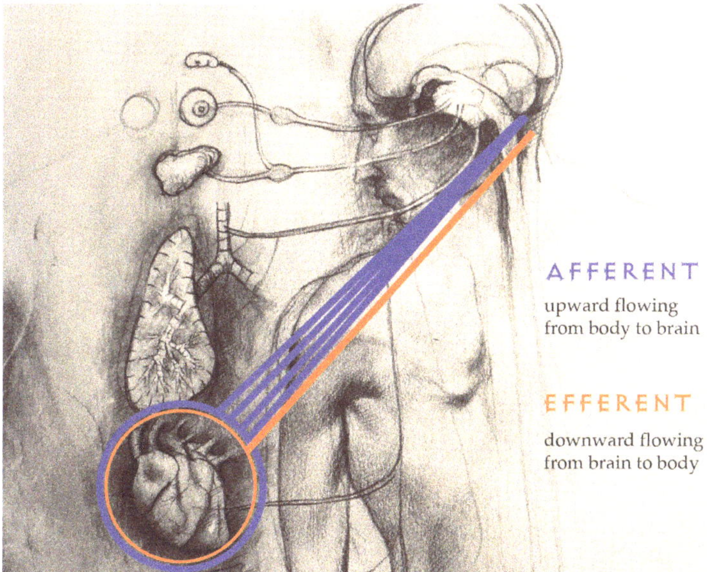

Do you know when that system activates?

When you are stressed out. When you move into a defensive state, your brain takes over your heart and your health degrades.[2]

Schematic diagram of HRV shift from ventral vagal (connection state) to defensive state (fight response)

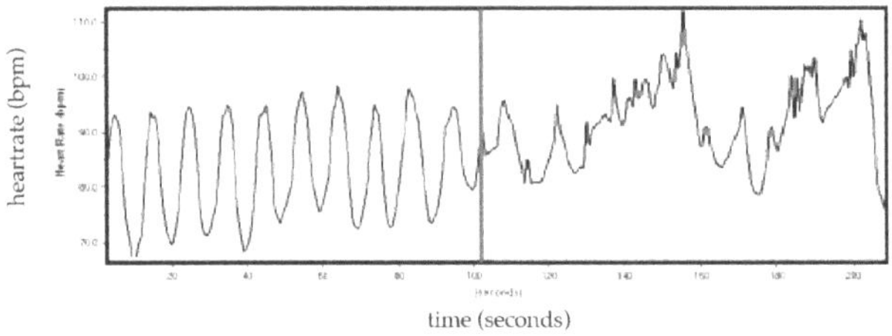

neuroception of danger at moment of (red line)

When you feel safe, and are able to activate the neurobiology of connection, you heart has sovereign control over your brain.

Since your Autonomic Nervous System begins in your body and ends in your brain, and its primary function is to synthesize information from the internal milieu, then this points us to an embodied field of awareness; the neurophysiological foundation of the intelligence of embodied feeling.

A nervous system whose crucial componentry is woven through the entire fabric of your body.

What we in neuroscience call implicit memory is not implicit: it is non-brain-localized. It has nothing to do with your autobiographical sense of self. It is the memory held in the body: in

2 We can see, in measurements of vagal performance, including Heart Rate Variability traces, the moment when this directional shift happens, from the heart running the brain to the brain running the heart. When this happens, the HRV trace shifts from looking like a sine wave to looking like the measurement of an earthquake. Chaos breaks out. Your health degrades at this moment.

sensation, emotion, and autonomic state. Your body isn't localizing identity, narrating a story about it, and presenting that curation to the world as who you are. Your body is interacting directly through the felt language of sensation and emotion.

This points us in the direction of both entire epistemological categories of alternate, verifiable ways of knowing (non-cognitive ways of knowing), and explains why the Lakota language, and other verb-based Indigenous languages have no word for 'I'. It explains why Buddhist teachers talk about 'no-self.' It explains why mystical traditions around the world talk about the importance of dissolving an egoic structure. If you are grounded in these other ways of knowing, constricting and concretizing an autobiographical self, which is such a deep part of the project of individuation undertaken in modernity, seems completely beside the point.

Rather, we open a channel to let experiences flow through this feeling intelligence. Sense of self is not an object. Not a noun. It is the living texture of the river of experience flowing through. *I feel myself, therefore I am.*

You are the feeling-in-the-body: the weather passing through the inward forest. The river running through autonomic banks. Notional construction of identity as a necessary or useful project rather falls away. We become *beings*, not *doings*. Human beings.

Verbs not nouns.

One of the reasons that this is practically relevant is if your primary autonomic states are defensive, these modes of awareness are not available to you, and your embodied experience is not going to be gifting you a lot of pleasure and enjoyment. If you are not centered in connection (your primary autonomic states are defensive) accurate non-cognitive ways of knowing (intuitional modes) are not readily available. If you are not centered in connection, your brain is not in its proper evolutionary relationship to your body. The brain, you see, is not an end in and of itself. It is a contextual organ, the integration end of signals coming from the body. In defense we organize a brain-centered cognitive self. In connection we organize a flow of feeling.

Shortly I'm going to help you understand that safety is the necessary doorway to open the door to connection, and connection then opens the interoceptive doorways that grant us the access to these ancestral ways of knowing-through-feeling.

A primary takeaway from this section is that your autonomic nervous system is not centered in your brain. She is centered in your heart and your guts and the whole of your body. This is not metaphor: it is neurologically factual. To center connection in your life, to enter into the kinship worldview is to restore the heart and deep belly and the body entire to their proper places in our lives, and their original meanings. For heart, in Aramaic, the language Jesus spoke, means *the center, the pith, the best part of anything*. The root of courage, the seat of the soul, not merely metaphorically, yet neurologically.

Once we do this, we can begin to recognize what we are talking about when we explore the neurobiology of connection is developing and refining the intelligence of the heart, and the deeper body, which are in fact the doorways to our original feeling and knowing-through-feeling capabilities, the ways of knowing of our endogenous/ indigenous bodies, which are much broader and more wonderful than you know.

Your Autonomic Nervous System fills the entire interior of your body, a vast aboriginal forest, woven like cotton candy or clouds through your body. Like the oak in summer, the neural receptors in your body form a complete clothing of leaves, so enwrapping to your organs and organ systems that the trunks and branches of your nerves are not visible, wrapped as they are in beautiful volumes of leaf.

You are a forest creature, through and through.

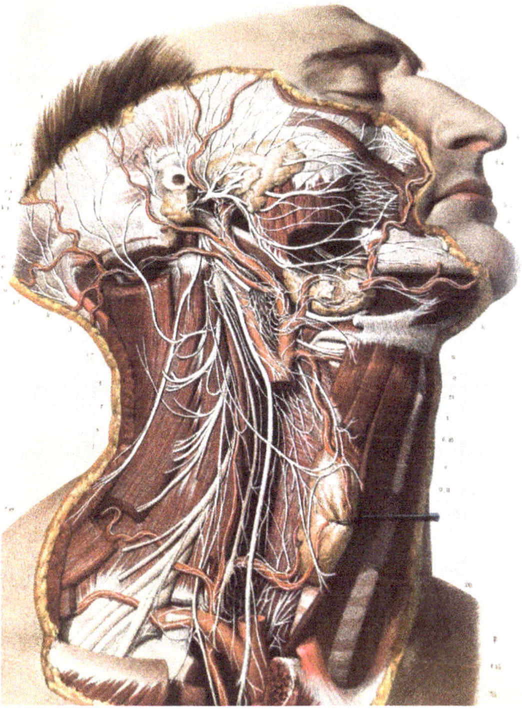

Facial, Trigeminal, and Accessory Nerves (cranial nerves 7, 5, 11)

Autonomic neurology of the pelvic cradle

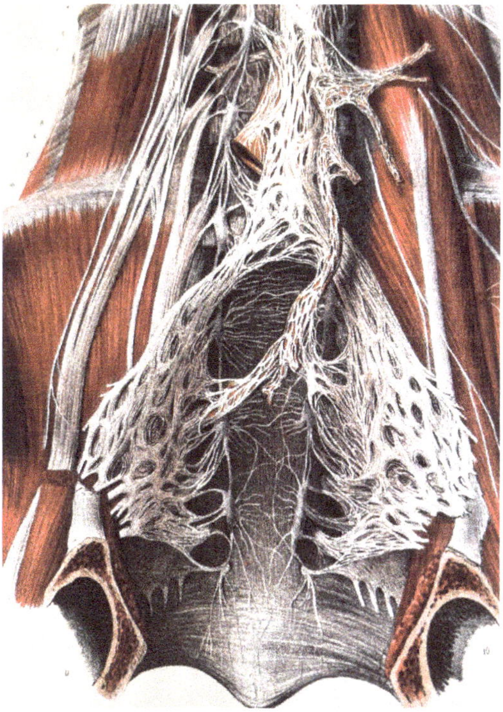

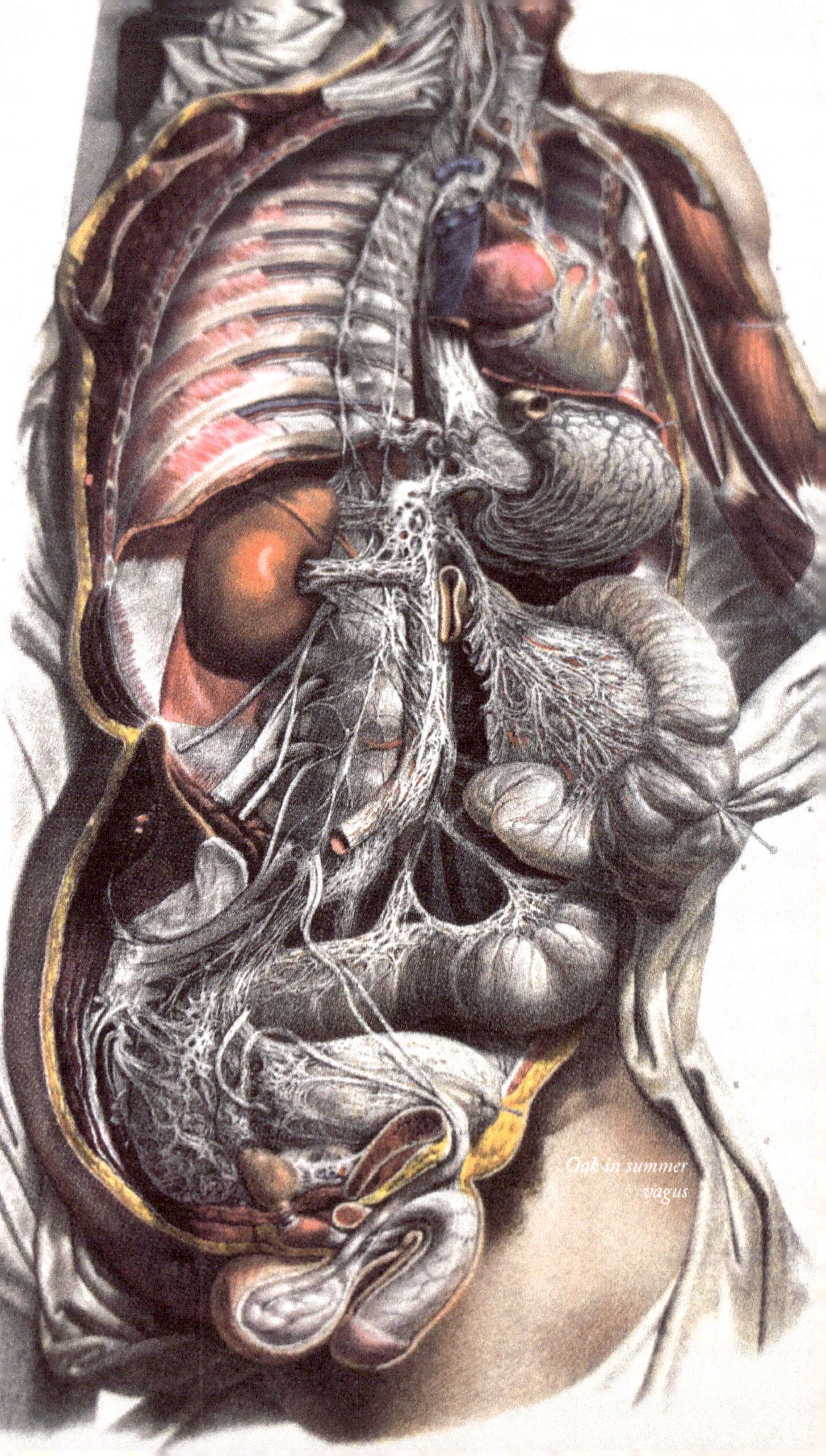

Oak in summer
vagus

Behold the cotton candy webbing of your autonomic physiology. Not telephone wires, not computers, but an inward ecology.

We have been de-storied by de-natured mechanical metaphors and it is time to reclaim our wild animacy. You are much wilder and stranger than you know, which is good, because it means that you have abilities much wider and stronger than you realize.

Your Autonomic Nervous System was born in your guts and heart, not your brain. The Autonomic Nervous System begins everywhere in the body, flows into the heart, and feeds up into the brain. Now after five hundred years, we can put the heart back in its proper place neurologically: the center.

Quod erat demonstrandum, by the neural architecture itself.

I am gendering the heart female in this framing, not because it has intrinsic gender at all, but because it is usefully disruptive to using the pronoun 'it', which pushes us away, and because I like this notion of the heart, whose primary intelligence is of feeling, as the primarily active agent in reaching out to the brain. Our neuro-embryology, in this regard, strikes me as matrilineal: led by feminine biological decision-making.

What I want you to take forward from this chapter is the primary songline of this book: your neural architecture is designed in such a way, centered in your heart, that the entire way that your physiology functions is derivative of whether or not you feel safe. This is because when you feel safe, your heart can run your brain (think 4X the neural refinement), which is what your physiology is optimized to do. When our bodies, hearts, and minds are operating like this, we experience the deepest wellbeing, the greatest flourishing, and our greatest capabilities. Running this circuitry is the foundation of what many indigenous cultures refer to as *becoming a real human being*.

It is worth considering that much of the assemblage of modernity has been put together by people who were not operating from this foundational intelligence at all, and that much of the operation of modernity will attempt to thieve it from you should you find your way back to this sovereign home.

Wellness in a sick society is a revolutionary act.

CHAPTER 11

I TOUCH THE GROUND, THEREFORE I AM

When we speak about the heart being the center, when we talk about reconfiguring the center of gravity in the mindbody system so that what is centered is feeling awareness and not cognition, an *I feel myself, therefore I am* rather than an *I think, therefore I am*, we are shifting identify from being organized by the left hemisphere of the brain to the right. We move from nouns to verbs, from thinkers to feelers.

We are changing the locational sense of identity in the organism from an identity that is experienced in words and internal images, an identity that localizes who we are in our heads, to one that shifts that experience into feeling, and registers identity primarily through the body: identity through the salience of embodied feeling.

Although I am speaking about the physical heart, the English language does not have a word that can properly contain the conceptual terrain we are entering. In Japanese, the word *kokoro* means heart, but it also properly means the emotion of love, the distillate of spirit. The reorganization I'm speaking about doesn't mean that your awareness is centered or located exclusively in the physical heart. It means that we re-inhabit the body, and give primacy back to feeling as way of knowing.

The shift at the heart of this book is in giving primacy back to feeling as the ground of primary ways of knowing.

In giving primacy back to the felt, we come back into the present moment. And we feel through the body. This shift in identity means that we process our experience viscerally through the animal body, that we come back into feeling and reside here. We occupy *ourselves*.

Last night, as though the dreamtime were the Goldberg Variations, I dreamed the same theme twice. The first dream was the more intense version, yet the second dream was the identical flavor, rendered with lesser intensity. Both dreams were about being deviated from the Source. I cannot say that is properly what they felt like, but that is their *meaning*.

Both dreams were of incredible suffering, but the intensity of the first was so great it was almost overwhelming. My awareness, my sense of my self, was locked into a container so extraordinarily small, under such great pressure, that I experienced a knowing and a simultaneous inability to know that the greater part of myself was missing. Void, vacated, absent. From within the framework of the dream, which I experienced as a fight to escape a sort of prison of awareness, without success, I awakened briefly, it must have been 2 AM, the visceral imprint of this feeling still animating my body, and lying there in bed, still hungering for contact with myself, which had suddenly returned, I felt a sort of opening in my diaphragm at the cardiac sphincter, and I felt myself pour back down through it, into my guts. I experienced this, even in the liminality between wakefulness and sleep, with total certainty as the reclaiming of my Grounding System, the inmost layer of the autonomic architecture, known in some circles as the enteric brain, the source of gut feelings, known to Polyvagalists as the Dorsal Vagal system. I then fell into the second dream, on the same theme, yet milder.

Polyvagalists do not realize that the Grounding System exists–they know it only in its lifethreat configuration. They do not realize that it is absolutely crucial to our embodied sense of groundedness, *is* the neural architecture of our connection to the Earth, is valent in this salugenic mode when we are deeply in connection, coordinates with the ventral vagal systems, and that when people talk about feeling grounded, talk about feeling their feet on the earth, as a visceral experience, it requires availability of this system.

For thousands of years we have been trained to worship sky-gods, and the chthonic and Indigenous attachments to place, to the underworld, to the burial grounds of the ancestors, to their mnemonic whisperings that make their way up from the depths of soil, through mycelial earthen networks of remembrance, have been cut off and outlawed with the same alacrity that Empire forbade mourning.

The result of this is that most modern people are dissociated at a magnitude that we would be hard-pressed to understand if the very scaffolding of our entire modern civilization so-called was not premised on this level of normalized alienation.

Many modern people I know cannot feel their bodies, they cannot feel their feet on the ground. They are popsicles sticks with big-ass heads, balloons untethered, that might at any moment go freaking off into the ether, possibly attempt to relocate to Mars.

Modern people's feet do not touch the ground because the sub-diaphragmatic system required to feel our own deep bellies is inaccessible, frozen solid as blocks of ice through the release of endogenous opioids, multi-generationally transmitted. The effect of this visceral shutdown is that it cuts off our visceral experience of contact with the Ground. And most people who are like this–void in their deep centers–*do not realize they are void.*

When White people got to Hawaii the Indigenous people named them *Haoles.* The etymology of this word is *hā* (breath) and *ole* (without). White people are *the people without breath.* If you go do a bit of research on this, you will uncover the assertion that this refers to the different kind of handshake White people gave, as the traditional Hawaiian greeting is to share breath with foreheads touching. But this thesis is, to put it technically, absolute horseshit. The reason Hawaiian people called White people *without breath* is because we do not breathe correctly, which means we do not breathe from the belly. And the reason that White people do not breathe from the belly is because the guts are physiologically frozen, due to trauma. It is hard to breathe if the place breath needs to originate is solid ice.

The reason Polyvagal Theory does not know that the dorsal vagal system can exist in a configuration other than the traumatic, is because the collective soma of modernity does not exist in a configuration other than the traumatic, so you could study modern people as your baseline and never discover that there was another way to neurologically occupy this vagal system. Bluntly put, when traumatized people develop a physiological system with their own bodies as the physiological baseline, their own trauma disappears into the baseline.[1]

Yet all we need to do is look at wisdom traditions from around the world, and the etymology of various gut knowings, ways-of-knowing that are surely referrant to the enteric nervous system, to begin to see a new picture emerging.

Hara, in Japanese, means belly, but it also means 'true nature.' The Japanese phrase *hara o waru*, for example, which literally means to split or crack open the belly, means 'one who speaks the truth'. It is a way of talking about someone who is sincere and speaks the truth: says what needs to be said openly and frankly. The Japanese word *hara* is linked deeply with the Japanese word *tanden*: *dantien* in Chinese. *Tanden* means 'elixir field'. It can be understood as the place where the elixir of life is made. This comprehension is directly connected to the Sanskrit word *agni*, which refers in Ayurveda to the digestive fire: the metabolism of food, thoughts, and breath. The *hara*, *dantien*, or deep belly is the center of life; the place where metabolism originates. It is the place the digestive fire begins. It is also the autonomic center of shutdown when we are pushed into lifethreat. Being pushed into lifethreat extinguishes the

1 This is incidentally, a similar problem to what is going to happen with AI. When people who are only using half of their brains insantiate these half-brains in silicon (Artificial Intelligence) they replicate their own alienation as the baseline of machine intelligence. This is likely to make the machines as dangerous and disconnected as the people who made them.

digestive fire, so to speak.

Take a moment to find and feel it in yourself, the area just beneath the belly button. Place your palms flat against this lower belly, dominant hand against your skin. Take a moment to acquaint yourself viscerally with the place we are visiting. It is a few inches below your belly button: the deepest part of the viscera. Take a moment to quiet your mind and simply attend to the experience of the deep belly in this present moment. Breathe into and with it.

- Can you feel yourself there?
- Can you feel this deep belly?
- Can you feel it as a conduit to the ground?
- Or is it unavailable?

If you do not feel safe here in the deep belly, you do not feel safe, full stop.

This is the origin of digestion in the body, and the place from which authentic movement originates. Under lifethreat it is the place from which immobilization emanates. Can you feel the truth of that in your own body?

Both Japanese and Sanskrit, and the traditions that arise around them, understand that we have to digest our experiences. We don't digest merely food. And we have to be able to digest our experiences in order not to choke on them, in order for them not to get stuck in us and putrify. If the digestive fire goes out, we cannot alchemize what happens to us. We cannot process it; it cannot move through. Experiences stay unintegrated and we are stuck in liminality: a dream-like (often nightmare-like) surreality fueled by the chemistry of dissociation and shutdown, high on our own endogenously released opioids.

Terror comes, like a gust of freezing wind, and extinguishes the digestive fire. Shock blows it out. This hearth at the center of the body, the furnace of digestion, the alchemical seat that boils the elixir of life: it can be brought to a standstill by trauma.

From autonomic physiology, and Polyvagal Theory specifically, we recognize this immobilized dorsal configuration, the one entered through lifethreat. This traumatic dorsal configuration, which is the one that the field of somatic trauma healing has developed largely to address, could be regarded as what happens when we do not have adequate safety to allow the *tanden* to make the elixir of life. This is why shutdown is always accompanied by digestive difficulties. This shutdown entered through lifethreat is the opposite of authentic movement: it is immobilization. This shutdown of the Grounding System through lifethreat is the doorway to hell realms.

The *hara*, is the expression of the Grounding System that arises when we are safe. Consider, for a moment, what this means. At the center of your body– just beneath the place that you breathed and ate through as an infant in utero– is the center of metabolism, known to ancestral and indigenous cultures around the world, and that can yet also become the center of immobilization when we cannot digest our experiences. Becomes the seat of ice when we are exposed to life-threatening danger.

Inhibitions that constrain the hara can occur in utero if there is sufficient stress chemistry in the mother, and then begin occurring for most of us with regularity when we are infants. Certain movement expressions that our bodies wish to make are suppressed. The more this happens the more the center of metabolism and authentic movement degrades into the center of immobilization, and the less we feel ourselves.

Many somatically-oriented trauma therapies focus on immobilization that happens as the result of developmental or shock trauma exposures without giving adequate recognition to the ways that the suppression of early infant movement patterns weave immobility into the neurological baselines of the body.

When we experience immobilization as infants– when we shut down in the *hara*, we cannot feel ourselves.

Viscerally, we do not know where we begin and end. We do not have a clear experience of our edges, we do not have boundary clarity. We do not feel solid.

Many people have gone into lifethreat responses before they experience what we ordinarily conceptualize as developmental or shock trauma. When our full repertoire of movement cannot express itself, when we attempt, as infants, to express our sovereignty by saying no, by turning our heads away from food that we don't want, for example, if this kind of innate, *No!,* is not respected, we learn very early that we have to inhibit the expression of our authentic movements. This is the same as learning that we have to inhibit the expression of our authentic selves, as the expression of movement is the animal expression of identity.

Let's say that mom holds out a spoonful of delicious spinach and it looks as appealing as eating pond scum, so I turn my head away and pucker my mouth up in refusal. Mom can either respect my sovereign *No*, or she can scold me and tell me to eat my spinach. Most of us were instructed to eat the spinach. The way that we learn our authentic boundaries is by resisting.
We learn who we are by pushing against.

That is how we learn in our bodies where our edges are. Indigenous and ancestral childrearing techniques include nearly constant touch compression on the baby for the first nine months, which is important because touch compression neurally instantiates the edges of our bodies. Feeling the boundary, feeling our edge, and pushing against this edge is how we learn to differentiate self from other. (This is why babies in traditional cultures were swaddled most of the time for the first nine months.) Saying *no* to the spinach I don't want is part of my knowing of myself.

Most parents do not like it when their babies do not comply. And so, as parents, we cajole, we invite, we insist, and if that does not work, we force. We push the spinach in through the puckered up mouth. We scold the baby. *Bad baby! Eat your spinach!* What this teaches baby's body is that it is not safe to express a *no*. When baby's body feels that it is not safe to move in resistance, baby's body moves into immobility. Deprived of safety when we authentically express, we immobilize. With the release of shutdown chemistry comes the dissolution of boundaries. We lose our clear edge. We become permeable, diffuse. Modern people are much more dissociated than most of us realize. Our reservoirs of ice are much deeper than we know. We are much more frozen, down in the core, down in the bowels of things than we are aware. This is because most of us, from a very early age, have been socialized into dissociation and shutdown because our sovereign *no* was not respected.

Being downstream of domination, which all of us moderns are, causes people to immobilize. Dad is dominated at work, he comes home and dominates mom, mom dominates children, children dominate dog. I know I'm painting a patriarchial feverdream even in this tiny sketch, but this is what we have inherited. If we are going to trace the full arc, it would be that humans attempted to dominate nature and subdue her, men attempt to dominate women and subdue them, colonialism attempts to dominate indigeneity and subdue it, whiteness attempt to dominate brownness and blackness and subdue it. This architecture of domination creates immobilization. And all of us are downstream of domination. Humanity has been seduced by domination mind.

He or she who is dominated seeks to dominate. When parts of me are turned to ice, I will attempt to turn others to ice. This is the passing of the wound. We harm others in the ways that we have been harmed. We want to bring others into our woundedness in ways that we are not often even consciously aware of.

When they are wounded as we are, others feel more real to us, more familiar. The parents who have, as babies, been forced to eat their spinach force their children to do so. To watch their children spit out the spinach, to watch them exert a sovereign NO elicits rage. *How dare you say NO when I was not allowed to say NO? You will eat this goddam spinach, baby.*

This is often the real power struggle between parents and children. Of course, it is not often one of which the parents are conscious. As parents, it would be good if we asked ourselves why the baby's resistance enrages us. Why the baby's innate knowing about what she wants drives us out of our minds.

If I have been beaten into submission by my parents, by the educational system, by the factory, the job, the fight to maintain dignity in a world intent on stripping it away, I will be damned if my own children are going to defy me. At least in the sphere of my own private home I am in charge. I am the dominator. And in this way before they can talk, we have often de-natured the most authentic parts of our children.

I am taking some darkly poetic license here, because clearly it is not always like this. There are other ways to raise children. Yet this is how the Grounding System, the deepest and most authentic parts of ourselves, gets shifted into its traumatic configuration: turned to ice. And it is this configuration that trauma therapy in its current incarnation understands. It knows nothing about the sovereign intuitional ways of knowing that come from humans whose deep bellies are not frozen: for whom the Grounding System operates as the bassline in the symphony of autonomic systems coordinated by the pulsation of safety.

Aloha is comprised of two words:

Alo ... one's face or countenance, essence; and hā .. one's breath, breath of life. Aloha is the loving, willing and reciprocal exchange of one's Alo or spiritual essence via one's hā or inner breath of life without expectation of return.

-Bill Rezentes, PhD
Hawaiian Licensed Clinical Psychologist

CHAPTER 12

I MOVE, THEREFORE I AM

I am laying out for you here the three primary autonomic systems. Although in classical neurology, and in Polyvagal Theory, they have names that you may have encountered before, we have renamed them because prior nomenclature problematically conjoins neurological systems with their neurochemistry.

This chemistry takes various forms, but it is typically comprised of hormones. When people talk about the sympathetic nervous system, for example, they are typically speaking of both the neurological Movement Systems mediated by spinal sympathetic ganglia, as well as the neurochemistry of activation, which is adrenaline and cortisol-based, and released by the Hypothalamic-Pituitary-Adrenal (HPA) Axis. If I use the words sympathetic nervous system, or fight-and-flight response, it invokes this conjunction of neurology and chemistry, and you visualize a system involved in responding to threats.

Yet the neurology functions completely differently in the presence of safety than it does in the presence of danger or life threat. It is in fact crucial for us to distinguish the neurological systems themselves from the interpretive context in which they are being utilized. Both of these systems (the Movement System and Grounding System) have ways of knowing– *vivid embodied intelligences*– that can be accessed when we have sufficient safety in our systems to come into contact with them in the presence of connection chemistry. Although they are often co-opted by threats, to view them simply as distress response systems fails to recognize the degrees to which they can (and must if we are to reclaim our deepest humanness) become functional parts of our awareness in the context of adequate safety, bringing particular facets of embodied intelligence online.

The philosopher, Ludwig Wittgenstein famously noted that, '*the limits of my language mean the limits of my world.*' Although the English language systematically lacks words to denote non-brain based intelligences, other languages are replete with them. The Yiddish word *kishkes*, for example, denotes a knowing that comes specifically from the guts. In German, the word *bauchgefühl* speaks to the same kind of knowing. In English we too speak of having a *gut feeling* about something, which is as close as we get to naming this deep belly intelligence. To understand these neurological systems only in their defensive (threat response) configurations is to miss out on half of their roles in our lives, and to fail to lay claim to parts of our Indigenous intelligence.

Orthodox neurology has placed the separation between the brain and the spinal cord in the wrong place. It is not until evolutionarily developmental biology that the proper distinction becomes more obvious, but the division is between the mid and forebrain, and the hind brain and spinal cord. The hind brain and spinal cord work as a functionally integrated motion unit in concert with the vestibular system and the eyes. These brain stem structures and the spinal cord house a stack of Central Pattern Generators (CPGs) that coordinate complex rhythmic movement patterns. At the cranial end of this column are the central pattern generators associated with breathing, sucking, and swallowing. As we move tailward down this neurological conduit, we encounter central pattern generators that govern the movement of the hands and arms, full spinal wave movements, peristalsis and digestion, micturition (urination) and defecation and orgasm, and the central pattern generators that govern the movement of the legs.

A central pattern generator is typically a cluster of interneurons tasked with organizing a complex patterned movement. Swallowing, for example, requires the coordination of twenty-six different pairs of muscles in a particular sequence. If you swallow really slowly, you can begin to feel the coordinated movement complexity required to complete that simple motor movement that we generally take completely for granted. (Sneezing and vomiting also require centrally generated pattern coordination, some of the same muscular systems operating in reverse order.) Most of us do not remember how difficult it was to learn to walk. A lot of what we do with our bodies requires incredibly refined coordination. Again, it is possible to take this for granted until it is interrupted. When we work with movement disorders clinically, or people in whom proprioception has been compromised, we often see the effects of disruptions to these coordinated movement and balance systems.

To be able to reach out for an object and grasp it without hesitation or tremor, and without falling over requires not only coordinated movement but the activation of our vestibular systems that allow us to keep our balance. We don't notice this because we do it automatically and without thinking. Yet when the vestibular system is compromised, people have trouble reaching out for an object without tipping over because they do not counterbalance the extension with muscular tension that pushes the body in the direction opposite their reaching. To conduct our bodies through space, in the field of gravity, and remain upright requires this incredible coordination of systems: to be able to bring a fork to our mouths with precision so we can eat, to be able to swing a tennis racket or a golf club, to be able to climb stairs or to run… all of these activities require tremendous coordination of multiple muscular groups with our vestibular and proprioceptive systems. If in order to perform any of these movement actions, we had to send neurological signals all the way up to the brain and back, we would not be able to move quickly and with efficiency. So rather, the complex coordinated rhythmic movements are governed closer to the limbs in which they happen. This coordination happens through central pattern generators. They are responsible for the coordinated control of these kinds of complex patterned movements.

Central Pattern generators are housed in the brainstem and along the spine. What this means is that when you are walking or running, the neural circuitry that governs the pattern of this motion is closer to the hips and legs than it would be if we had to refer these neurological signals all the way up to the brain. Central pattern generators are extraordinarily interesting and under-documented in the literature. Central pattern generators are called this because they can create a pattern in the absence of signals from the cranial brain. Many of them essentially function as oscillators. They have their own intrinsic rhythmic cadence. That noted, they can also be governed by the brain, governed by sensory inputs, or governed hormonally by neurochemistry in the blood.

Neurologically, and specifically with respect to Autonomics, the spinal cord organizes this set of central pattern generators along a single column. This body plan harks back to the origin of vertebrates. What we have historically called the Sympathetic Nervous System is in fact the entire vertical arrangement of these central pattern generators. The antique conceptualization was that the sympathetic nervous system activated under threat, but this isn't actually correct. This spinally-mediated Movement System is active any time you move. It turns on simply when you transition from sitting to standing. What characterizes its response during threats is the addition of activation chemistry. When we feel in danger, this system's role is amplified and modulated through the release of adrenaline and cortisol. Yet just as the Grounding System centered in the lower dantien has an aspect that expresses itself when we feel safe, as well as a different aspect that expresses when we are under life threat, this Movement System is the same.

In the absence of the chemistry of activation, and in particular when it is being organized by the orienting rhythmicity of the pulse of safety, the Movement System is what helps us to move the limbs gracefully, and to coordinate the internal rhythmic movements of the body and organ systems. Having this system appropriately calibrated, available, and online, is part of what helps to govern the homeostatic rhythms of a healthy body, and allows us to coordinate crawling, walking, running, swimming, complex coordinated arm swings, coordinated full spinal movement patterns, sucking, swallowing, vomiting, breathing, digesting, peeing, pooping, sexual thrusting/ hip rotational range, and orgasm.

If you examine the neuroanatomy of the spinal cord, as it descends from the brain stem through the cranial nerves of the spine, and between the shoulders, you see widening in the cord at the lateral level of the arms. Part of this widening is associated with the generation of central patterns of our movement. You also see widening at the base of the spinal cord in the cauda equina, or horses tail, where the central pattern generators for the pelvic cradle, including peeing, pooping, and orgasm are organized.

With the addition of this system, we have now articulated an autonomic triad. Although this is an oversimplification, we can say that the Connection System is centered in the heart, the Grounding System is centered a couple of inches below the belly button in the *hara* or *lower dantien*, and the Movement System is organized along the vertical length of the spinal column.

CHAPTER 13

EPIPHENOMENA OF AUTONOMIC STATE

Is it possible that the first lesson on the first day, from the moment you emerge into the earth situation, this having a body, the lesson that you will spend your life learning, is the difference between safety and danger? Between what it makes sense to move toward and what you need to move away from? That this discernment will continue to grow more and more subtle throughout your life?

Even single-celled organisms are endowed with the chemistry required to approach and avoid. Is this the original feeling? The wellspring from which all others rise? The trunk from which all others branch? Attraction/repulsion. Approach/withdraw. Turn toward/ turn away.

Our mammalian nervous systems are organized in such a way that the parts of us that are most human–most open, connected, fluid, flowing, creative, generative, and relating–are also the most vulnerable. Our deepest availability is most easily harmed. The infant reaches out to her caregiver so nakedly needing to be picked up. There are skylights in her skull and she is totally undefended against loss. The gesture is pure, without calculation, and she holds nothing back. Our deepest availability, which is the wellspring of our deepest humanity, is most easily harmed.

Our mammalian nervous systems are organized in such a way that the parts of us that are most human are also the most vulnerable.

And so the doorway to opening these parts of ourselves is of safety. We require safety in order to open. Opening cannot be forced. Like flowers deciding to bloom, safety is a sovereign determination made inwardly. Force the bud and it will not blossom.

Watch a snail pulling back into its shell when threatened. Then watch the slow re-emergence, the gradual way it comes out again, tentatively, re-inflates its rubbery antennae. The gate on the snail shell, that threshold across which the soft body either blooms or suddenly withdraws? It is called *neuroception.*

I don't prefer to use a technical word. We could call it t*hreat detection*, except that it also is assessing for safety.

Here is the continuum of vulnerability, by way of a snail.

Our first snail is secure. Closed in on itself, it cannot be harmed. Yet neither can it live.

Here, our second snail extends a bit. He can still pull back, but some of him is out of his shell. More life, and more vulnerability.

Our third brave snail? Fully extended. Fully alive. A snail living with vibrancy. And yet, for a snail this is deeply vulnerable.

Life– the fullest expression of our capabilities – happens with our courageous snail. The undefended snail heart, if you will. Yet what is most available is without defense.

Neuroception is the perceptual mechanism that shapes our bodily decisions about how open we will allow ourselves to be, in light of the vulnerability that arises when we become available.

This conceptualization of neuroception developed by Dr. Stephen Porges is one of the major conceptual/physiological breakthroughs in autonomic physiology over the past fifty years. Neuroception is the moment-to-moment bodily detection of safety or threat that governs the surfacing of platforms of behavior in the autonomic physiology. When we neurocept safety, connection states are able to arise. We blossom into the fullness of our potential, become our most human. These states are, functionally, the physiology of the deep ancestral baseline. These states are where we access our truest, deepest, highest, most creative sense of self.

In modernity, however, what we often see as normed baseline states are deviations from this ancestral physiological baseline due to the deflections of chronic, toxic, and traumatic stress. Non-resolving stress, which is underpinned by the enduring neuroception of danger, re-tunes the physiology in characteristic ways.

Neuroception governs interoception, and shifts neural setpoints in the physiology, altering our visceral state, inflammatory responses, emotions, cognitions, interpretations of the world around us, and our behaviors in predictable ways.

Consider for a moment that the polycrisis of modernity is the predictable result of a world where most modern humans have drawn their soft bodies back into the defensive perimeter of their shells. We can only come out of the retraction of our defensiveness–into our full humanity– when we feel safe enough to do so, yet our soft bodies know that in coming out of our defensive fortresses we are vulnerable. Look around and as far as the eye can see there are cues of danger. Social cues, economic cues, ecological cues, epidemiological cues. Is it any wonder that modern people have doubled down on the politics of defense? On authoritarian regime? Stranger danger? Xenophobia? Our body politic cannot but be a reflection of our physiology. The gated community is an allegory for the defended psyche.

The problem is that the absence of danger is not the same as a sense of safety. Building defensive walls may prevent the intrusion of the stranger, but these walls also defend us against joy. Against curiosity. Against life itself. Defense excludes life.

How shall we coach the modern snail out, when so many times they have been wounded?

CHAPTER 14

NEUROCHEMISTRY

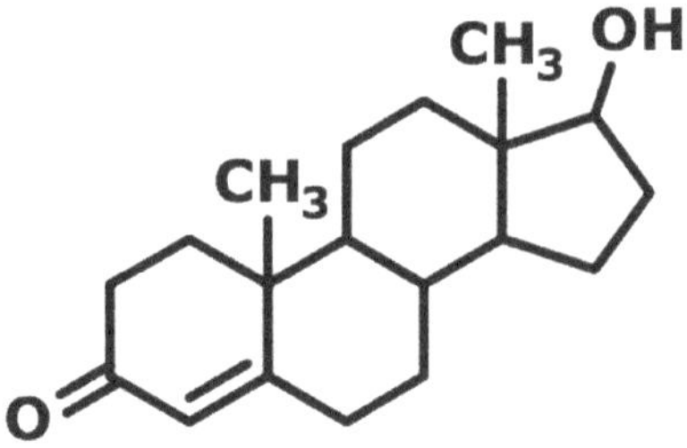

TESTOSTERONE

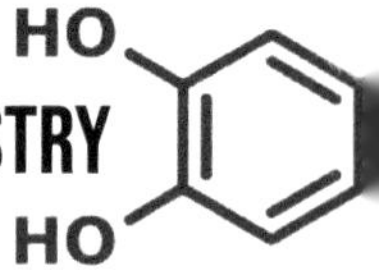

DOPA

OXYTOCIN

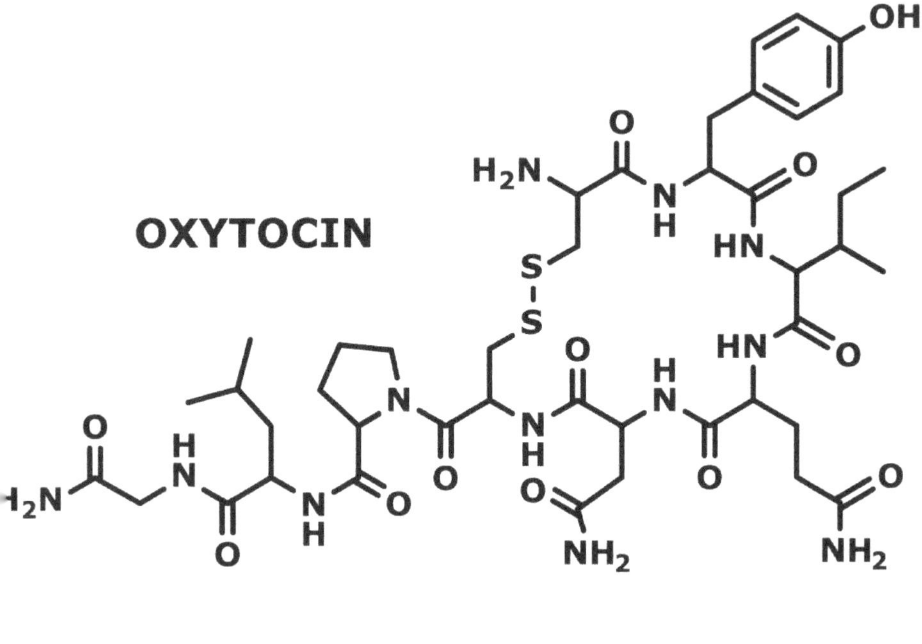

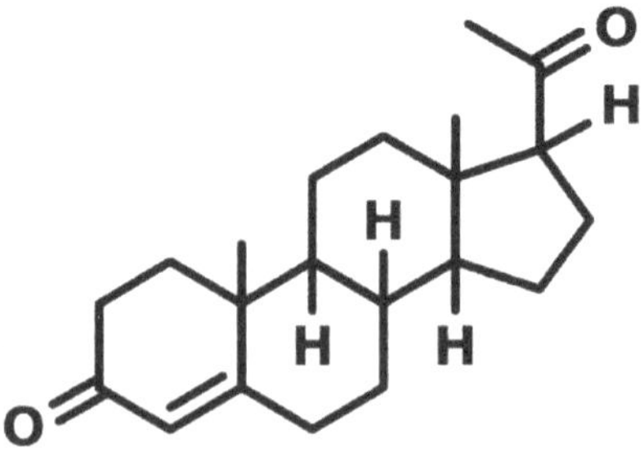

PROGESTERONE

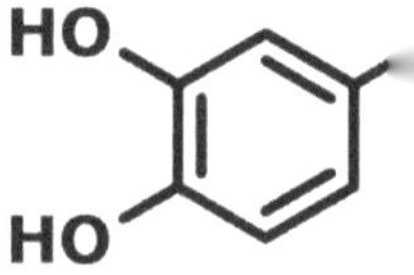

ADREN

Your nervous system is comprised of neurons, and the primary mode of signal tranduction that we've been focused on so far is electrical. For your nervous system is in fact an electrical system, and it propagates signals through your neurology as electrical impulses.

This neuro-electric transmission is quite precise. Neural conduits, like most transportation infrastructure, come in various grades to accomplish differing tasks. Neurons are classed into three categories of fiber, depending on their thickness and degree of myelination, which govern the velocity of signal transmission.

A fibers are like neural super-highways. They transmit at 10 times the velocity of B fibers, and 100 times the velocity of C fibers. This velocity is a function not merely of the thickness of the conduit, but also its insulation. Part of the reason the telephone wire analogy has been so enduring in neurology is that certain neurons are ensheathed in myelin, a fatty substance that functions in much the same way that the rubber insulation around a telephone wire keeps the signal tightly constrained. Faster connections, which are those that are more important, and more used, are both physically larger, and they have a more robust insulation. Much of the Connection System is conducted with insulated neurology (think of the precision of micro facial-expressions), whereas the Grounding System has unmyelinated neurons (when you feel sick to your stomach it is global in the guts). Insulated wires provide tighter control and precision, uninsulated wires support broader field awareness. And so within this neuro-electrical system there is a level of functional complexity that organizes the way signals move, from trunk lines, to branches, to leaflets: just like a tree.

There is, however, another mechanism in the body for exerting neurological influence. Neurogenesis is always accompanied by angiogenesis. Neurons, as they grow, call into being their own blood supply.

Because of the degree to which these two systems are interwoven, we might properly think of the nervous system as a neurovascular system. [1]

This addition of the element of blood is important because it points us in the direction of the second way that neurons are influenced in the body, which is through chemistry.

If we take, once again, a very long evolutionary view, we note that we are creatures that came up out of the ocean, and brought that ocean inward.

I became intrigued several decades ago by the health benefits of trace minerals. These soluble ionic minerals–which include magnesium, chloride, sodium, potassium, sulfate, lithium, boron, calcium, silicon, selenium, phosphorus, iodine, chromium, manganese, iron, copper, molybdenum, zinc, and vanadium–and which have extraordinary restorative effects on the body, are the naturally occurring ionic trace minerals in seawater.

You don't have to analyse the ideal health in a human population, find robust specimens of vitality, cycle their blood down and measure its trace minerals constituents, measure those amounts, and then formulate a supplement with precise quantities of the aforementioned minerals. All you have to do is find pristine seawater and boil it down: the perfect and precise concentrations of trace minerals are already there, because the inward circulation in our bodies *is* the ocean. Nature did it for us. Nature is you.

You are an inward forest and an inward ocean. Are you starting to notice something about how we are related to the living Earth?

It is no accident that making sure our inward ocean contains

1 This is important for the energetic geometry of autonomic state (energy-processing templates) because the circulatory system is required to create an electro-magnetic field.

these minerals is salugenic. We are, literally, the ocean, enclosed in a sack. Our bodies, as you're probably aware, are 70% water.

In this inward ocean, communicating through it by exhibiting specific effects on target glands, organs, and neurology, are hormones. These cellular messengers are a primary manner in which the body communicates with itself. In order to act on a part of the body, its cells must contain a specific receptor for the hormone in question. Some hormones, such as Thyrotropin-releasing hormone (TRH), which is sent from the hypothalamus to the anterior pituitary gland, and tell it to release thyroid-stimulating hormone (TSH) and prolactin, are extraordinarily specific in their targets. TRH is a message for the Pituitary, no one else. Other hormones bind to receptor sites throughout the body, and have much more globalized impacts. Cortisol, a glucocorticoid hormone, which has a very complex set of anti-inflammatory effects on the body, and is involved in the fight or flight response, is uptaken by receptors all over the body. Every cell type in the body has receptors for glucocorticoids. Cortisol, when released, is a message broadcast everywhere.

Hormonal communication in the body is quite different than neural communication. It works through the medium of the blood, and in the same way that when you add a drop of ink to water, you can watch it slowly spread out and percolate, the circulation of hormones through our system is a diffusion process.

The fact that we can shift into a fight-or-flight response, activated by the neuroception of danger, and the sudden release of adrenaline, shows us that hormones can act quickly. In light of the necessity to diffuse this chemical messenger throughout the body, elevated heart rate, which essentially stirs the inward ocean more quickly, makes a lot of sense.

What I'm referring to here in this book as neurochemistry are classes of hormones that engage in specific ways with our neurology. The scope of this interaction is in fact, extraordinarily nuanced. Our bodies manufacture an incredible number of hormones. They control everything from regulating our circadian rhythms, to metabolism, to the female ovulatory and male ejaculatory cycles.

I'm going to constrain our conversation about hormones here to a discussion of five primary chemicals. I want to say at the outset that this is highly reductionist.

I would like to acknowledge a couple of limitations before we proceed. While I have been studying neuroscience intently for thirty years, my study of neuroendocrinology is more recent. It is my sincere hope that colleagues will expand and deepen this area. The interaction between neurology and the endocrine system is an absolutely crucial dimension of Autonomics. While the broad contours of what I am about to describe are accurate, there are nuances missing.Another very salient limitation of what I'm about to describe is that it does not include a comprehensive understanding of the way that the female ovulatory cycle (Gonadotropin-releasing Hormone (GnRH), Follicle Stimulating Hormone (FSH), Luteinizing Hormone (LH), Estrogen, Progesterone, Testosterone), and male hormones such as Testosterone, Estradiol, LH, and FSH interface with these chemistries and neurological systems. In this aspect, the present presentation of Autonomics is incomplete.

With these caveats noted, in this chapter, I would like to bring our attention to three specific neurochemistries, and five chemicals. The neurochemistry of love and pair-bonding. The neurochemistry of activation. And the neurochemistry of shut down. Oxytocin, vasopressin, adrenaline, cortisol, and endogenous opioids.

It is possible that neuroception is not a discrete neurological phenomenon in its own right, but rather a vagal gestalt interpreted by the hypothalamus. Could neuroception be the aggregate surveillance of the entire embodied autonomic systems, all three of them, summed in a single equation that sets off the production of neurochemistry in the Hypothalamic Pituitary Adrenal axis?

The hypothalamus is located above the brainstem, and beneath the thalamus. It is on the body's midline, about the size of an almond, and links directly to the pituitary gland just below. Although the vertebrate forebrain generally exhibits the greatest evolutionary diversity in the structure and function of the brain regions– the forebrain of a fish and a chimpanzee are radically divergent– the hypothalamus stands out as a stark exception, with nearly identical structure across the entire vertebrate line. It seems to be so ancient that it predates the vertebrate body plan: hundreds of millions of years old.

The purview of the hypothalamus is the regulation of the deepest physiological substrates of the body. It is responsible for

- Fluid and electrolyte balance
- Feeding and energy metabolism
- Wake-sleep cycles
- Thermoregulation
- Stress responses
- Sexual behavior and reproduction

These are the foundational physiological balances. It seems to me that in the same way that the hypothalamus is responsible for thermoregulation of the body (our body temperature) based on continuous surveillance of autonomic inputs from receptors, neuroception is another way in the which the hypothalamus takes the body's temperature. What it is measuring here is not warmth, but rather the relative continuum from safety to danger to lifethreat.

From deep within the brain, the hypothalamus takes the temperature of the body across numerous fundamental physiological dimensions, and one of the primary manners in which it responds is through directing hormonal release in the pituitary and its targets in the adrenal glands, thyroid, and the sex organs.

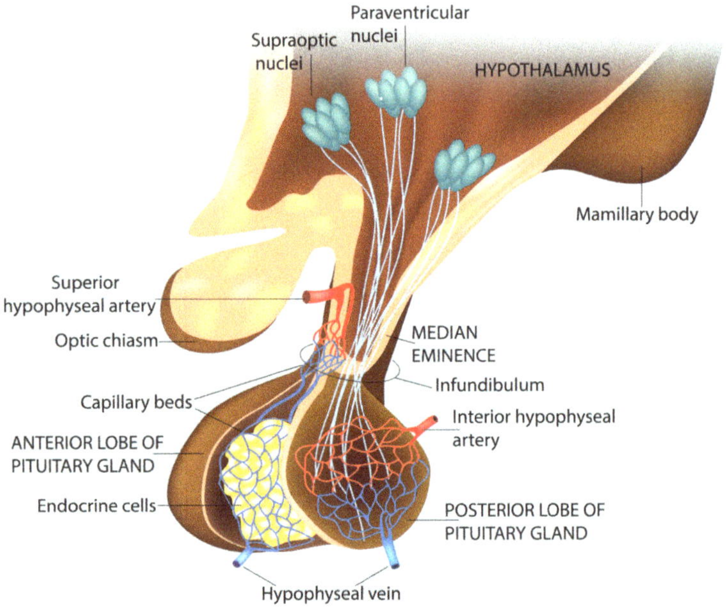

This nexus of the hypothalamus with the pituitary and adrenals is specifically called out in the stress response, where the HPA axis is understood to be intimately linked to motoric components of what is known in orthodox neurology as the Sympathetic Nervous System (the SNS is this aggregation of the spinal neurology with this neurochemistry). As I've noted numerous times, this naming, which conjoins the neurological systems with their associated chemistry is problematic, because the neurology can function with different chemistries to different effect.

This HPA axis is implicated in the stress response (adrenaline, cortisol, etc.), yet the hypothalamus also produces both oxytocin and vasopressin (aka Anti-Diuretic Hormone). Oxytocin, to the chagrin of Dr. Sue Carter, the internationally acclaimed expert in behavioral neuroendrocrinology, has become known as *the love molecule*. She finds this problematic because it is involved in an intricate dance with its similarly ancient molecular cousin, vasopressin. Together they are the primary chemistries of Connection.

Vasopressin and oxytocin are both nine chain amino acids. These ancient neuropeptides have a complex geometry, like a three dimensional molecular key. They are almost structurally identical, with variations in the third and eighth amino acid chain only. This nearly homologous structure allows them to bind at the same receptor sites, which are located throughout the body: perhaps the reason we can toggle so rapidly between feelings of connection and feelings of aggression is undergirded by this chemical phenomenon.

Remember, we are creatures who came out of the ocean. We drew it inwardly. Vasopressin, which is the older of the two hormones, is anti-diuretic. It essentially allows us to change the inward osmotic balance of the body, regulating free water in the body by changing the amount of water reabsorbed by the kidneys, and constricting blood vessels, which alters blood pressure.

Vasopressin is synthesized in an androgen-dependent manner. Men produce more vasopressin than women, and since its effects are synergistic with testosterone, experience its effects more strongly. Vasopressin is strongly implicated in territoriality and aggression. In its diuretic properties, it is connected to urination and scent-marking. When your male dog is marking his territory, this is vasopressin-mediated behavior.

In the prairie voles that Dr. Carter has studied for decades, pair-bonded males, after conceiving, become aggressive with strangers. This is part of the neurochemistry of affiliation.

Prairie vole babies are vulnerable to being eaten by just about everybody: other marauding rodents, cats, etc. And so part of the chemical preparation for prairie vole fathers is to activate aggression towards strangers. This aggression toward strangers is coupled with partner defense. So the balance between vasopressin and oxytocin in pair-bonded prairie voles is part of the neurochemical substrate that causes prairie vole fathers to defend their families.

Oxytocin is the connection chemistry, and it induces pulsation. The contractions of birth are pulsations organized by oxytocin. When Robert Naviaux, one of the world's leading mitochondrial researchers, talks about safety being pulsative, while danger signals are constant, he may be talking about the pulsations induced by oxytocin.

The half-life of oxytocin is somewhere between 1 and 6 minutes in plasma. You can test this half-life experientially for yourself by basking in the afterglow of an orgasm, and seeing how long it lasts. Both men and women produce oxytocin and vasopressin. Both release oxytocin at orgasm.

This pulsatile nature of oxytocin is fascinating. When I am working manually with somebody, and the body shifts out of a defensive state and back into connection, I can feel this as the return of the pulsation of blood to the tissues. In the contractions of defense, there is no pulsation through the tissue. It is without tone, and feels inert. With your hands, gently cupping the area of someone's body where a localized shut down has occurred, the transition autonomically back into the rhythm of regulation is accompanied by a return of this pulsation. The pulse literally returns to the tissue. Is this oxytocin binding to receptor sites in the localized area? Is it the metabolism of immobilization chemistry outflowing? Is it the co-occurrence of both?

The pulse is not something abstract, or metaphorical. It has a distinct cadence. A rhythm. Safety is a pulse whose neurochemical heartbeat is governed by oxytocin.

It is probably not accidental that males of the species, who have more vasopressin, and more experience of it because it combines synergistically with other male hormones, tend to shift in the direction of fight when they experience danger.

The neuroception of danger activates the HPA axis, releasing adrenaline into the bloodstream. Adrenaline

- increases your heartrate to send more oxygen to your muscles
- increases your blood pressure to send more blood to your heart and muscles
- improves the efficiency of breathing so you don't have to breathe as deeply to harvest the oxygen your body needs
- opens your pupils so you can take in more light
- increases your alertness
- raises blood sugar levels
- increases your sweating
- reduces pain
- increases strength and performance

When I study this list, I tend to think that with the possible exception of sweating more, none of these phenomena are unpleasant. Who doesn't want to have more oxygen and blood available for the muscles, more efficient breathing, more strength, less pain, and more alertness? In fact, when I think about sitting down to write this morning, and the cup of coffee I poured myself, this list sounds a lot like why we drink coffee. And indeed, caffeine increases adrenaline levels in the body.

The thing I've always found paradoxical about the fight-or-flight response is that fighting and fleeing are subjectively such entirely different behaviors, it has never fully made sense to me that they were classed together. But if you look at the effects of adrenaline, you can see that these same attributes– alertness, increased heartrate, blood pressure, muscle tone– can be applied to confront a threat or get away from it. The tilt then, either toward the threat or away, is not mediated by adrenaline, or the cortisol that typically follows to sustain the response if it endures (cortisol, which is released secondarily in the fight flight response, is anti-inflammatory, and sustains the metabolic activity required for exertion), but rather the vasopressin-oxytocin continuum.

Vasopressin dominates and you will fight. Oxytocin dominates and you might turn toward the threat (tend-and-befriend, accommodate, appease, placate). Without vasopressin or oxytocin, we probably flee. These chemistries are not gender-determined, but they are gender-correlated, because men and women generally make different amounts of, and are differently sensitive to, vasopressin and oxytocin.

Let's just take a moment to dwell, subjectively, on the experiential difference between feeling our own neurology– the various configurations of the Connection, Movement, and Grounding systems– versus experiencing chemistry. Neurology tends to be more targeted in the body, and is accessible to us through sensation, whereas chemistry, which plays out through the inward ocean, tends to be more diffuse.

Could we experience neurology as sensation and chemistry as mood? Neurology as feeling-tone and chemistry as weather? A more global and yet somehow background phenomena?

Take a moment to see if you can make this discernment in your experience. When you imagine the various autonomic states, what component of them is neurology, and what component is chemistry? What is the sensation of being connected versus the chemistry of it? The sensation of an orgasm as compared to its emotional tonality? I'm inviting you to attend to this with precision because being able to feel the ways that these elements combine directly increases your self-awareness. Have you ever eaten something and tried to pick out a flavor that you cannot name? You dwell on the dish, able to isolate but not identify a particular flavor note. *What is that?* You ask your dining companion to take a bite. *Can you taste that?*

This ability to discern, at a subjective level, what parts of your autonomic experience are originating neurologically, and what parts are originating chemically, is part of knowing yourself through an autonomic lens. And this will pay dividends in your self-understanding.

The third chemistry, which I am referring to as shutdown chemistry, refers to endogenous opioids that are released under life threat. The motoric impact of endogenous opioids is to take the Movement System offline.

Opioids, as you may well be aware, are painkillers. When we move into a lifethreat response, endogenous opioids are the chemistry that gives us the grace of not feeling it when we get eaten. In extreme cases of shutdown, the release of indigenous opioids leads to both immobilization, and numbness. In life threat situations, the uptake of endogenous opioids is very localized and specific in the body. They bind to receptor sites in interneurons in the postganglionic Movement System (sympathetic ganglia in the parlance of orthodox neurology), taking central pattern generators off-line.When this happens, posture is deeply impacted. Someone in a shutdown state will physically look collapsed. Curvature goes out of the spine. The head falls forward. The deep abdominal muscles that bound the core lose their tone. This posture is a result of the endogenous opioid cascade, removing tone from the Movement System.

A secondary effect of the release of endogenous opioids is the dissolution of boundaries. In the life threat responses, this boundary dissolution is typically accompanied by a violent ejection of the sense of self from the body. This is why victims of sexual violence will often report watching their bodies from above.The body immobilizes, boundaries dissolve, and our subjective sense of self vacates the body. The state of *union*, intriguingly, combines endogenous opioids with oxytocin. It is possible in this state to experience immobilization without fear. A pedestrian example of this occurs during orgasm. The body is immobilized, but there is no fear.

You will notice that one of the possible byproducts of this experience is a sense of boundary dissolution with a romantic partner. If our hearts are connected to someone else, in the moment of orgasm, with immobilization and boundary dissolution, we can experience Union. This is the focus of tantra.

It is also a pretty good description of a spiritual experience. If we are not under threat, but in a state of centered connection, and boundaries dissolve, rather than being ejected from the body, we experience a sense of connectedness with everything.

I want to invite us to continue thinking about chemistries in terms of primary colors. In the same way that the Connection, Movement, and Grounding systems organize and align in different ways in different autonomic states (*Union* is lots of Grounding and lots of Connection, *Play* is lots of Connection and lots of Movement), these foundational neurochemistries also blend.

We want to understand the individual signatures of them, and we also want to understand how they combine, and the way that feels. At an experiential level, you are already familiar with this, even if you didn't realize that was what was happening. Anytime you are playing a team sport, and the team gets into some kind of flow state, you are in the realm of both oxytocin/vasopressin, and adrenaline/cortisol. Oxytocin is what delivers the sense of connectivity. It's the chemistry of the heart connection you have with your teammates. Vasopressin is part of what gives you loyalty to your own team, and feel aggression toward the opposing team. Adrenaline is the thing that allows you to race around the court. Cortisol sustains this elevated activation. Both of these sets of chemistries need to be online, and both of them need to be in a certain kind of balance.

Carlos Alcaraz is my favorite tennis player in the world to watch. Just thinking about watching him play puts a big shit-eating grin on my face; it is happening right now as a write these words. I've played tennis for nearly 40 years, and Carlos' game style appeals to me as much as anyone I've ever seen. Part of what makes it so extraordinary is the unfettered joy he brings to playing. This creates a positive feedback loop of oxytocin release, and when he's really on, playing at his best and beaming after points, it can ignite a stadium. It can, at times, even ignite his opponents, who will at times smile and laugh in disbelief. That is how good he is.

In the quartfinals of the 2022 US Open, in front of a stadium packed to the rafters, Alcaraz was playing Jannik Sinner, a brilliant Italian player who is the current world number One. At the time, both of them were relatively new on the professional tour, and people were already sizing them up as the rivalry of the next generation. Alcaraz would win this match, and go on to win the tournament, the first Grand Slam of his career, but at the midpoint of the match it did not look like this was how it was going to go. Sinner was playing at an extraordinarily high level, they were battling it out, there were 15,000 people screaming, and it got so stressful for me, simply watching it on TV, that I turned the match off and went to sleep.

Before I turned it off, at the end of the second set, a set Alcaraz would ultimately lose in a tiebreak, there was a point played that I will never forget. Alcaraz is trying to close out the set, he is up 6 games to 5, and Sinner is serving, trying to hold him off. Sinner has the advantage, and needs to win one more point to get into a tiebreaker. This is, for both players, the most significant match of either of their careers, on the world's biggest stage, at one of the four most important tournaments in men's tennis. It could not be a more high-pressure moment.

Sinner hits a huge serve, like a laser, to the outside of the court, Alcaraz' backhand. Alcaraz is returning from 15 feet behind the court, and he lunges for the return, it is so far wide that he can only get one hand on it– he looks like a fencer lunging sideways with his rapier. He returns the ball at full stretch, he gets it back, but the shot is totally defensive, and Sinner gets a high floating ball that is an easy put-away. He closes to the net with all the time in the world. The ball floats up in front of his forehand, which is deadly. Alcaraz has, meanwhile, somehow anticipated where Sinner's next shot will go, and after scrambling back toward the court's midline he breaks into a sprint sideways as Sinner makes contact with his forehand. The ball comes off Sinner's racket at 90 miles an hour, but Alcaraz, who has already sprinted, is waiting for it. The problem, however, is that springing sideways across the court, he has overrun the ball, and so now it is behind him. There is no way he can turn around, no way he can spin, the point is over.

Except that is not what happens. Instead, and watching the replay it is still unbelievable that he is able to do this, Alcaraz returns Sinner's ninety mile per hour strike behind his back. He simply swings at the ball behind his back– he cannot see it. Not only that, the behind-the-back return is so good that it dips below the net and Sinner has to volley it upward, Alcaraz closes on it, and hits a clean cross-court backhand winner to end the point. It is a staggering display of dexterity. Between the eruption of the fans, you hear the commentators in the background, who say...

Oh! No Way!!! They are drowned out by screaming from the fans. *In your life...*, says the announcer. (More fans screaming) *Have you ever seen anything like it?* The camera cuts to Alcaraz' coach, Juan Carlos Ferrero, himself a former world number One– we watch him standing, smiling, shaking his head in astonishment and appreciation, and from his mouth, simply, the single word, *Wow*.

If you're playing tennis at an elite level, you are returning serves that move at 125 miles an hour, and you're hitting forehands that approach 100 miles an hour. You are simply obliterating the ball. In order to even stay on the court at this level, you have to be able to race around and you have to have incredible endurance. Adrenaline and cortisol undergird the high-powered Movement System foundations of this performance.

Yet Alcarez is special in the degree to which oxytocin activates his Connection System, bringing a level of touch to his game that is perhaps entirely unprecedented. He is so strong that he can stand at the baseline and inject incredible pace into a forehand or backhand. He is fluent and deadly off both wings, striking hard enough to push any player on the tour back. In the middle of a rally like this, sometimes opponents are 20 feet back from the baseline. But the thing that Alcaraz can do better than anyone I've ever seen is to disguise a drop shot. After hitting two or three shots at 100 miles an hour, he will abruptly cradle the ball, completely changing the arc of it, looping it high over the net and dropping it three or four feet beyond the net with so much backspin it doesn't move forward. In order to do this with a ball hit at him at 100 miles per hour he has to remove the pace from it; to soften into it with exquisite touch. To hit this dropshot he swings forward at the ball while falling away from it at the same time. Sound hard? It is almost imposible to do well, particularly with a shot struck at you at ninety miles per hour. Of all the shots in tennis, this is the one that even the most high-level players are apt to screw up most often.

This level of touch comes from access to his Connection System, and oxytocin. I love using tennis as an example. People who don't play do not realize the degree to which it combines two very different activities into a single sport. Tennis is sometimes a dance, and sometimes a fight. When players are rallying, exchanging shots back-and-forth, this happens within the framework of a call-and-response pattern. Although players don't talk about dancing, this is really what's happening. You are in call-and-response with a partner, in fairly neutral territory on the court, exchanging the ball back-and-forth. This is highly social behavior. Since tennis is also gladiatorial, what happens at some point in the dance is that one of the players decides to kill the other one. The second aspect of tennis, it's competitive aspect, is the fight. No one gets paid to lose a tennis match. So at the end of the day, it's usually the person who's a better fighter who takes home the trophy.

On occasion, the degree to which Alcaraz is running oxytocin through his system, and enjoying the dance of connection, costs him matches. Unlike some of the other players on tour who seem to thrive exclusively on domination, who delight in crushing and dismantling opponents, Alcarez enjoys the social aspect of the game so much, thrives so deeply on entertaining the crowd, that he will sometimes stay in a point or in a rally simply because it's so beautiful, and he's enjoying it so much. I've watched him lose several significant matches, often to players who are much lower ranked than he is, because he's having too much fun dancing. It's quite strange, until you look at it through an autonomic lens. The thing that makes him such an extraordinary player – at 21 years old, he's already won four Grand Slam tournaments, including winning Wimbledon twice, is this level of touch and spontaneity that comes from his ability to stay in connection, and to play with oxytocin coursing through his system.

The downside of this, if you will, is that he's really playing with a social architecture, and so there are times when that tilt causes him to make choices based on dancing rather than fighting. This is an example of what happens in a sport, when there is a little too much connection chemistry, and you're supposed to be focused on winning. Your supposed to beating people, not hugging them, Carlos– yet as I say it I cannot help smiling.

DOORWAY
OF SAFETY

CHAPTER 15

SAFETY IS A SOVEREIGN DECISION

We cannot make anyone feel safe. Safety is a sovereign determination of the individual embodied nervous system. The part of us that knows we are safe, bone-deep knows this, marrow-deep knows this, or does not know this, is not our ordinary autobiographical sense of self. It has almost nothing to do with thinking, though thoughts stream up from it like water from a spring. Follow the thoughts down to their source, and you start to get to the places in the body where this awareness is rising up from.

I want to spend time in this chapter acquainting you (or re-acquainting you) with the bodily experience of safety, and some ways to invite more of it into your life. I am often surprised, though I should not be, at how little we humans understand about what makes us feel well. I have watched many intelligent people undermine their own wellbeing because they do not understand at all where it is coming from.

In this world we are living in – filled with conflict, war, violence, aggression, transgression, transaction, uncertainty, exploitation, normalized extraction– all the byproducts of a world-view birthed from domination, translated through trauma, and requiring us to spend much of our time defended: armored up against the onslaught of systems perpetually seeking to entrap us – it is of particular importance to understand what the visceral experience of safety can do for us.

I'm calling it a visceral experience, an embodied experience. I think this is an appropriate word to use because visceral shares the same root as viscera, which means your guts. And the kind of safety that I am talking about really begins in the guts. The deepest belly. You may know that you are safe or unsafe by the talk in your head, or the pounding of your heart, but if you follow these roots down to their source, they end in the guts. Or, better said anatomically, they begin in the guts.

The roots of your safety begin in your guts.

This is partly because your guts, embryonically-speaking, are the inmost layer of you. They are your endoderm, or inner skin, folded upon itself through the ornate infolding that organized your interior (first a primitive streak, then a turning inward). Your guts are the deepest inwardness of you, and from this deepest inwardness grew up, neurobiologically, in the direction of your heart, and then your brain, your deepest (deep belly vagal) circuitry of embodied safety assessing.

The kind of safety that I am speaking about, the embodied experience of it, is physical, yet also *intrinsically* psychological. You can be physically safe yet not be psychologically safe. If, for example, you work for an unpredictable boss, it may be the case that your physical body is never in danger, but that your psychological self is constantly so.

We can also know that we are safe intellectually, know that we are loved, but the physical animal body may not feel this. Raised by parents who were not themselves held, did not receive the physical expression of affection, we can know that we are loved and yet our bodies can still hunger for this physical expression and in its absence not know the depth of safety of which we are speaking.

The kind of safety that I am speaking about permeates both. It is an enveloping safety: a surround. To experience this kind of safety is to be embraced by it, both physically and psychically. The touch compression in Indigenous childrearing, which takes the form of swaddling and being carried skin-on-skin by our caregivers, mimics the compression in the last trimester in the womb, and *is* this physical enveloping safety.

Most of us modern people did not experience this enveloping sense of safety. We do not know it in our animal bodies, in our bones. If we were not held, touched, and cradled, we do not hold, touch or cradle others.

And so as you read this, you may find yourself grieving as you recognize that you did not receive the neurological inputs, as an infant, that would allow you to experience this level of embodied safety.

And yet, *and yet*, simply becoming aware of this is hopeful because part of the beauty of the plasticity of the autonomic nervous system is that we can convey this sense to the deepest parts of ourselves if given the right context. Part of the success of well-administered psychedelic treatments for trauma, from MDMA to ketamine to psilocybin is the ability of entheogens to penetrate into the deepest layers of our psyche and conduct us back into this primordial experience of safety.

Many of us do not experience safety regularly in the world, though we may experience it in our most private moments, with those most intimate in our lives. And even then it is often imperfect, as we are often imperfectly witnessed, imperfectly held, even by ourselves. From the sovereign seat of safety we can be uprooted by difficulty, danger, abandonment, disconnection, grief, tumult, violence, disaster, illness, uncertainty, sarcasm, immorality, lying to ourselves, lying to others, the unexpected. When it is thieved from us, we have to find our way back. This is part of the great and necessary work of our lives. If thrown from the horse, we must find our way back to this seat.

Safety is the throne from which the seat of sovereignty arises.

When I think about safety, what calls to me is this notion of surrender, the notion of letting down our guard, of being undefended. And since one very visceral way of understanding modernity is that we are walking through a world where everyone is armed – if not with a literal knife, then with ways of taking up space that are armed, even language that is used as a weapon – becoming undefended is something that we must be very careful, very precise, about how we do.

When our guard is down, we are vulnerable. When our guard is down, we can be hurt. We can be hurt anyway, hurt with our guard up, but at least then we are braced for it, we were expecting it, in a way. If your shield is up, and you are struck with a blow, you were at least prepared, no? Yet if you were undefended, heart open? Hurt with our guard down? Hurt when we are undefended? The babe reaching up to her parents with open arms? These are the kind of harms that cause people to lose hope: harms that dim the inward lights. The kind that break hearts and shatter psyches. So we have to be careful, not in the sense of caution, but rather of discernment. We have to exercise discernment in when and where and with whom it is safe to let down our guard.

But what I want to tell you, from the standpoint of your health, is that finding places to let down this guard is essential to your wellbeing. Finding places to let down this guard, to surrender, is required if you are to be well.

There is a sort of posture to becoming undefended. I want you to think about it. If we imagine the archetypal gesture of defense to be hand's up– hands out - (the superhero on the right), (the tiger on the left)

Then the gesture of surrender is to allow the hands to drop without collapsing. This surrender is not a collapse. Collapse is a form of giving up. That's not what I'm talking about. What I'm talking about is removing the barrier. Removing the boundary guarding us from the world.

It is a very worthwhile exercise to ask yourself when in your life you are able to do this. When can you let your guard down? The answers to this question are as varied as humanity. In my previous book, *Restorative Practices of Wellbeing*, I wrote about 300 practices that might allow someone to do this. This book was produced based on 700 trainings that I ran with about 5,000 people, where I asked them the question: *How do you come home to yourself?*

Where in your life can you be undefended? For some it was in solitude, for some with trusted others, for some in Nature. Some people found it through creativity, some through movement, some through working with their hands, some through listening, some through their ancestors, some through meditation, some in or on the water.

Sometimes people think about these things as hobbies. After university, when I lived in San Francisco and was waiting tables at night and struggling to live in a city where our rent was six times higher than the place I'd lived before, I drove the beat-up 1992 Toyota Camry we had bought from a friend for two thousand dollars to Pacifica to surf. On top of the car I'd strap an eight-foot surfboard, pink and purple, clearly designed for ladies, but the size I wanted, and purchased on Craigslist cheap, to a Pacific Ocean break called Linda Mar State Beach. I would pull into the parking lot, shimmy into a wetsuit, and paddle out. A couple of hundred feet out in the ocean, bobbing up and down on a fiberglass plank, rising and falling with the lilt of waves, I found my breath again. I had to literally get off the land to find this space.

I am a very novice surfer. I kinda suck at it. At a certain point I realized that the reason I came out to surf was not to surf. In a two-hour session, I was probably up on the board for four minutes. I came out to find myself. To come home to myself. To sit or lie on the board and surrender. To watch the seabirds wheeling through the sky, behold the emerald beauty of the coastal hills, the jewel-tones of the sunlight on the water. Sometimes I would climb off my board and just float in the water, letting it pull my body, rising and falling with the rip and lap of the waves.

This train of recollection leads me to thinking about one of my favorite beaches in the world, in Napili, on the island of Maui. There is an unusually steep beach, and a wave that comes in and picks everyone up, such force, and slides them fifty feet up the hill, as if mother ocean has placed her children in a giant swing. One of my favorite things– and I'm a grown-ass man– is to lie down in this wave, close my eyes, and surrender to her. I have to be somewhat careful in this, because I'm not the only one–it's a sea of people often doing this–but when on Maui, I have sought out this primal sea-swing. I want to let go, to be held, to be rocked by the Mother. Usually, I cannot stop giggling. The wave is very strong. I watch it knock people over all the time. If you struggle against her she will overpower you. But if you just totally relax, she will carry you, very quickly, fifty feet up, and then swing you back fifty feet down. It is mother nature's amusement park ride. She has tremendous force, this wave, but if you settle into her just right, she is also very very gentle.

Candidly speaking, it is easier for me to trust Nature than people. I'm just being honest. When I go camping, I always notice its the sound of unfamiliar humans that tenses my body like a board. My wife, God bless her, will grab me during the night and go, "What's that?" It might be a skittering outside the tent, a growling, a call in the night. But as long as it's not human, I generally feel quite relaxed. Strange human voices, though? I'm on full alert.

We have to learn the language of our own bodies, what makes us feel safe. As we start to understand this, we can begin to arrange our lives so we have more of these experiences. I know people who go to church very consistently not because they are studying the scriptures, but because something in them is able to surrender in that place, and this is foundational to their wellbeing.

I like to make things. I write, I sculpt, I paint. I make furniture, buildings. There is a kind of state I can enter into through creative work that gets me to this place. The breath finds its own natural rhythm. The mind begins to unwind, begins to travel and spool in interesting and non-linear and felt directions. My hands awaken, become autonomous beings, and take me with them. The canvas begins to paint itself, some vehicular dialogue between hand and eye to which cognition is a mere passenger. At times I realize that my thoughts have gelled or crystallized upon something, and I either behold it closer, or unhook the thoughts. The mind is a hunter. This is a way of finding my way home, finding my way back to the surrender of the wave.

Of course, there are depths of safety. The first time I drank what is commonly known as Ayahuasca, in Brazil, in 2006, I spent the entire night fighting with myself. I was outdoors, with about 120 people I did not know, in a dimly lit outdoor room, in a language I barely understood, and I would begin to relax, to be able to feel my heart, and then something in me would shift to terror, and I would shoot up into cognition and my mind would travel like a run-away train. And suddenly I would notice this, realize I was freaking terrorized through the ether, and have enough presence to be able to slow my breathing, and find my body again beneath me, fall back into it, feel my feet, and begin to relax, and then something would shift as I got closer to a ground state, and a switch would flip and I would go to terror again, and find myself ricocheting around in my thoughts. This experience is etched in my memory because of the battle it was. During this ceremony, which lasted for about four hours, I brought my attention down out of my thoughts probably five hundred times. It kept happening over and over and over. I could not find and sustain a continuous sense of safety at all.

Now, just to give you the full picture– I was in another country on a two month artist residency, alone, with a large group of people I had just met, strangers effectively, in a forest with which I was unfamiliar, in a language I barely spoke, in the middle of the night, using a psychedelic substance I did not know much about. It did occur to me, more than once, that if something went wrong there was very little in the way of a safety net. Yet this is not why I kept shooting into terror. And this is what is interesting. It became clear to me that this reflex of terror had almost nothing to do with anything external, despite the apparent or possible dangers involved in what was happening. What I was fighting with was a habitual contraction into terror in my own mind and body. The layer of terror I kept encountering was *intrinsic.*

It was my own terror, which I followed deeper and deeper down into my body. I was, at heart, scared to death of *life.* I want you to notice that when you move into a defensive state, you lose contact with your heart. Defense often shoots us up into thinking. This is the result of a physiological reversal that happens during threat detection that we spoke about in chapter ten (I Feel Myself, Therefore I am), where at a certain level of baroceptive pressure in the heart, in the carotid arteries, the direction of information flow reverses. When we shift from the heart running the brain to the brain running the heart.

When we are open and connected, our hearts are running our minds. When we move into defense, our minds are running our hearts, and we find ourselves in our thinking. Ejected. In exile. *I think, therefore I am.* No, you think– *therefore you are defended.* This is the meaning of the Fall, is it not? This ejection, this exile, from the heart, from feeling, from relatedness with the Source. This seizure of defense.

After several hours of this, in the ceremony, as it wound to an end, I started throwing up. And I could not stop. I threw up five, ten, twenty times. My entire body would convulse, and I would throw up everything inside me, down to my toes. I was wracked by the convulsions of this. It was two in the morning, and I was surrounded by a hundred and twenty smiling Brazilians, eating fruit and soup in the middle of an Atlantic forest, and I would have to step aside and excuse myself and go vomit outrageously for five minutes.

I couldn't imagine how they could eat. Putting something into my stomach seemed like possibly the worst idea in the world. I couldn't even hold down water – I tried. I also couldn't understand how my extremely loud and egregious vomiting wasn't destroying their appetites.

We left, and I sat in the passenger seat of the car, and for the entire ride home, which was nearly an hour, every few minutes I would say, *Disculpe* (sorry), and they would pull over by the side of the road and I would step out into the screaming wall of insect noise, the fecund smell of the soil, and vomit everything inside me, probably waking up anyone in a 500 foot radius of our car. I was a homing beacon of barf, moving through the pre-dawn silence of Salvador retching every half-mile.

What was I experiencing terror of? What was I vomiting up? Back at their house, near 5 am, I finally fell into a dreamless sleep. When I awoke the following morning I felt measurably lighter.

Terror of relating, it came to me. *Terror of actually being touched. Terror of an undefended heart.* The ego is a contraction to avoid relationship.

Four days later I drank the tea for the second time. It was as if the first night I had spent just unclogging all the pipes on the toilet, vomiting up all the crap – all the experiences of unsafety I had accumulated over the thirty-one years of my life to that point. The second time was like a jetplane taking off. I was able to stay in my body, without contracting, to stay in my heart. Suddenly I was being operated on by tremendous primal forces, surgery by a cadre of dragons, then I was perceiving a coherent spiritual landscape opening all around me, the very Sacred Heart of the Universe palpable; available to direct experience.

There is an experience of safety at the heart of an experience like this – a ceremonial experience – where we begin to experience the safety of belonging. The safety of feeling, in our hearts,

that we are part of the Universe. That we are her children. That we can be at home here- in our bodies, in our lives. The relief of this is so existential it is far beyond the ambit of words.
In this second ceremony, I was re-homed into safety. Re-wombed. Rewoven into the surround of safety. Listen to any mystic worth their salt, anyone who has had a genuine mystical experience, and you will hear some version of this.

In 2005, the organization that I directed at the time held a retreat at the Institute for Noetic Sciences (IONS), near Petaluma, California, which was founded by the American astronaut Edgar Mitchell after he had a mystical experience in space during the Apollo 14 mission in 1971.

Our retreat happened to correspond with the IONS annual board retreat, and on the second morning of our stay, I was up early getting breakfast when I ran into an elderly man wearing a plaid shirt with suspenders, a pocket protector, and a pocketful of pens. It was Mitchell himself. We were the only ones there, and I turned to him and asked, "What was it like being in space?" He responded with a kind of child-like glee. He told me about a moment during the Apollo 14 mission, where they were flying in the ecliptic, which is the plane of the Sun, the Earth, and the Moon. The craft was spinning to avoid accumulating excess heat, and he was sitting in front of a giant window. Outside the earth's atmosphere, starlight is amplified by a factor of 10X. So he was confronted by a portrait of our earth, our moon, our sun, arrayed against the brilliantly backlit blackness of celestial space. He said that he knew, cognitively, that we were made of star stuff. Recombined elements from stars. In that moment, he said, his body began to know it at a cellular level. I am beginning to shake just telling you this story.

He described an experience then, of a kind of cosmic hum coming into him, a state he would later describe as *savikalpa samādhi*. You'll notice that is not an English word, because we don't have a word in English to describe this state. Neither did any of the other countless scientists to whom he tried to describe the experience. It was not until he met with an Indian spiritual guru who looked him in the eyes and described the state to him with precision that he knew what had happened to him. He had experienced Union with the Creation. Sanksrit has an entire vocabulary to speak about these refined states with precision.

This kind of an experience provides us the safety of belonging. Now we are getting into the terrain of the possibility for safety that exists. And here is the deepest wellspring of wellbeing. Safety is the deepest medicine, full stop.

THE SAFETY OF WHICH I AM SPEAKING

The safety I am describing, even the most blessed among you can barely fathom. You know it not.

It begins in the bottom of your heart
At your deepest navel
A place you have not touched
You will have to dig down deeper than you know you exist to get there
clear away what has overgrown the hallowed ground
to this place in umbilical center
this Original flame untroubled
by any breeze any hint of danger
with no flicker
/

to get there you will have to find your way home,
through millenia of hurt
Pain smuggled like contraband through lineage/
everything that has been done to your ancestors
to your Sacred body
every blessed damn time
you have been here wearing a skin.

And even that is only the beginning,
for once you have found this place in yourself
once you have located this unperturbed flame
you will have to fan her
coax her out upon the lacelines of the nerves
until she spreads and consumes your body /
immolates you entire.

Only then will you know,
in your body, the safety of which I am speaking.

It is hard to be you
in ways you barely understand

so much has been thieved from you
you are the victim of human sacrifice
over and over and over again

sacrifice upon the mountain, the steps of the altar
your body strapped down and your chest opened by knives
and you don't remember this
because you can't bear to.

your depression, your contraction is the veil,
the sheet wrapped around
what has happened to you
what the body knows that you cannot know/
The reach of your infant arms towards Parents who did not pick you up
Because their parents did not pick them up
Because...

We are the grandchildren of the humans
sacrificed on the altar of the empire of domination.
Our bodies have been broken into
our heads separated from us
we have been hung broken
quartered in four ritually
the violence enacted upon our sovereign bodies is unthinkable.

We know this
in the bodies we are wearing and yet
we know it not
/
You will have to let all of this go and yet
only by feeling it can you do so.
/
And how will you have the courage to feel this
when your heart today is wracked by all the pain it bears
just trying to make it through the battle
of the aggressions in the unkind world we find ourselves in?

The kind of safety I am speaking of– it comes from the center
of you
the silence beneath the voice
you know as your truest self
the you beneath the verb of you
the ground out of which
that figure-in-motion emerges.

From safety like this a new world is born.

TEN PRACTICES FOR EXPLORING EMBODIED SAFETY

Now, I am going to walk you home through five doorways, and safety is the first doorway. Let us start with the recognition that safety, a moment-to-moment felt embodied sense of safety, is the deepest substrate of wellbeing. Safety is the medicine.

Since this is so, augmenting your moment-to-moment experience of safety is a direct route to experiencing more wellness. In addition, because of our biological tilt toward noticing 'what's wrong', called negativity bias in science, most of us do not have acute tracking of our own embodied experience of micro-increments of increased safety. We simply do not attend to them because we do not realize they matter. This is food we leave on the table because we don't realize it is food.

Let's look at our lives through the lens of safety to see about increasing access to this foundation resource.

SIMPLE NOTICING

Do you notice when you feel safer? Can you build the capacity, in the present moment, to notice when you feel more safe? Since our bodies are constantly, as in every single gosh-durn moment, neurocepting, there is a crucial wellness skill to build here. The first part of this is noticing, in your own body, what happens when you feel safer. Like, how do you know? Does your breathing change? Do your shoulders drop? Does your belly soften? Does your voice change? Does your mood lighten? Do you get emotional?

What are the particular micro-adjustments that your body makes when you feel safer? How does the base of your spine change angle? How do your hips open and your leg bones socket into them? How does your chest open? How do your shoulders relate to your spine? How does your head get carried differently on top of your neck? How does your belly soften?

How does energy flow through your genitals? How do the soles of your feet open to the earth? If you aren't aware of any of these shifts, it could be that you have a significant amount of residual shutdown - no judgement if so. If that's not the case, it could be that you simply haven't trained yourself to notice these adjustments because they didn't seem important to you. The foundation of this kind of noticing is mindful awareness, dwelling in the present moment. There are lots of places that you can learn mindfulness, so I won't explain it here. Yet once you have a bit of this foundation, begin to direct your attention at noticing your bodily response in different environments. For example, you can experiment with simply witnessing how your body reacts to being in a crowded subway station versus deep in a forest. I assure you, the bodily experience of your sense of tension, boundaries, space, and emotion will be different. The tilt of your hips, openness of your chest, softness of your belly, rootedness of your feet…all of these adjust. This difference pivots around your embodied felt sense of safety. You can begin to notice how the body adjusts around different people in your life. Parents, partners, children, bosses, colleagues, etc.

Begin making it a practice to notice the bodily needs for safety, and allow yourself to adjust to this. Most of us do this without thinking. We use our phones as shields, for example, pulling them up in moments of vulnerability or to deflect someone else's inquisitive or intrusive energy. You couldn't do this if you weren't neurocepting danger. The invitation here is to make this more conscious and intentional. To bring it up above the threshold into conscious awareness, so that you are noticing the shifts in the felt sense before you simply adjust. How can you calibrate your gesture, posture, and felt sense to respond to the different degrees of safety you may experience as you move through your life?

And what if you notice, in particular, the moments when you feel increased safety? When you can let down your guard. What if you begin to organize your days around feeding yourself more safety, recognizing as you are beginning to that it is the first and foundational doorway to greater wellbeing?

NOTICING IN TIME

Choose an interval of time, for example- the last day, the last week, the last month. Take a moment to mentally and emotionally scan the overall texture of this time period with regard to safety. Examine the felt texture of this time interval noticing when you felt the deepest sense of safety. Ask yourself–*When did it happen? Where were you? What were you doing? With whom? And, How did you know that you felt safe? Or safer?*
In your first pass through the time period, just look at this broadly and loosely. The time period may have occurred during an epoch when you have a greater or a lesser sense of safety for many reasons. Yet even within this, there will be variations, different amplitudes, if you will, of safety. Take the time to focus on when you were feeling the most safety, and how you know this in your body. There is often a correlation between experiencing greater safety and experiencing greater wellbeing. If you are dealing with symptoms of illness, the periods of feeling greater safety may correlate with alleviation of symptoms. Greater safety will likely correlate with greater ease, access to positive emotions, creative thoughts, etc.

As you remember spans of time, bring conscious attention to surfacing moments of greatest safety in the texture of your life. You will remember moments of greatest challenge, difficulty, and defense by default. Augment your memory to include safety. Scan your days for the moments of greatest safety.

NOTICING IN TONE

What are the reliable and repeatable signals in your particular body that you are feeling safe? Or safer? Are there certain things that you do? Say? Activities you engage in? Ways that your body feels? Ways you hold your posture when you are feeling safe that let you know that you are feeling that way?

Facial expressions you make?
For example, I don't sing if I'm not feeling safe, nor do I often swim, nor do I have a sex drive. (Just keeping it real.) When I'm not feeling safe my face doesn't feel relaxed. I don't smile. If I am feeling unsafe, this generally takes the form of anger (fight response) but sometimes of shutdown. These have very different energetic signatures. When I'm feeling safe (enough) I want to play tennis. If I'm not feeling safe I don't. If I notice the kind of music I want to listen to, I can tell whether or not I'm feeling safe. (Music is often how we create a felt soundtrack for our lives, so this is an interesting way to observe the tilt of our bodies obliquely.)

When we are not feeling safe, we generally do not drink enough water. Take note of this. It is a physiological response to stress. Stress and dehydration are a viscious self-reinforcing cycle.

See if you can approach your own felt sense of safety like a researcher who was studying you, helping you to seek to elicit this response. Take note of the things you like to do when you are feeling safe, because when you are not you will not remember them, but they can serve you.

This living room, with its simple stone hearth and ancient wooden bench sends me cues of safety.

Make a list of these. Notice when you are not doing them. Ask yourself, when that happens, if doing them could help shift you back into feeling safer.

NOTICING IN PLACE

Where do you feel safest in your home? I mean, physically and in an embodied manner. Is it the kitchen? The bathroom? The garden? The bedroom? Why? See if you can figure out what, at an embodied level, is generating cues of safety for you. What are the poetics of space as they generate or reduce your felt sense of safety? If there are places in the house that you do not feel safe, why is that? Attending to this helps to build your conscious awareness of neuroception- the moment-to-moment embodied feeling of safety and danger, and of constructing environments that serve our deepest needs. Our neuroception of safety can be activated by feeling like we are in control, not vulnerable, able to perceive our surroundings, as well as by acoustics, textures, and the look and feel of space and color and architecture. If you made it an explicit goal to have a home that granted you a deeper felt sense of safety, what would you do to make the space different? How could you make it warmer, more inviting, more restorative?

This can come down to very specific details, such as where a chair is positioned in relationship to a window. If you can be seen from behind by someone on the street, for example, your body is likely to feel more vulnerable. Attending to the spatial aspect of this– like rotating the chair so that you won't be watched from behind– can change the way it feels to be in your space.

More broadly, what spaces that you move through in your life do you feel safest in? As you move from home to work to activities, etc., what are the spaces and places that elicit a bodily feeling of safety? Which do not? How can you invite more safety?

NOTICING IN NATURE

If you have access to the Living World (I hope you do), what qualities of the Living World convey a sense of safety to you? There has been considerable research done, for example, on two types of primal view situations. One is a vista view, which tends to be from elevation, looking out, and provides you a broad view of who might be approaching. This view situation tends to convey an experience of spaciousness, and give you clear views of approaching predators. This is why real estate is often more expensive 'in the hills'. The other view is known as a 'refuge view'. Refuge views tended to be tucked in, nested, cozy, and convey a sense of security that is more womb-like. When kids build tunnels, or tuck themselves into small nooks and crannies, they are seeking a refuge view. From a refuge view, we can often look out on a broader landscape while experiencing a sense of protection, and controlling how visible we are from outside. What are the attributes and qualities of places in nature that you feel safe? Do you find yourself drawn to the expanse and height of mountains? The tucked-in qualities of refuge? Notice and take note.

NOTICING IN YOUR BODY

If you think of your body as a living architecture, everything about the way that you carry your structure - tension in muscles and joints, openness of chest, verticality of spine, balance of head on neck, is connected to the neural patterns of the Autonomic Nervous System. The body you wear when you feel safe is different from the body that you wear when you feel in danger. For that matter, the body you wear when you are angry is different from the body you wear when you are anxious is different from the body you wear when you are shut down. Can you feel the way your body wants to move, open, stretch when you feel safe? Can you give it this permission? Can you, like the cat, spend more time each day stretching, opening the hips, opening the chest, freeing the joints so that your body can respond fully and kinesthetically to life? If so, what does that feel like?

NOTICING THROUGH CLOTHING

What kind of clothes do you feel safest in? Does some of your clothing feel like armor? Does some of it feel like a nest? To what degree does your personal sense of style, and your comfort with that style, provide you a felt sense of safety? How do color choices, jewelry, ornament, hairstyle, tattoos, piercings, etc., play into this? I can wear a black leather motorcycle jacket that says 'back off', or a camel-colored dad cardigan that says, "I'll help you re-tie that shoelace." Think Mr. T versus Mr. Rogers. The clothing that I wear, often without me fully realizing it, reflects the level of safety or threat that I am feeling. Like music, it can provide an oblique reflection. So too do our choices of ornament. We are wearing our physical bodies, and we are also dressing and styling them in ways that broadcast social (and or protection) cues. If you consider safety through the lens of dress, what do you notice? No judgment.

NOTICING THROUGH OBJECTS

A certain Pentel pen feels comfortable in my hand–I don't think of it as a 'safe' pen, but safety arises when I hold it. The silverware we have includes a fork with a particular shape. In our silverware drawer there are also a smattering of other forks. If someone sets the table and I end up with one of them, I'll swap it out. These are objects that talk to my hands, give me a feeling of safety. My tennis racquet. A certain necklace I wear. A couple of rings. Objects imbued with something relational. Do you have objects like this? Do you *know* that you do?

NOTICING THROUGH RELATIONSHIPS

With whom do you feel safest and why? Think about who you share personal stories with, and why? What are the qualities or attributes in others that invite your sense of safety? What can others do, or not do, that helps you feel safe being vulnerable or intimate with them? If you are not feeling safe with someone, what, if anything, can they do to help re-assure you? Where are your redlines, e.g., those places that if someone crosses will break your trust in them? How do you navigate boundaries with people you must work with, or communicate with, but with whom you are not able to feel fully safe? How do you take care of yourself in these relationships, be they with co-workers, bosses, or others who might have structural or positional authority that can impact upon your life?

NOTICING THROUGH TRIBE

Each of us exists in a variety of social locations. These include our gender, sexual orientation, ethnicity, geography, class, religion, nationality, etc. Each of these locations, which structure a facet of our embodied experience, exists somewhere along a continuum from what is centered in our culture(s), e.g., considered normal, normative, or privileged, to what is more marginalized, e.g., de-centered, considered impoverished, or even ascribed deviant. The more our locations are centered, e.g., reenforced as normative, the less we have to be aware of this, because the less our identity locations cause friction for those in power, and therefore the less they evoke threat response from power, structurally diminishing our access to safety. The United States, where I live, centers Whiteness, Maleness, Christianity, Hetero-sexuality, traditional gender identity and presentation, Affluence, and US Exceptionalism. We could label this centralizing as White Supremacy, Patriarchy, Christian Supremacy, Hetero-supremacy, Capitalist ethics, and US Nationalism. Each of these structures social and relational gravity in the collective soma (the collective body). The degree to which my embodied identity aligns with the templates of this Supremacy gravity is the degree to which I am aligned with the locations in this society that confer unearned safety.

If I am a white, rich, hetero-sexual, cis-presenting Christian Nationalist man, I basically occupy the social location of rule-writing in this country, to the degree that I may be astonished, if and when I invade the seat of Federal Government, that they don't just open the doors for me. If you find it strange that I am saying this, you haven't listened to interviews with people who came to Washington DC on January 6, 2020 to disrupt the Federal certification of the Presidential election in the US House of Congress. People were like, "We traveled all this way to take over the government, and they wouldn't even let us in." From this location, it's like–*Wait a minute, it's called the damn White House- we made that house. That house works for me!*

When you occupy a centered social location, you don't feel the gravity. You are standing in the location from which the lines of gravitational force emanate. Yet step out of the center, out of maleness, out of whiteness, out of straightness, out of gender binary, out of wealth, out of Christianity, etc., and it's like, *Whoa, there's a lot of force moving through here.*

The farther you get from the center, the less safety is granted. Centered locations are generally invisible to people who occupy them. They constitute 'normalcy'. Part of the necessity of Michel Foucault's analyses of power are that they provide us the conceptual tools to visibilize these structural relations, which, without critical awareness, are often experienced as either unearned elevations (unearned privilege) or unearned diminishments (unearned deprivations), depending on where you are situated in relation to the center of a particular axis of social location/ continuum of power.

The reason that I'm drawing out this analysis is to help you put a conceptual framework around what is essentially an embodied experience, because each of us is welcomed/ excluded, granted safety/ denied safety, broadly speaking, by the collective soma (the collective neural gravitational architecture) structured by what is centered in our societies.

Lest this seem theoretical to you, consider that your teenagers are navigating this landscape of inclusion/exclusion every single time they navigate friend groups, clothing choices, what it means to be seen with this or that person, this or that social group, what music they listen to, what movies they watch, which sports teams they are fans of, whether or not to be emotionally vulnerable or sarcastic. Adolescence is an embodied experience of navigating how to find safety and belonging in a shifting landscape of embodied power relations. Popularity is the word for this aggregate social capital that ebbs and flows, creating gravity differentials among teenagers based on their social locations, appearance, choices, attitudes, activities, etc. The expansion of civil rights, the movement to grant greater freedoms, is the movement of enlarging the center to include a broader ranges of subjectivities and locations. I will note here, for pedagogical purposes, that this whole construction of centers of gravity is predicated on a domination paradigm. Where relations are horizontal, it is not necessary, and is in fact in-advisable, to center a particular location, because that location will be blind, i.e., architected to be without consciousness of itself, i.e., structurally vacant of self-awareness. In plain language, if you exist in the center, everyone is looking at you and you can't see yourself. This is part of why celebrities are often so unhappy.

All of this duly noted, this 10th exercise, around safety in tribes, has to do with understanding where the different aspects of our social locations feel more or less safe, and with whom, and under what circumstances, and why.

Let me give you an example. We moderns are people who like to use adjectives in front of our nouns to qualify who we are. Depending on mood and context, I might introduce myself as a father, a CEO, a connection phenomenologist, a writer, a person of Jewish ancestry, a tennis player, a competitive tennis player, a tennis player with a 5.0 USTA ranking. Each of these adjectives, or adjective combinations, has a particular set of valences and connotations and meanings in certain cultural subsets. In some contexts, such as a professional context, introducing myself as a competitive tennis player is irrelevant. So I, like most of us, calibrate the adjective(s) I use to describe myself based on with whom I'm speaking and about what. If you don't know me, you might not know that I have Jewish ancestry. I can selectively choose whether or not to highlight this attribute. (My bodily awareness that in this fraught political moment I am less interested in highlighting this is evidence that I can feel the lines of force pushing through this less-centered location.) My maleness, and my whiteness is more public-facing.

My friends who are women and/or Black may choose to veil their geography, class, or religion, but generally are seen as women and/or Black. These locations are more or less visible, and we can, in our curation of self, either exaggerate or diminish or ignore them.

So we wear our social locations in different ways. We curate them, consciously or less so. What I'm inviting you to do here, in this 10th mode of listening for safety, is to consider the ways that you are curating your identity, and what parts of you, to use Master Mindfulness and Diversity Trainer Lee Mun Wah's words, *you are leaving at the door.*

In order to experience safety, there are times when we leave parts of ourselves at the door. To experience more economic safety, there are times when have left my class background at the door. Times when I have left my Jewish ancestry at the door. Times when I've left my intelligence at the door. Do you understand what I am saying? I have Black friends who are executives and run organizations who have to leave parts of their Blackness at the door. Friends who are professional women who have to leave parts of their femininity at the door.

The body I personally am wearing is conferred a lot of unearned structural privilege. Yet there are still times when I turn up or down the volume on aspects of self, in pursuit of various kinds of safety. Most of us do this. We don't wear the same thing at home on Sunday morning that we wear to the office. We also don't necessarily wear the same body.

My invitation is that you inventory your life, and the tribes you are part of (by which I mean the social groups that you identify with– friends, colleagues, hobby groups, religious groups, affinity groups) to look clearly at the types of safety you derive from your affiliations, as well as what you might be giving up to maintain those associations.

I'm not asking you to DO anything with this awareness. I simply am inviting you to consider sociality, and the landscape of your social life, through this lens. Where, if anywhere, do you trade one category of safety for another?

There is no judgement in this– it is a simply a kind of mirror you can look into to better understand the choices your body is making to navigate a world that doesn't prioritize keeping us safe and connected. A way to catch a glimpse of parts of ourselves that can be hard to see directly.

Whatever you notice, be gentle with yourself. You may find that this exercise is emotionally evocative or distressing. You may discover that you've given up parts of yourself along the journey to where you are. Most of us have.

This sacrifice is the shape of your illbeing.

As your guide here, my objective is to gently, tenderly, and carefully help you restore your sovereignty: the complete beauty of who you are, not the version curated for consumption by a society motivated by domination. If there are parts of self that you have buried, disowned, disaffiliated, dismembered–and they are genuinely part of your spirit–the journey to wellness will including resurrecting them.

You might want to pause at this natural fulcrum of the book and let it sink in. If you are feeling grief, it is sacred. Once you've been present with this deeply, we will move to the second doorway: the doorway of connection.

CHAPTER 16

SAFETY IN A WORLD OF EXISTENTIAL THREAT

One of the crucial contributions to the modern discourse on wellness is surely Dr. Stephen Porges' conceptualization of neuroception, with its concomittant awareness of the role that an embodied felt present moment registration of safety, danger, or lifethreat has upon the variable recruitment of Autonomic systems.

Part of what Polyvagal Theory has shown us, elegantly, is that we have access to different neural circuitry when we feel safe, as compared to when we are in danger, and different neural circuitry still when we are in danger as compared to when we are in lifethreat.

Porges had to invent a word to describe the neurological faculty of this discernment, since there is no word that exists in English adequate to the task. He chose *neuroception*, from *neuro-*, for neurological, and *-ception*, etymologically originating in the Latin, where it means layers, and could be taken to mean 'nested inside itself'.

So this inward neurological perception, that drives our sensory and interoceptive engagement with the environment as we check for safety inwardly and outwardly– the same way your dog, when introduced to a novel environment will sniff, walk the perimeter, and circle before marking territory and settling down to rest– governs the autonomic systems that center our experience.

What we then know is that a bodily felt registration of safety is required for flourishing. It is pre-requisite to wellness. To be perfectly clear, safety is the frequency whose pulsative cadence organizes your physiology, and allows your Connection System to coordinate the activities of your other autonomic systems.

Many traditions, including ancestral and Indigenous healing lineages from around the world, are aware of, have conceptualized, and work with this pulsation of safety. It is not abstract; rather it is a coordinated waveform that can be assessed physiologically and the coherence of which can be measured quantifiably. It is not Heart-Rate Variability, but is reflected in heartrate variability. It is not a function of breath, but is reflected precisely in breath. While signals of danger are constant, signals of safety are always pulsatile in nature.

Learning autonomic physiology is an embodied learning process, and it takes time because it begins to reorganize your experience of yourself. What do I mean by this? As we, for example, begin to understand lifethreat responses, and begin to review our own pasts for places where we have moved into shutdown, and begin to do the healing work to transform these places, our systems begin to heal. As this process unfolds, endogenously, it begins to change our physical bodies, the way we occupy them emotionally, the degree of our embodiment, how much time we spend in our heads, our understanding of our own histories, how we feel in our skin, and who we understand ourselves to be.

For people who have been using the maps of autonomic physiology as frameworks to help them transform their experience *of their experiences*, which is the most useful way to relate to these materials, there is an experiential confirmation of its reality that begins to seep into your bones. Learning and embodying these maps, particularly in community, allows us to directly grasp the primary levers of our own transformation. And this process takes time to unfold.

One of the students in my Autonomics seminar, an activist, aware of the degree of existential threat all around us – endless war, ecological catastrophe, economic turmoil – simply stated that 'Safety is an illusion.'

I think we can boil this down to a single question: *What do we mean by safety in a world of existential threat?* I contend, in agreement with this person, that we are in collapse at the end-stage of capitalism. All you have to do is look at the news cycle to be reminded of the encroachment of existential threat.

So what do we mean by safety in the midst of all of this? As I grapple, it becomes apparent to me that safety is an approximation for a territory that we would do well to spend some time naming in ways that enrich the associational matrix of the meanings we would like to accrete as it relates to our wellbeing.

Each of you reading this is held within the gravitational field of a planet hurtling through the void of interstellar space at unfathomable speeds, spinning on an axis, where at the present time human civilization so-called is engaged in the projects of undermining the integrity of the only biosphere in the known universe, blowing one another up, and binge-watching Netflix on smartphones to dissociate from all of this. For those of us making an effort to unplug from the machine and soberly assess how to live in a grounded and loving manner amidst the plunderings of the extraction engine of end-stage capitalism, while attempting to feed our families, and stay sane, it is an important and legitimate question. Does safety even exist at an existential level? Which is to say nothing of the more granular negotiations of the everyday- from engagements that are strained because of money, gender, ethnicity, age, religion, family, etc. Does safety even exist?

And while I cannot answer this in an absolute sense, beyond wondering aloud with you– in our several million year long species heritage, when we were still wandering in small bands of hunter gatherers as we did for 99% of our lineage history, still held within the ambit of ancestral human normalcy evolutionarily speaking– gathered around the fire on the savannah, at night, with apex predators loping through the dark, would we not have wondered the same thing?

And yet, in this archetypal formulation, down in the very roots, the toes if you will of our humanness, comes a possible response. Out there, in the night, *No, safety does not exist. Alone, solitary, in the dark– No, safety does not exist.* Yet here, in the glow of the fire, little ones held close, their tiny fingers and toes enclosed by adult bodies nearest the warmth of the fire, and eyes looking out, watching, *Yes, there is more of it. Some possible temporary island of it.*

Here I propose some alternatives to the word safety, because the language of the deep belly is not a word, it is a feeling. What is the feeling we are pointing at?

What if, as Gabi Jubran proposes, we think of this in terms of home? At the dawn of western culture, as I was taught it representationally through the canonical books that delineate a meta-narrative trajectory of the modern mind, is *The Odyssey*, Homer's epic poem. It tells the story of Odysseus, that man skilled in all ways of contending, who is stranded far from home and hearth. He is filled with yearning for home. In the Greek, *Nostos.*

This yearning animates him throughout the text. Adam Johnson, Pulitzer-prize winning fiction writer and my creative writing teacher, taught that yearning is the great governor of character. Not just for Odysseus, but all of us. And the great yearning of character, the archetypal yearning at the root of the story of the modern world is a yearning for home.

The Odyssey, to my understanding, yields most readily to interpretation as a text that grapples with the question: *How do we get home from war?*

This is the literal line of the story– our hero is stranded in the wake of war, but *The Odyssey* is a psycho-symbolic narrative, I would propose. The war is inward. The question is how do we get home, inwardly, from war, which is itself a palimpsest for alienation.

Distillation of the Odyssey: *How do we overcome separation and re-unite with the Source?*

And so what if, instead of safety, we mean '*At-homeness*'?

Can we access a relative version of that? Can we feel 'at-home' enough to let our guard down, to bring all of our autonomic systems online? To coordinate our breathing up from our deep bellies in an inhalation of belonging?

The first Tibetan Buddhist retreat I ever attended, prior to each session of discourse with the Rinpoche, there was an invocation amongst the assembled to generate the energy of *boddhichitta*. This translates roughly as *awakening-mind*. We could also call it the yearning-to-awaken. It is a verb-based feeling. A reaching, as towards the Sun, even if we cannot see it.

If, at some level, safety is an illusion, can we awaken within ourselves the *yearning-for-home*?

Can we make our way down, out of our heads, out of the cranial certainty, out of the depthless loss that burbles and rants as incessant naming, and home to the deep belly center of ourselves, a silence in subterranean caverns of self, and from here can we yearn-for-home? Can we reach for it?

And in the proper context, with the proper support, can touching into the possibility of this ignite the re-organization of our physiological systems to draw us back into the coherent pulse of life?

Can we prostrate ourselves before the difficulty of it all, link arms, and find our way home?

DOORWAY OF CONNECTION

CHAPTER 17

CONNECTION IS THE FOOD

There is a genius waiting to come through you: the genius of Nature. Yet this genius, to strike, must come through the spark of who you are: the snowflake signature of your innate vitality. It must pass through the chamber of why you are here. In the Tzutuji'l Mayan dialect of Santiago Atitlan, in present-day Guatemala, the word that is translated as to live is *N Kasea*. Yet what N Kasea really means is 'the ember yet burns.' This ember is who you are: your particular genius resides where this ember is fanned by the flames of Nature.

There is a Native American saying that reminds us that the farthest and most significant journey a human can ever take is the 18 inches from the human head to the human heart. And this journey, its consummation, is where your sovereignty and your destiny await. What guides you on this path is yearning. It is the same force that guided Odysseus, man skilled in all ways of contending, on his journey back to home and hearth over 3,000 years ago, for Odysseus, even then, was stranded by alienation. Odysseus is modern in this regard. We say that home is where the heart is, but to configure this notion in a slightly different way, which puts the emphasis in the proper place, find your heart and you have found your true home.

Find your heart and you have found your true home.

And let us be clear that your heart resides not in another person, though another person can spark it into flame, certainly. Your heart resides at the intersection of your unique vitality, your particular ember, and the flames of Nature: the Source context. Where your particular ember meets the Original Fire. This journey that we can take as humans is one that cannot be guided by cognition. Thinking cannot bring us home to our hearts. Thinking is, in actual fact, useless here. The journey home to the heart takes place *beneath* thinking. You know that you are on the journey when you feel the gravity of forces guiding you, but don't understand what is guiding you. A homing instinct: intuition.

For modern people, accustomed to worshipping sky gods, trained to look to the heavens above, you may be surprised to learn that the most fruitful direction you can turn, should you wish to come home to your heart, is down. My friend Tiokasin Ghosthorse, of the Cheyenne River Lakota people, laments that people talk about inspiration as receiving a download. "What about an upload?" he asks. "What about an upload from the Earth." And this is, indeed, a marvelous question. *What about an upload?* What happens if we turn our attention not up, not out, not to the stars, but to the good earth beneath our feet? What happens if we turn our attention down, to look at the ground?

Most of us, directing the gaze downward, staring at our own feet, will notice that there is rubber interposed between our feet and the earth: the soles of our shoes. Take off your shoes and let your feet themselves rest on the good earth. And now take a moment to consider the fact that all of your life, you have been cut off from rootedness. This is part of what it means to be modern. Modern life is a conspiracy to teach you that you are not an earth-based creature. Yet you are a terrestrial mammal, not an untethered brain. This is the deviation of modernity. The notion that we are above the earth, outside of her laws, not four-legged mammals walking upright.

The story of modernity is the story of supremacy, the story of power *over*. It is the story of the human transition from laterality (kinship spread across the earth), reciprocity (I and thou), and horizontality (the circle) to domination and verticality. The fever dream of domination: the story of empire. You and I are the orphaned children of Rome. Descartes, writing, "I think therefore I am," is writing as a dutiful imperial citizen, whether he knows it or not.

And here, in this simple and profound error– for you *be* not simply because you *think*– is the organizing principle of what has pulled you up out of your heart. The inheritance of millenia of trauma. For time, and thinking, in the way modern people do it, began the moment that we left our hearts.[1]

And this leave-taking happened, in all probability, some sixty-five or seventy thousand years ago, with the first diasporas of the San heading out from the Kalahari, our ancestors in deep time who had turned war-like, fractured relatedness through murder, and developed the genetic modification known as the war gene (the adrenergic a2C receptor (ADRA2C).[2]

It is appropriate that the Odyssey begins in the wake of a war. We enter the scene *en medias res*, with our hero, a stand-in for modernity, marooned by war thousands of miles from home. If we can see through the particulars of the moment, see Troy as simple stage decoration, the situation that we are being shown is the roots of the world we have come to inherit. Here is someone skilled in all ways of contending, someone gifted in all the arts of culture and of life, who is still alienated. Odysseus is filled with longing to return home, but it is not simply to his wife, Penelope, his son Telemachus, or even his native country, his kingly hall. Here is a man stranded on the far side of war, and although the formulation in the Homeric text configures the obstacles to his return as Sirens, as sea monsters, as

---

1 It was our friend and mentor, Kuuyux Ilarion Merculieff, of the Unangan people, who first explained to me that time began the moment that we left our hearts.

2 The war gene– a genetic modification that modern people carry, and that the San people do not– is a concept that I originally encountered in a series of lectures by connection phenomenologist Jon Young that were given at Schumacher College, and are available on Youtube. *Young, Jon. Nature & Genius: Awakening our Ancient Potential for Health, Happiness, and Creative Emergence, July 10, 2016* Further research into this directs us to the Adrenergic a2C receptor (ADRA2C), which is a genetic modification on the neuronal control of the fight-flight-response and increases aggression. *https://journals.plos.org/plosgenetics/article?id=10.1371/journal.pgen.1007311*

a one-eyed Cyclops, these are again, seen clearly, the artifacts of trauma. The nightmare obstacles that stand between him and the reclaiming of his heart.

Odysseus, the sovereign in exile, is far from home and hearth. He is alienated, lost in the fog of war, and yet filled with yearning. And it is this yearning, the yearning the Buddhists call *boddhichitta*, an active, burning desire to awaken. To transform. To come home from exile. In from the cold, out of the alien, back to home and hearth. *Bodhi*, from the Sanskrit for 'awakening', *citta* from the Sanskrit root cit, meaning 'that which is conscious.' Translated as *awakening mind*.

And yet this awakening mind, this fire of yearning, longing for home, *nostos*, that burns in Odysseus, it is not a *mental* fire. It is not in his brain. It is not a thought, not a cognition. This fire is in the heart.

The Indian spiritual physicians of 3,000 years ago, writing at the time of the Rig Vedas, knew so well the neurocardiology of the heart that they described, in minute detail, the sino-atrial node that is her pacemaker. This tiny structure, they explained, the size of a grain of rice is lit with a blue flame (the venous architecture), and governs the yearning of the heart. And it is true, neuro-physiologically. The sino-atrial node receives inputs from all autonomic systems. It produces, effectuates in the cadence, rhythm, and contractile pressure of the heart, the translation of your bodily sense of safety or danger into the heart rate variability that expresses it in realtime.

What then feeds the heart? What are the portals to the heart whereby we can journey in our quest to find our sovereign selves, to heal from the hurts of the world, and to transform? And let us be clear that *transformation* is the word, not simply healing. Healing implies the return to some state that was before; and yet we cannot go backwards. Time doesn't work like this.

Time doesn't exist, in truth. The spiral we feel of her is rather the perturbations of eternity that come from having left our hearts. We cannot return to what was before any more than we can become younger. Our only hope is alchemical. What we are seeking is *transformation*. To melt down the experiences of grief, of loss, of alienation, in the crucible of the heart, to allow them to transform into something else. To earn, from the combustion of these base metals, from all the things that have hurt us, the gift of growing into the fullness of our hearts.

There are portals to the heart that are both spiritual and neurophysiological. The face, the eyes, the ears, the voice, the breath, the hands, the feet, the skin, rhythm, gesture, posture, the genitals, and language itself. Through each one is a pathway toward the reclamation of who you are. Come now, let us continue on a journey in the direction of home. You have passed through the doorway of safety; let us now open the doorway of connection.

A BRIEF NOTE ON THE RELATIONSHIP BETWEEN ATTENTION AND AUTONOMIC STATE

Before we dive deeper into the portals to connection, I would like to clear up a confusion with respect to the relationship between mindful awareness and Autonomic state, because lack of clarity about the ordering of this relationship is systematically causing awareness practices to be applied in ways that are not useful, and sometimes harmful. What I am going to teach you with the portals of connection are a series of approaches to working with attention. And before we do that, I want you to understand the relationship between attention and Autonomic State. The way attention functions, in the present moment (which is the only place attention functions) *is a derivative of present moment Autonomic State.*

When I say derivative, I mean it in the mathematical sense. A derivative in calculus is the instantaneous rate of change of a function with respect to another variable. In the human nervous system, the controlling variable that determines how attention functions, e.g., its depth, scope, shape, directional orientation (inward/outward), is the present moment state of the ANS. Whatever the dominant autonomic state in the present moment, this will control the flow/shape of attention. This is the energy-processing template through which the river of your experience is flowing.

If we simplify this, we can say that the primary governor of attention is neuroception. There are three primary neuroceptions: safety, danger, or lifethreat. Traditional mindfulness practices were developed under the assumption of a neuroception of safety. The physics of attention mapped by these system (Buddhist inner science, Taoist, etc.) presume a baseline in the neuroception of safety.

Traditional mindfulness practices assume a baseline neuroception of safety.

If we are neurocepting danger, e.g., our center of gravity in the present moment is an experience of threat, this will alter the shape of attention in predictable ways.

Attention potentiates experience. It is a form of magnification. This means, therefore, that when we are neurocepting danger, traditional inward-oriented mindfulness practices are contra-indicated, because they will potentiate (magnify) neuroception of danger. Again, practically, what this means is that unless someone is a very advanced meditator, having them engage in traditional mindfulness practices when they are feeling under threat (angry, anxious, etc.) will predictably make them feel worse.

Mindfulness practices developed for the baseline neuroception of danger are called martial arts.

If someone has moved into lifethreat, across the threshold into a dorsal vagal state while in a neuroception of threat, traditional mindfulness practices will incresae their dissociation. E.g., in this nervous system state, traditional mindfulness instruction will make someone more dissociated. This is the opposite of what is useful.

There are no mindfulness practices developed for the baseline neuroception of lifethreat.

I am spelling this out here because it is so widely misunderstood, and because as you begin to work with portals to connection, it is important to understand that coming to them through the doorway of a felt sense of safety is necessary to open them in ways that lead to the deepest embodied sense of connection.

First safety, then connection.

CHAPTER 18

THE BODY YOU ARE WEARING

Scientific revolutions often arise from challenge: something inexplicable that doesn't seem to make any sense and yet is true, which requires us to re-examine our fundamental understandings about how reality works. The origins of the Polyvagal Theory were in a paradox that Dr. Stephen Porges refers to as the 'vagal paradox'. At the time he formulated the paradox, Dr. Porges was a psychophysiologist running a laboratory studying heart-rate variability in infants, and published a paper in neonatology about the protective features of the vagus in preterm infants. The Vagus is the 10th cranial nerve, and the primary conduit of what orthodox neurology calls the parasympathetic nervous system. Porges' lab had discovered that a healthy vagal system was protective to newborn wellbeing, but after publishing the paper received a letter from a neonatologist noting his admiration, but explaining that he had learned in medical school that the Vagus could kill an infant. This 'killer' vagus was a reference to sudden drops in infant heartrate that could be lethal. Porges was left confronted by a paradox: how could the same nerve be both protective and lethal? He spent several months with the letter in his briefcase, and then began a deep study of comparative evolutionary neuro-anatomy.

The method by which he eventually arrived at discerning the social engagement physiology or Ventral Vagal Complex, the mammalian neurological system that unites the neural regulation of the face, voice, sucking and swallowing, with the neural regulation of the heart and the lungs into a single functional unit, was based largely on comparative evolutionary neuroanatomy. The development of an embryo is a window into deep time, and deep time is a window into how we develop. Somewhere around 200 million years ago, a second (ventral) cardio-inhibitory (heart-slowing) vagal pathway began to differentiate itself from the oldest (deep belly/ enteric) cardioinhibitory pathway. These evolutionary studies showed a 'ventral migration' (a movement forward) of a group of cardio-inhibitory (heart-slowing) neurons that moved from their original location in the Dorsal Motor Nucleus (DMX) of the brainstem forward into the Nucleus Ambiguus (NA). Through the long spans of evolutionary time, these ventrally migrating neurons begin to link up, through interneurons, with a set of neurons that innervated the ancient gill arches (the breathing apparatus) of bony fish. This migrating complex evolved into the neural regulation of aspects of the mammalian face and voice: specifically an anatomically defined and functionally integrated Social Engagement System.[1] The Vagus is not merely a nerve. It is a conduit that holds two distinct neural systems, both of which slow the heart. An ancient unmyelinated system that is primarily sub-diaphragmatic (it originates in the guts) and flows into dorsal areas of the brainstem (Porges call this the Dorsal Vagal System, we call it the Grounding System to denote that it can be harnessed by safety), and an evolutionarily more recent myelinated system that innervates the face, voice, and connection systems and flows into ventral areas of the brainstem (Porges calls this the Social Engagement System, we call it the Connection System).

The Social Engagement system is a face/voice complex that unites the neural regulation of the face and voice, synchronizes sucking and swallowing, and integrates the heart and breath. This complex unites the neural regulation of the face, voice, larynx, pharynx, tongue, sternocleidomastoid muscle, upper trapezius (both muscles necessary for the baby to find the breast, turn the head, and nurse) with the heart and breath into a functional unit. It activates to coordinate sucking and swallowing with breathing in a rhythmic pulse. Part of Porges' certainty that this complex functioned as a neuro-anatomical unit (the face-voice complex) came from the observation that

---

1 Porges, S. *The Vagal Paradox: A Polyvagal Solution.*
"As the cardioinhibitory neurons migrated ventrally, regulation of the structures that emerged from the ancient gill arches (facial and head structures in mammals) appear to have developed interneuronal connections with the ventral cardioinhibitory neurons. In mammals, the product of this brainstem neuroanatomical integration links the ventral vagal cardioinhibitory nucleus with nuclei that regulate sucking and social cueing via facial expression and vocalizations. Functionally, this neuronal circuit provided reliable pathways (e.g., vocalizations) to communicate autonomic state to conspecifics (e.g., others of the same species). Developmentally, this is easily observed in humans because the social engagement circuit is active in full term newborns, creating an adaptive portal for co-regulation between mother and infant."

unlike any other voluntary muscles in the newborn, this system is myelinated at birth. This is actually quite strange, because it gives the infant voluntary muscular control over the face, voice, eyes, and middle ear, which seems normal– because it is how we are born– until you recognize that the infant does not have voluntary control over any other skeletal muscles.[2]

Think about it. A human infant is helpless at birth. It cannot hold its head up, sit, stand, or walk. It is also unable to articulate its interior state to the caregivers on whom it is utterly dependent for survival. And so this system of face (all facial expressions, eye gaze), voice, tuning of the middle ear, turning of the head and neck, uniting with the heart and breath, is the physiological portal whereby the caregiver is able to perceive the interior of the infant. We say that the eyes are the windows to the soul, yet the Face-Voice Complex, we could say, is the original doorway to the heart.[3]

Identifying the Social Engagement System is one of the most significant neurophysiology breakthroughs of the 20th century (another would be identification of the Grounding System). It is not simply the anatomy that is important, but its living functionality, something astonishingly difficult to deduce from the neural circuitry. These breakthroughs combined to scaffold the framework of the Polyvagal Theory, which heralded a revolution in our understanding of the human Autonomic Nervous System.

Yet the Face-Voice complex is not the full extent of the Con-

---

2 There are three types of muscle in the body. Smooth, cardiac, and skeletal. Smooth muscle, found in the walls of hollow visceral organs (such as the liver, pancreas, and intestines), is involuntary and controlled by the Autonomic Nervous System. Cardiac muscle is striated and involuntary. Skeletal muscle is the voluntary muscle that you think of when you imagine muscle. It is attached to the skeleton and is under conscious control. When we are born we cannot move our legs or arms because they are not yet myelinated. Only as they myelinate do we gain control of their function.

3 We say idiomatically that some people wear their hearts on their sleeves, but this is not accurate. We all wear our hearts on our faces and in our voices.

nection System. It is the central aspect of the Connection System that *is myelinated at birth.* Fortunately for you, your Connection System continues to myelinate as you grow. The Social Engagement system that Porges identified is a central subset of the Connection System we are elaborating in this book, but does not comprise its full extension.

Why does Polyvagal Theory constrain the Connection System to the face/voice complex? In order to understand this, it is necessary to understand the method whereby Porges arrived at delineating this system, and its criteria for inclusion. To set the theory on a tenable foundation, because it was assembled across disciplinary boundaries, he had to ground it in both in comparative evolutionary neuroanatomy, and embryonic development. Therefore part of the criteria for inclusion was that the system be myelinated at birth, when embryonic development was 'complete'. Yet this gives rise to an unusual problem, another paradox, because humans are not actually neurophysiologically complete when we are born.

Humans are unusual with respect to our gestational period in the following sense: human infants are born with only 25% of our mature brain volume. This is because our brains are so large that if our gestational period was longer, our heads would be too large to fit through the birth canal and we could not exit it. The gestation period for elephants is 24 months. Baby elephants arrive fully baked. Humans, not so. We are parbaked at birth – cooked less than halfway. We have to be finished outside the womb. Developmental theorists have conceptualized the first 18 months after birth as an external womb period. Ashley Montagu calls this, most artfully, "A womb with a view."

What this means, neurologically, is that our foundational neurological systems are not completely built until we've completed this womb-with-a-view period.[4] Our full gestation is

---

4
Consider that a horse can walk within five minutes of being born, yet that a human infant takes nearly a year to accomplish this task. One could argue further, with merit, that we are not completely built until the pre-frontal cortex is finished, which

not 9, but 27 months with the first nine inside, and the next 18 outside.

The full human gestational period is not 9 but 27 months, with 9 inside the womb, and the next 18 outside of it. [5]

If we want to understand the extent of the Connection System we need to look at its instantiation not just at birth, but when our neurology is completely myelinated. And that is at a year-and-a-half old. By this point, myelination of the Connection System is complete, and it includes a number of systems that are not online at birth, and are not part of the face/voice complex: the ventral surface of the hands, parts of the soles of the feet, the genitals, deep postural muscles, and the skin.

By this time, the developing human, now as a toddler, has fully arrived in a body. And ideally we have successfully completed a sequence of autonomic motor movements, in the context of our dyadic relationship with our caregivers, that are organizing our developing sense of self, boundaries, and experience. If we study Indigenous and Ancestral childrearing practices transculturally, looking at what elements of childrearing transcend time and place, we notice a variety of clear trends. One is that during the first 9 months after birth, the infant is typically swaddled and carried around on the body of a primary caregiver, front or back. *Infants typically are not placed on the ground.* After 9 months this shifts to the infant riding on the caregiver's back, looking over their shoulder. From 9-18 months the infant is crawling and developing into standing and walking.

This modeling shows us three distinct developmental phases coordinated with the three 9 month intervals of infant devel-

---

doesn't happen until women are in their early twenties and men in their late twenties. Completely built though, the way I'm using it, doesn't mean that our neurology is done transforming. That continues to happen throughout our lifetimes.

5 I encourage you to take a moment to reflect on what this means. The first 18 months of life are actually crucial neuro-developmental periods that will profoundly sculpt the rest of the life. How would society be different if we understood that for the first year-and-a-half of life, children were still being born?

opment. The first is accomplished in the internal womb, and lasts 9 months (womb). The second period happens after birth for about 9 months, with the infant swaddled against the caregiver's body (swaddled). The third happens from 9-18 months (observing). Developmentally, there is a sequence of priority in establishing neurology. The most crucial developmental period is the 9 months in utero. Second is the swaddling period (0-9 month). Third is the observation period (9-18 months). And then there is the rest of your life.

- 9 months in utero
- 9 months swaddled
- 9 months observing
- The rest of your life

All of us who are reading this are in the fourth stage. Neural templates and patterns set in the first three stages will occupy us for our entire lives.[6]

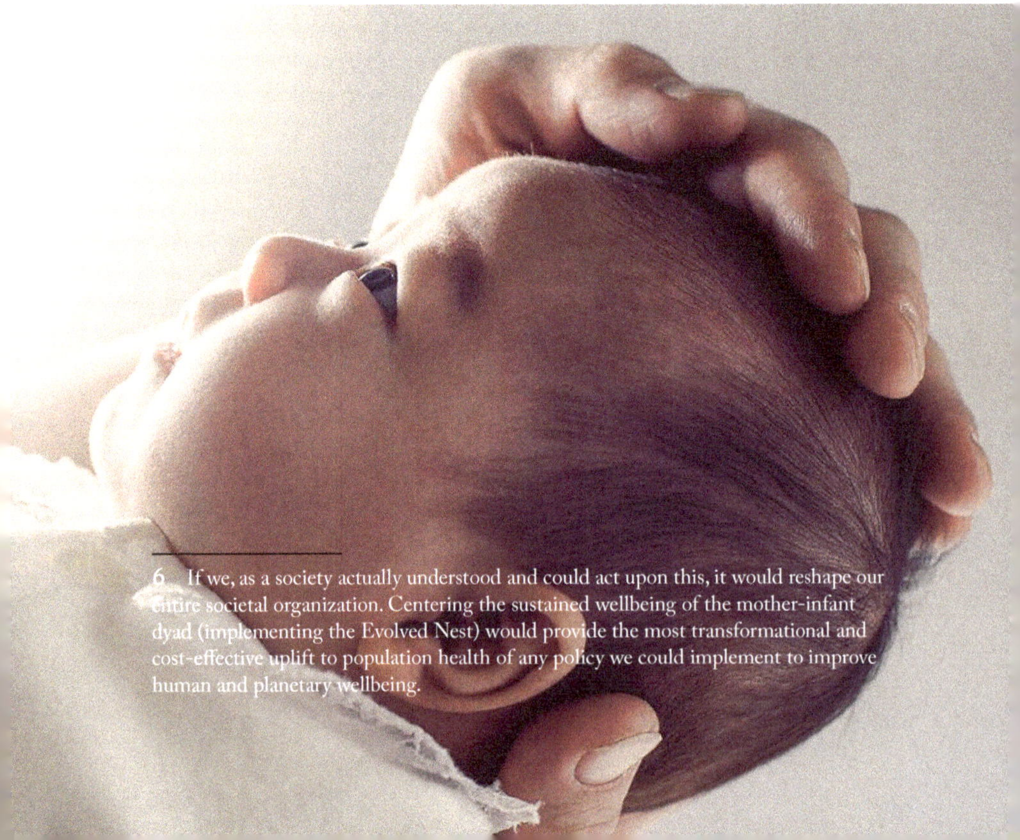

6 If we, as a society actually understood and could act upon this, it would reshape our entire societal organization. Centering the sustained wellbeing of the mother-infant dyad (implementing the Evolved Nest) would provide the most transformational and cost-effective uplift to population health of any policy we could implement to improve human and planetary wellbeing.

FONTANELLE

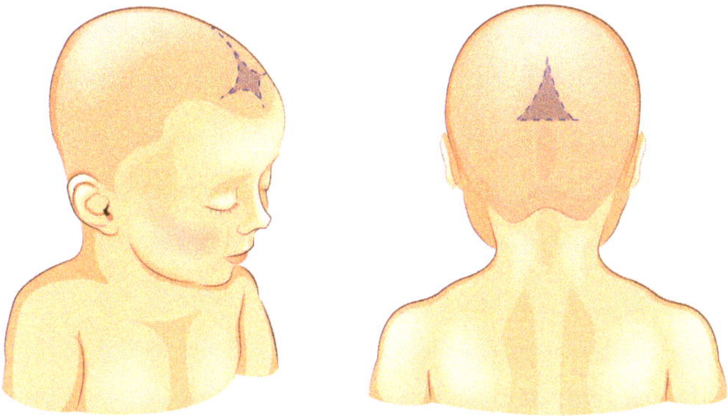

The Face/voice Complex is architected in utero during the 9 months of gestation. But our Connection System is not complete at this point.

We are born with compressible skulls by virtue of the anterior, posterior, sphenoid, and mastoid fontanelles, the soft spots in a baby's skulls where bone growth is not complete and the sutures in the skull have not solidified. This spacing allows the skull to compress slightly as it is passing through the birth canal, one of mother nature's elegant design accommodations to help us come into the world.

Until the fontanelles close, and you can witness this with a newborn, there is a soft spot on top of the head that is not protected by bone. The brain is just beneath. You can see this spot pulsing, at times, with the heartbeat.

In the beautiful image on the facing page, Dad's left thumb is over the anterior fontanelle. It is not visible in the photo, but if you look closely at baby's eye gaze it is apparent baby is staring into dad's eyes. Part of the reason I am drawing this out is connected to Indigenous and ancestral childrearing, and is the following. *Babies are extremely permeable to energy.* We arrive very sensitive, with very transparent (functionally non-existent) boundaries. The permeability is anatomical at this stage. The brain is not fully encased. The two primary fontanelles, the anterior and posterior, are along the brain's midline, essentially windows into the skull. The posterior fontanelle will close first, but the anterior may not fully knit closed until 18 months of age.[7] This is the anatomical completion of the 27-month infant developmental period.

With respect to the development of the ANS, these post-birth stages are crucial. For the first 9 months, ancestral childrearing techniques focused on several things. First, giving the baby consistent touch compression, similar to what we had in the womb. This uniform compressive touch, generally given through swaddling, is part of what is required to help baby develop embodied awareness. Compression is part of what gives us the neural feedback required to localize our awareness into the body. Baby, on arrival, is not localizing consciousness fully in the body. Awareness is not tethered in this way.

Not only is the brain itself more accessible to vibration because there are skylights in the skull, but the body itself has not solidified with respect to sensory inputs. So part of the role of compression is in continuing to help us find and localize our body. Skin plays a crucial role in this. The C-tactile fibers, which permeate the skin at a density of 9 square meters per inch, are cutaneous autonomic receptors. They are touch-sensitive. We learn to occupy our bodies, to feel ourselves from inside, to feel where we end and the world begins, because we experience touch. And compression is a form of uniform touch that give us a sensory experience of where the body ends. This

7 There is some variability to the timing of these closures and they do not exactly correlate with the neurodevelopmental schedules. In a review by the American Academy of Pediatrics from 1950, the age at which the anterior fontanelle closed in 1677 infants varied between 4 and 26 months, with 90% between 7 and 19 months.

This is what is happening neurologically for the first 18 months after birth. Baby is leafing out, and connections that are being used are myelinated; encased in protective sheathing to increase signal strength, precision, and transmission speed.

information is conveyed to us through receptors in the skin. During this period, post-birth, skin is being integrated into the Autonomic Nervous System.

In the deep analogy of the inward aboriginal neural forest, of the vagus as oak tree, touch compression is the sensory input that helps your Vagus leaf into a summer oak.
Broadly, we can say that skin is an under-utilized portal to the Connection System. Of all the clinical interventions of which I am aware that have the goal of increasing regulation and connection, using the skin as a portal is under-documented in the literature.

Yet if you think about how global inputs to the skin impact upon your autonomic experience, it becomes clear that you can use the skin consciously to change neural inputs. Think for a moment about getting in the water. Part of why taking a bath, or going for a swim, or taking a cold plunge can have such a deep impact upon autonomic state is because the skin is the largest organ in the human body, and when you change the inputs to the skin in broad and uniform manner, you generate tremendous novel neural inputs to the ANS.

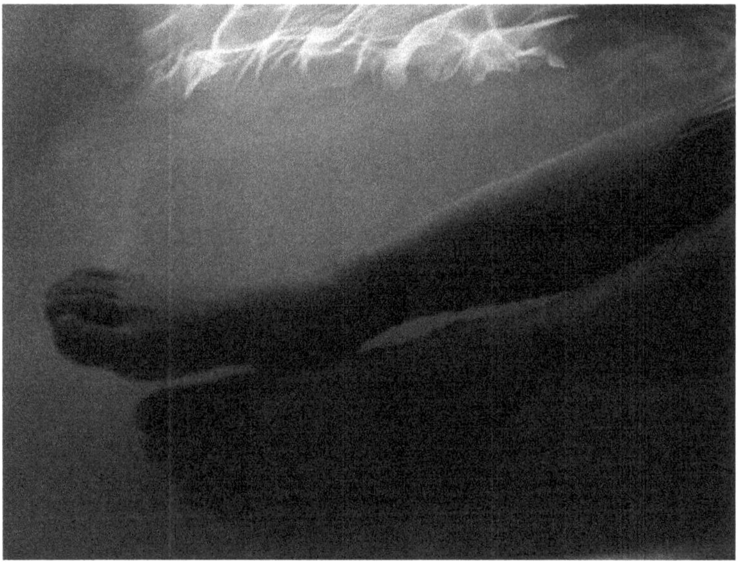

Much of the daily texture of our dysregulation– moments when we feel out of sorts– not severely distressed, but just off– happen because we are not conscious of the degree to which we are feeding ourselves the neural equivalent of candy, rather than neural inputs that are properly nutritious. Your iPhone is the neural equivalent of a Snickers bar. Scrolling Facebook is like tossing handfuls of M&Ms into your mouth. TikTok is a waterfall of Skittles. Sitting inside, in an office with blank walls, typing on a screen, which is what many of us are doing a lot, is the neural equivalent of feeding yourself a slow drip of morphine. It is deadening to your ANS. So after doing this for eight hours, we are likely to feel out-of-sorts.

The ambit of modernity is not designed to give your ANS the kind of juicy, vitalizing neural inputs that we crave. Social media, which has very effectively hacked our social biology, has moved into this barren landscape to provide a simulation of connection– yet it is in many ways worse than the desert, because although it looks like connection, it is not. It is like trying to eat those little food displays you sometimes see in old-school restaurant or sushi bar windows. It looks like real food until you put it in your mouth. Then you realize it is made out of wax or plastic. A simulacra.

Plastic food for restaurant display

Your connection system, which you want to continuously nourish so that you reside in a Connection state, is not cognitive at all. It is embodied, and the portals to it are the bodily conduits through which we relate. What this means is that providing pleasing inputs to this system, and attending to them (*savoring them*)– because attention is required to allow these inputs to actually register– is the recipe for maintaining yourself in a state of safety and connection.

In my humble opinion, everyone with a nervous system (that's all of us) should have a Brazilian- style hammock (in Portguguese this is called a *rede*, (a net). Unlike American-style hammocks, the Brazilian hammock has no cross-piece so it wraps around you, providing touch compression across your body. This combination of full-body compression and rocking is one of the most primal settling experiences for a mammalian nervous system, mimicking the rocking we received (hopefully) as a baby. This is one of those investments that will pay dividends for your wellbeing in multiples very quickly if you use it.

When you are in your hammock, to really make good use of it as a restorative practice, deeply attend to the sensation of being held, and allow yourself to feel and receive the sensation of being rocked and fully supported.

You can find this style of hammock in sporting goods stores where it will be labelled as a 'travel hammock'. These will generally be made from a compressible synthetic material. You can also find these on the internet by searching for 'brazilian-style hammock' or 'rede'. These will generally be made of cotton, and are heavier and more substantial.

If we really understand that our earliest infant experiences of support are laying the groundwork for our experience of regulation the rest of our lives, and we really take seriously the notion of the Evolved Nest, looking trans-culturally for the practices that have been used across cultures, since the dawn of time, to support the development of healthy humans, we see that swaddling an infant and placing them on the back of a regulated caregiver is one of the first and most central.

Infants are swaddled to help them develop a sense of boundary, and because safety is connected to containment. As adults, a sensory experience of compression is not something that we ordinarily receive. I remember marveling at animal behaviorist Temple Grandin's account of placing herself in a modified cattle squeeze chute to emulate this experience of being squeezed.

Temple discovered that being squeezed uniformly– uniform touch compression– was deeply settling to her nervous system. I would be out on a limb here, yet probably one that will bear my weight, if I observed to you that children born in the last twenty or thirty years are less likely to receive swaddling, and uniform touch compression, than generations before. Since these are ancestral techniques, they tend to be carried on through lineages and handed down from elders to youngers. But in the modern world, where children consistently move away from their parents to start their lives and their families, fewer and fewer parents are around experienced elders. We learn to parent from Youtube and TikTok.

My wife and I learned to swaddle our daughter from my mother-in-law and my daughter's godmother, who were both a generation older. Learning this felt like a form of magic. Swaddling instantly brings a neural sense of safety to baby. Our daughter would generally relax immediately. I had some resistance initially because it bothered me that my daughter's arms were strapped down. But at the time I didn't know anything about infant development, and I couldn't deny that despite how strange it looked to me, she would settle right away.

What I want you to understand here is the principle, neurologically. Because the skin is a conduit to the ANS, the way that we are touched (or the lack of touch) is a relevant input to our Connection System. The skin is the body's largest organ. To fail to utilize it to support regulation is to ignore the largest, most global, and potentially quickest way to transform inputs to the ANS.

Now, all of this said, you could simply drop $8000 on a cattle squeeze chute for your living room– they do make modified versions for humans. Alternatively, you could understand your ancestral neurobiology, get a Brazilian hammock, and figure out how to get more hugs and snuggles from people with whom you feel safe. I think you have a sense what direction I'll be encouraging you in. But if you do buy the squeeze machine, please invite me over because I'd like to try it.

TEN WAYS TO USE THE SKIN AS A SOURCE OF REGULATION

1. Get a *rede*, like I said.
2. Take a cold shower. (Not very pleasant, but watch what happens to your ANS.) In this category: splash cold water on your face, cold plunges, etc.
3. Swim. When you do this, intentionally allow part of your attention to anchor into the sensations of water on the skin. Make this a habit when you are immersed. Note that the crawl activates a sequence of central pattern generators in the spine.
4. Spend time outdoors in clothing that permits your full body to feel the ambient air temperature. Notice transitions between indoors and outdoors in this regard. In winter, for example, when you go outside, really allow your full body to respond to the change in temperature. Expose yourself to more temperature variation during the day. Practice being slightly cold. e.g., dress for slightly less warmth than you need. Do this as a practice of allowing the Autonomic Nervous System to adapt to temperature. Notice what happens.
5. Dr. Jeff Rockwell, a pioneer in autonomically-informed manual therapy, describes the kind of touch that nourishes your Autonomic Nervous System. He says, "For the social Autonomic Nervous System…very importantly, the sensory portion of these nerves exist in abundance in the skin and its adjacent superficial fascia, which makes them very available and amenable to gentle, kind touch. Hilton's Law states that "the nerves that innervate a joint innervate the muscles and skin superficial to it." The converse is also true. By touching the skin a particular way, we can communicate with the nerves…A gentle gliding movement on the skin, like a caress, is particularly effective. So is a "static caress," as in a gentle tangential stretch of the skin and tissues immediately superficial to it. It has been referred to by researchers as "emotional touch."
6. The Brazilians have a word, *cafune*, which is the art of gently caressing someone's head or hair. I don't know about

you, but having someone wash my hair can be just about one of the most pleasurable things on earth, simply at a sensory level.

7. If you watch professional tennis players, you will notice them blowing on the palm of the hand that holds the racquet between points. There are several things going on here, but one of them is the blowing on the skin part. Try blowing on the palms.
8. Use a weighted blanket. There are therapeutically-weighted blankets, but simply wrapping yourself in a heavy blanket and attending to sensation is useful. If you wanna go down this rabbit hole, there are full sensory body socks for adults. Another limb that will probably bear my weight: the explosion in sensory-processing issues in children over the past couple of decades is directly related to the diminishment of the ancestrally-required neural inputs in the 'womb-with-a-view' period. Compression, swaddling, being held against a caregiver, rocking, etc. There, I said it.
9. Get a self-massager. They look kind of like a medieval torture device, but self-massage is a wonderful way to do this.
10. Because of Hilton's law, which states that "the nerves that innervate a joint innervate the muscles and skin superficial to it," giving yourself pressure around joints, specifically, e.g., squeezing and compression, transfers into the Autonomic Nervous System. There are a variety of Chinese self-massage techniques that include touch compression of joints. Take a minute to just squeeze the joints of your fingers, palms, wrists, and elbows, and shoulders on each side of your body, and drink the sensations. What do you notice? Powerful, no?

INFANT AS TADPOLE

I've had a somewhat mystical connection with tadpoles since I was a child. I think it's because a tadpole sort of looks like an egg with a face on it wearing a tail. Like something God made when she was a child and still learning to sculpt.

But then it's really amazing to watch tadpoles grow, because legs and arms emerge right out of their egg-shaped heads.Human infants, neurologically, are a lot like tadpoles. In neuroscience we go on and on about the somatic sensory homunculus, which is this sort of distorted vision of the shape of a human based on the amount of sensory area allocated to different parts of the body in the brain. This little person has giant hands and a giant face and lips. But before you were grown, this homunculus started out looking like a tadpole.

Before you were grown, neurologically you look an awful lot like a tadpole.

Functionally neurologically, when you come out of the birth canal, after your nine month gestational period, you are like a tadpole. You have voluntary muscular control only of the eyes, face, voice, middle ear, and turning of the head and neck. This is basically a tadpole. You're like a giant face neurologically, connected to a swallowing tube (your throat) stomach and anus.

And then, at about three months of age, you grow arms. The physical arms are already there, but your neurological arms don't grow in until you begin to be able to control them. So for all practical purposes, you grow your own arms at three months old.

At nine or 10 months when you start to crawl and then learn to walk, you are growing your own legs. It doesn't matter that the physical legs were already there. You couldn't control them. Our Connection System is this tadpole. We grow into our bodies neurologically. They are *apparently* there at birth, but they are not *actually* there. It feels useful to remember this when we are thinking about early infant development, and the development of the Connection System.

CHAPTER 19

RHYTHM & REGULATION

When we look at this womb-with-a-view period, and specifically the first 9 of the 18 months after birth, we notice that swaddling is closely coupled with another ancestral technology, which is that baby is being carried around on the chest or back of a regulated caregiver nearly all of the time.

What happens neurologically to baby when they are carried on a caregiver's back? Part of it is that they are in rhythmic contact with the parent's heartbeat. And breathing. And bodily rhythmic frequencies. This phenomenon of entrainment is poorly understood and under-emphasized in healing. When we talk about rhythm, people tend to think about dancing, but neurological rhythm begins with heartbeat. Drum, the primal rhythm, moves us the way it does because it emulates heartbeat. Music externalizes and augments rhythms that are within.

Our entire beings pulse, in fact, with rhythm. Dr. Bob Naviaux, one of the world's leading researchers on mitochondrial disease told me in a recent conversation, "Signals of safety are always pulses. It is danger signals that are constant."

Signals of safety are always pulses. It is danger signals that are constant.
-Bob Naviaux, MD PhD

Everything in the Universe, to use the Zen Master Joshu Sasaki Roshi's phrasing, is expanding and contracting. "There is Zero," says the Roshi (I'm taking some slight liberties here with the translation, via Shinzen Young), "But zero is intrinsically unstable, because zero consists of all positive and all negative taken together. And therefore zero intrinsically breaks apart. And the positive only knows how to expand, and the negative only knows how to contract."

The breath is expansion and contraction. The heart is pulsing. The cranio-sacral rhythm is a pulse. Every autonomic signal in our body is, in fact, a rhythm.[1] In vertebrates (phylum Chordata, to be technical), the brainstem and spinal cord contain a great number of central pattern generators. The central pattern generators control crawling, walking, swimming, and flying (birds). They also govern sucking, swallowing, vomiting, digestion, peeing, pooping, and orgasm. All of these are rhythmic pulses. We are in fact, comprised of sets of nested rhythms. Musics ensphered by musics, from the celestial to the subatomic. *Physical matter is music solidified*, said Pythagorus.

Because our culture places a mistaken premium on cognition, we do not recognize that cognition rests atop emotion, which is organized around regulation. The deep roots of these autonomic foundations organize everything that rides around on

1 Our colleague Samar Singh, PhD, working with mathematical models elaborated from the respiratory rhythm, has elucidated dozens of component autonomic signals that move in rhythm with each inhalation and exhalation. Illness is the breakdown of the coordination of these rhythms: the neuroception of safety is what creates the possibility of them uniting (re-uniting).

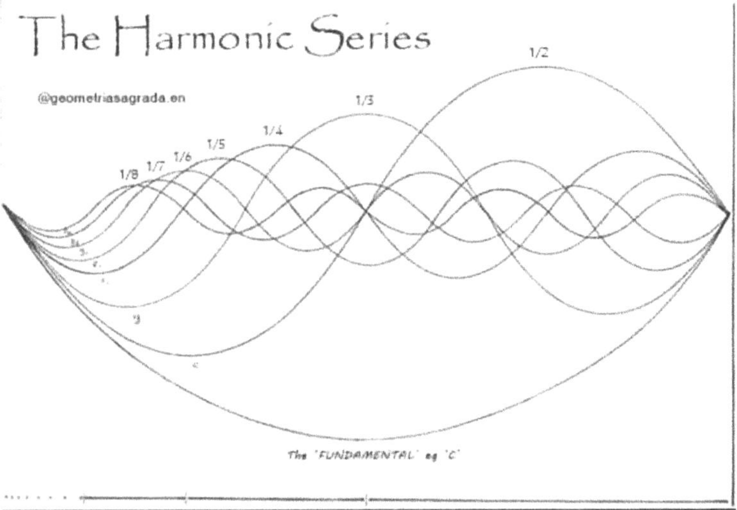

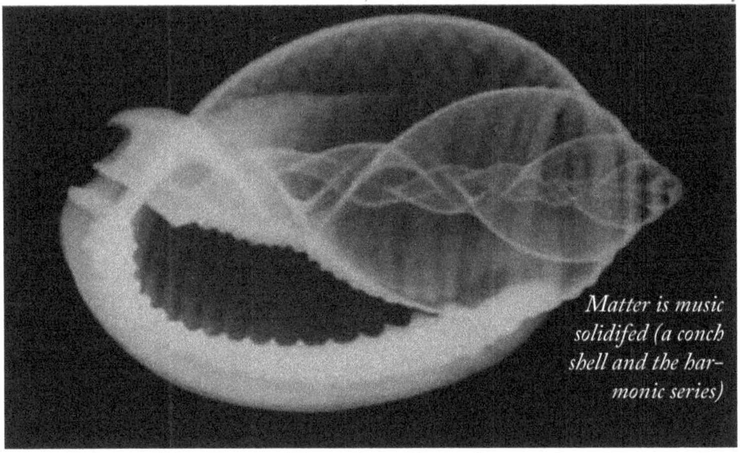

Matter is music solidifed (a conch shell and the harmonic series)

top of them. Rhythms of breath, of heartbeat, of sucking and swallowing, the peristaltic rhythms of digestion, of turning toward and turning away, expansion and contraction are the first and primal language of the body: the proto-language from which discernment arises.

Do you remember the first time you melted into a lover? I don't mean the first time you had sex– I mean the first time you allowed yourself to fall into rhythm with someone else's heart. It doesn't even have to be a lover. I am now remembering slow dancing with a girl named Hilary Oertli in 8th grade, a person I probably haven't thought about once in the past 35 years. It was middle school, and the second or third time we had a school dance, which was a form of initiation for me– the slightly awkward, slightly comedic entrance into my own awakening sexuality.

The DJ would play fast music, and we would bop and crash and jump around for four or five songs, crazed on hormones and Coca-cola and REM telling us it was The END OF THE WORLD AS WE KNOW IT, and I FEEL FINE…and then when we were sweaty and worn out there would be a lull, and the dance floor would clear out, and they would put on a slow song. There were a list of standards, old standby songs that still awaken in me a visceral recollection of middle-school dances when I hear them. This would have been 1989 or something. The song could have been *Stand by Me.* I can still barely listen to that song without closing my eyes and I'm a grown-ass man. I asked Hilary to dance, and she agreed, and we walked out onto the dance floor and time evaporated. It was dark, and there weren't that many people dancing so there was space around us. Occasionally we would be washed over by a spot of purple or blue or green light from the strobes above. I wasn't quite sure where to put my hands, and we weren't dating, and I'm not even sure that she liked me. But it was the first time I had ever really placed my hands on the hips of a girl, and we stood there for three or four minutes while the song played, and just swayed from side to side, her with her arms sort of loosely draped over my shoulders, and me listening to the pivot of her hips and breathing into some sort of moment that opened up between us.

We were both fully dressed, we didn't kiss, didn't hug, didn't speak a word to each other, nothing– yet it was probably the most sensual experience I had had up until that point in my life. My attention– all of it– fastened into the movement of her, the way it felt to sway, and my awareness expanded until I could feel us both as though we were one creature united on the pivot of her hips, the bones of which I could feel through her jeans.

The song ended, and she thanked me for the dance, and my heart opened and I said something, and she walked away, and I stood there for a moment in the smell of Drakkar Noir and the fog machine, watching her silhouette as she walked toward the door, my heart now beating out of my chest practically, thinking I was in love, then looking up into the darkened lighting standards with my whole body still alit with her, delighted to be alive in a way that might have been the real moment my sexuality was born.

A slow dance illustrates the rhythm of intimacy, right? Think about it. A highschool dance is a neural exercise swinging from play (connection & movement) to union (connection & grounding) and back. Slow dances are ballads. They are crooned the same way we sing to babies. They bring us into safety and intimacy. I find it interesting that sometimes they are sung by men known for their machismo– *hard rockers*, let's say. As if the harder you rock, as a man, the more you have permission to fall into the softest of ballads. (I'm thinking here about the band Extreme, very hard rock, whose most successful song of all time is a ballad called *More than Words*. Or The Red Hot Chili Peppers, who jump around spastically for hours harnessing mosh pit energy, Kiedis bare-chested always and never fatiguing, Flea wearing pants made of stitched-together stuffed animals, throwing his head around over his bass in a way that makes my neck hurt just watching him, yet whose most effective songs soften into something quite tender, e.g., *Under the Bridge*).
I think this is interesting, primally, because there is something about masculine strength that is associated with protection, and so when someone who has a strong protector archetype grants us permission to feel safe, it allows us to move deeply into this. It is like a baby sung to sleep by their father. There is something important about that. Something we are missing, as a culture.

Riding around on the back of your father or mother or someone who loves you is to ride around on that voice, that heartbeat. Is to be held in that ballad. As the baby, you are the one entranced by the rhythm of union. It is a rhythmic experience: *the music of being loved*. At that age, the size we are, we fall asleep against the chest or back of a caregiver and it is a warm wall to us.

You don't have words for this, but is deeply sensual because you are a creature with a skylight in your skull, a tadpole in the process of growing arms and legs (you really do grow them neurologically, despite the fact that they are materially already there), and you are riding around on the back of a whale (not a literal whale, but the size discrepancy between you and your parent makes it so). If you are riding around on their back, and your head is turned flat against them, your ear is pressed into the back some two inches from their beating heart. You awaken and sleep into the rhythm of the parent's heartbeat. The music of their body, as you grow your own.

Don't think for a moment that you weren't feeding on the sounds of your mother's living body when you were inside her. The swoosh and gurgle and the movements of digestion, the pulsing of blood, all of this was an acoustic environment conveyed to you in a fluid medium. And you spent 9 months immersed in this ambient sound. And so resting against your parent's back, ear to them, body not yet under your voluntary control (tadpole stage), is also a return to womb-time. Womb-music. The sounds of being inside, being underwater.

I cannot emphasize to you enough how profound this kind of training is. And it is a training to be sure. An *entraining*. As you grow a bit larger, a bit older, a bit more curious about the world, you will no longer be content to face inward on mom or dad's chest or back, but will want to look around. We'll start to carry you on our chests facing outwards so that you can see, look around, interact. So that you can receive smiles and notice what is going on. So that people can notice you and croon over you and smile and kiss you and play with your feet, while we, your parents, still have your back, literally.

And yet, as this happens, the music in the background, the slow dance underneath it all, is the parental heartbeat and bodily rhythms. Let's say that we are in the field at the edge of the forest together, and you are on mom's back. Seemingly out of nowhere a strange dog approaches, one unfamiliar to either of us. As mom notices it, she begins preparatory orienting, and her heart rate jumps up just a little, her body stiffens, her hearing becomes more acute. All of this happens to you, downstream of her rhythms, vicariously. You orient as well. Your heartrate picks up. *What just happened?* You are suddenly alert.

Mom is neurologically training your awareness. She doesn't say, "Look at the dog." You can't understand her yet anyway. She doesn't have to. You are sharing a rhythmic field, and all of this information flows straight through the bodies– it is the music of the moment.

If the dog is friendly, begins to wag its tail, mom knows enough about the world to interpret this, and her arousal will drop back down to normal, and you will feel this. You've just learned that dog is safe– not a single word exchanged. You will feel it in the arousal, and in the way she talks to the animal. Your own nervous system will settle.

By contrast, if the dog is not safe, you'll learn this too, because you'll feel her arousal increase. You'll feel the tightness in her voice. You might begin to cry. If she shouts for help, or starts making noise to chase the dog off– all of this you will learn too, again no words exchanged between you and mom about the dog.

This is part of the ancestral importance of keeping baby skin-to-skin with caregivers. For a baby in a stroller, none of this learning happens. Baby doesn't have a felt reference point, a visceral compass through which to measure experience. Mom on her iPhone doesn't notice the dog. The primary scaffolding of embodied awareness gets ignored. Baby in stroller is a baby in a cow catcher, pushed out at the front of the train into an indifferent and sometimes hostile world that she cannot interpret.[2]

2 I want to be careful not to shame anyone with this awareness. We are acculturated to push kids in strollers. Don't beat yourself up for doing this. Simply understand the ancestral neurotechnologies of why this is less than ideal.

CHAPTER 20

GRAVITY AS DANCE PARTNER

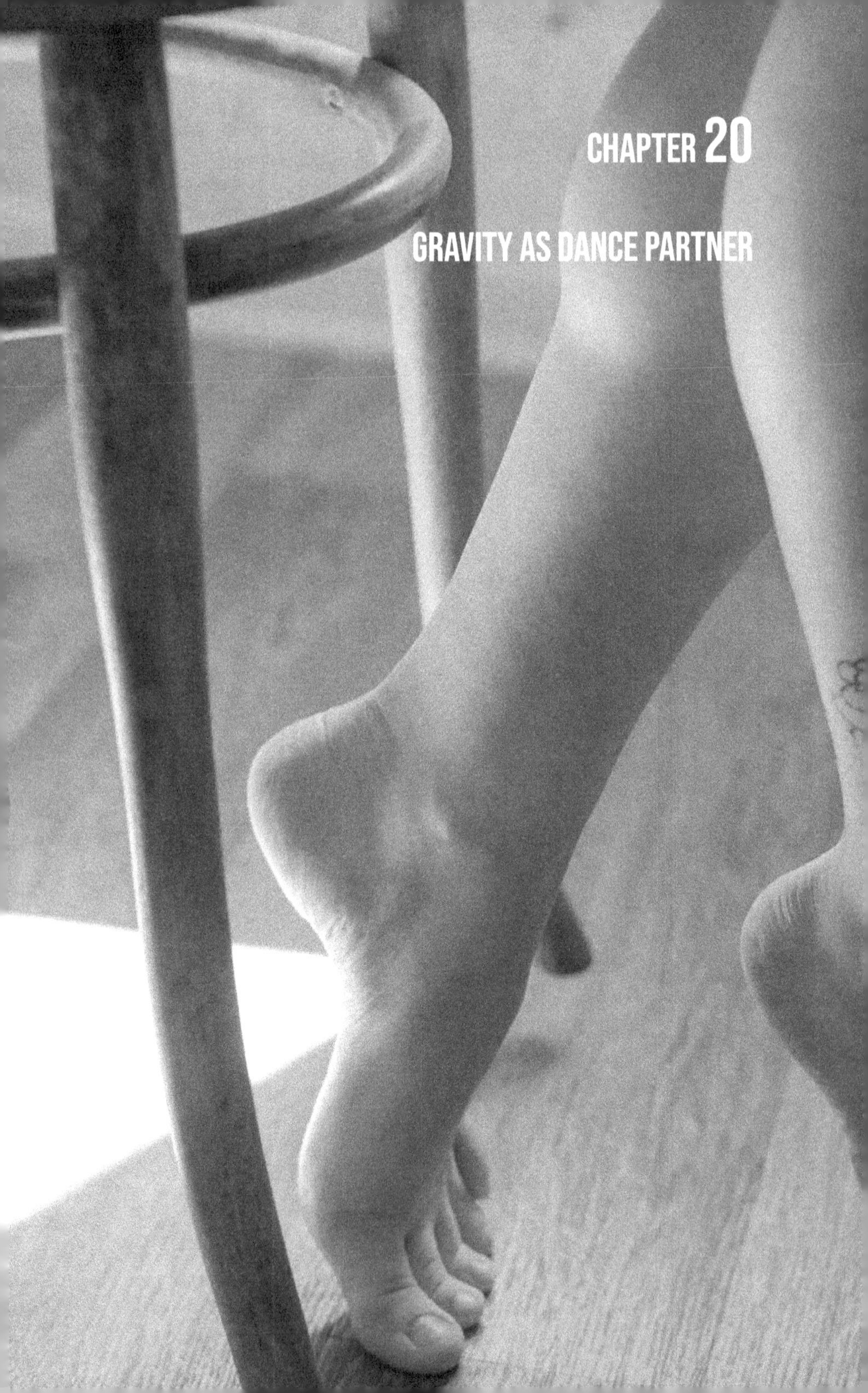

Most of us do not think of gravity as a dance partner, but it is to our detriment that we do not. Gravity gifts us deeper access to our Connection Systems when we are aware of her. Let me explain. As you are reading the book, right now, you are likely sitting. Maybe you are sitting in a chair, maybe you are reclining, possibly you are in your bed. I find it unlikely that you would be totally horizontal: that's not comfortable. So in some way your head is above the book, and you are looking down at it. Your body might be directly under you, or tilted.

Take a moment to experience gravity. Not abstractly, but the actual contact between your body and whatever is holding it up. If your feet are on the ground, start by feeling the way that they press into the floor: a proxy for the good earth. Is it simply a coin-sized disc of contact under the outer edge of your right foot, because your feet are stacked on top of one another tucked under your chair? Or do you have full contact between your insoles and the floor? You don't have to shift it, change it, or make it different, but feel it. Bring your interoceptive attention to the weight. Feel its pressure. And then, to make it your dance partner, as my friend mindfulness teacher Paula Rámirez reminds us – *Rest into gravity*. Let yourself feel her. Let yourself feel the weight, the pressure, the mass of you relating to gravity. Let yourself feel the way that your body translates itself into pressure through your joints, and the way that this force distributes as muscular tension and sensation across your foot. Feel the way the body bears your weight, and soften into this.

Same thing with your seat. Feel the contact between your bum and whatever is holding you up. Is it a chair? A couch? Again, where does the contact start? Just above the knees? Are you sitting straight, such that the pressure is distributed horizontally in balance? Or are you twisted slightly, because there is a beam of sunlight and you are holding the book off to one side? If so, the body is cantilevered slightly, turned just slightly, and in order not to collapse your core is tightened just a little bit more on that side. You are bearing the weight asymmetrically, and this is pushing your right side just a little bit deeper into the chair. See if you can feel this.

See if you can feel the way that this particular rest into gravity is deeper above the same foot bearing the weight, and that it will change if you alter the way your feet are arranged. We do this so artfully, with such little thought–arrange ourselves as sculpture cantilevered in air–that we don't attend to the sensation of this. But here, in this moment, let yourself feel the weight distributing in the back of your legs, and let yourself feel the press of your seat into the chair.

Learning to hold ourselves upright was so difficult. We forget this now, when it happens so automatically. But watch a toddler trying to figure out how to balance himself on the sand of a beach, or how to navigate the changing contour of a hill and you realize we have to figure out how to hold our bodies up. The way we do that is to feel them in relationship to gravity. And then, as with language, we become fluent in this dialog and stop noticing. Interrupt that now.

Can you feel the way the back of the chair supports your lumbar spine? If you notice that, if you feel into it, does the curve of your back in this moment support you, or it is collapsed? If you feel yourself from inside, nourish yourself with sensation, how does your posture change? And this is the inwardness of using posture to help us find our Connection System, and in this moment our Connection System feeds on the dance with gravity, on the weight of the body, on the way that we are balanced, like moving trees, me in this moment writing with my hands resting on these keys, two arms like branches hanging down at my shoulders, the palms placed squarely on the flat part of the laptop just beneath the keys bearing their weight–my forearms–for that is what they are, needing to catch that weight of the arms in the palms so that my fingers can dance: are free. But even the fingers, if I sense into them, flying over the keys at a velocity I will inhibit if I start thinking about it, are playing in gravity, and if I don't keep allowing them to rest into gravity will fatigue.

The body in motion, the body at rest, the way we find moments to rest into, and the way we breath into these moments– all of this is a neural exercise. Who are we kidding? The only reason we don't know this is because we are creatures who keep forgetting we are animals, keep forgetting we are bodies.

But the cat doesn't. She stretches religiously. The dog doesn't. Everytime he wakes from a nap he stretches himself out, vertebrae by vertebrae. Sometimes you can hear them clicking as he shivers. Every time I call him to a walk he, like a yogi, does that stretch they have aptly named downward dog. What is he doing? Resting into gravity. Finding his way back to connection through posture. And then shaking himself out.

Take as one final experiment with this, the human head. It is heavy. It rests on a pivot of the spine, aptly named Axis and Atlas. But sitting here, reading, set the book down for just a moment and attend to the weight of the head. Raise and lower the jaw just a few centimeters, just a fraction of an inch. And notice the cascade of sensation that radiates from this movement. Notice how many micro-adjustments of the body are required. Turn the head, slowly, left and right, and marvel again at everything the head is connected to. Can you feel the sensations in the back of the neck? The shoulder? Now tip your head back as you turn the head? Can you feel the way the chest comes forward, the way the spine is involved? How movement generates counter-movement in the body? Levers and cantilevers? How to turn this way the body must go that way? In this manner, gravity becomes your dance partner, for to orient in her, the whole body becomes involved.

Now do this with awareness of breath. Knowing that breath is the rhythm of the Autonomic Nervous System. Move your body, slowly, allowing the head to swivel on the neck in a figure eight, and relax until your body moves through this motion with the smoothness of a modern dancer. And once you've found that grace, bring attention to your breath. She has changed, shifted, moved into a different cadence. And this is the proof, the evidence, right here. Not far away, inaccessible, dwelling in some chart you'll need electronics to measure or interpret or fancy terminology to understand. Your Connection System is right here, available, just waiting for you to discover it. Just waiting for you to *recover* it.

A FEW WAYS TO REST INTO GRAVITY

THE HEAD

The head is very heavy. The balance of the head on the neck is deeply connected to our Autonomic State. Yet most of us– I'm doing this now as I write this– spend a good deal of time looking downward at something on a table in front of our hands. This tilts the head forward in a manner that requires the top two cervical vertebrae to tilt forward. Interestingly, one of the first postural adjustments when we get stressed out is a change in the placement of the head on the neck, and a tightening of the tiny muscles that connect the vertebrae to the base of the skull and anchor into the dura mater. When this rotation of the Atlas and Axis happens, it reduces blood flow to cranial nerves. So proper ergonomics are important for a variety of autonomic reasons. If you spend a good deal of time working on a computer, find a way to put the screen high enough that you don't have to put your head into a stress posture in order to do this. Also, spend more time looking up. In a forest your gaze is naturally drawn upward, into the canopy, and towards the sky, whereas in a city it is dangerous to let your gaze float up like this. More forest, less city.

SITTING

Sitting for long periods of time is highly deviated from our ancestral movement baseline. When we sit for long periods of time the lower parts of the body- hips, legs, feet, tend to drop offline. If you work in an office or in some other occupation that requires you to be still for long periods of time, get a standing desk, or experiments with alternatives to traditional chairs. My office stool wobbles by design, so that whenever I am sitting on it I have to engage my core. Your deep ancestral body, for that matter, has no idea what a chair is. We never ancestrally sat with our legs at 90 degrees. We were squatting lower than that. Notice how the back, hip cradle, hip flexors, and leg muscles reorganize when you squat. That is a position with which your body is ancestrally familiar.

FEET ON THE GROUND

Take time, periodically, to feel your feet on the ground. As in, notice the sensations moving through the legs and feet. Notice how you are bearing weight. There is much deeper exploration of this in later chapters, but feeling our contact with the ground is necessary in order to rest into gravity.

WHOLE BODY

Learn to take a moment, when you sit down, or when you lay down, to allow your body to rest into gravity. We tend to habitually contract just slightly, and not really set down our weight even when the body is in positions where we can.

Let your shoulders drop away from your ears. Let your hands and arms feel the weight of gravity. Particularly when we are using our hands often, we forget to fully release their weight back into gravity. When you lay down, spend a few seconds just releasing your weight into gravity.

LET THE BONES HOLD YOU UP

Let the bones hold you up. Bones are what should be holding us up, not muscles. See if you can learn to let your muscles drape around and envelope the bones. The Structural Integration work of Ida Rolf, and her lineage, addresses this.

CHAPTER 21

THE GREAT MYSTERY OF BREATH

Now that we've talked a bit about rhythm and regulation, and gotten oriented to the embracing field of gravity, let's bring our attention to the breath. If life is an ember, the ember that yet burns (*N Kasea* in Tzutuj'il Mayan), breath is the wind that fans the flames. The breath delivers, to each cell in the body, the wind that fans the flames in the form of oxygen that combusts in the cellular furnace. It also removes the accumulation of built-up gases, the carbon dioxide that we do not need, aggregates it, and expels it from the body with the exhalation.

You breathe in new life, you breathe out what has been combusted. Breathing is, in its primary manifestation therefore a mechanism of gas exchange. It tends the metabolic fire of life. It organizes the flow of energy through the body. And yet even this barely touches into the deep mystery of the breath.

If we approach it with humility, then bring an autonomic lens, we must wonder at this mechanism that sits at the intersection of the conscious and the automatic. Your breath will breathe you whether you remember it or not, yet what other autonomic phenomenon can you so directly modulate with awareness? The heart, while meditation masters have learned to steer it after prodigious feats of concentration, is not so easy to speed up or slow down intentionally. Yet even a small child can change the rhythm and depth of their breath. They can hold it when they jump in the pool. They can lengthen it when they blow out a candle, make it deeper by exaggerating the rise and fall of the belly. It is a mechanism through which we can regulate or dys-regulate ourselves on purpose or accidentally. And all of us know, whether we have done this by choice or had it happen to us, how deeply the breath governs the feeling of being in our bodies.

Whether it is discovering the relief that comes from a deep breath, or accidentally moving ourselves into panic by breathing too shallowly, if we are wearing a body we learn at a young age that breathing is central to how it feels to be in the skin we are wearing. We learn this whether we develop any sense of how to work intentionally with the breath or not, but keep in mind that if you did not already have exquisite breath control you could not talk.

The neural architecture of the rhythm of the breath is regulated deep within the brainstem, in areas that govern pattern generation in the body. These clusters of inter-neurons, that link together various aspects of the autonomic physiology, govern the central rhythms, the oscillations of our fundamental being. These pattern generators are technically, "neuronal circuits that when activated can produce rhythmic motor patterns such as walking, breathing, flying, and swimming in the absence of sensory or descending inputs that carry specific timing information." (Eve Marder 2001).

The final clause of this sentence is crucial because it suggests that these pattern generators can operate based on their own intrinsic timing. They don't require a directive from higher in the brain. That duly noted, the body's rhythmic pattern generators can be influenced by four factors:

- their intrinsic organization
- fibers from higher brain centers
- sensory information incoming (fibers coming from the body)
- chemical messengers such as hormones and neurotransmitters [1]

At the heart of the Breathing Pattern Generator, deep in the medulla oblongata of the brainstem, is a cluster of about 10,000 interneurons called the Pre-Bötzinger Complex (Pre-Bötc). Interestingly, and non-accidentally, in mammals this complex is located in a part of the medulla adjacent to the ventral vagal source nucleii in the Nucleus Ambiguus. See the two circles in the diagram below. The NA in the diagram, which you are viewing at right below in transverse section, is basically

1 These CPGs are directly impacted, for example, by endogenous opioids released with the neuroception of lifethreat. This is the mechanism whereby immobilization happens under lifethreat.

sitting on the PreBötc's shoulder.

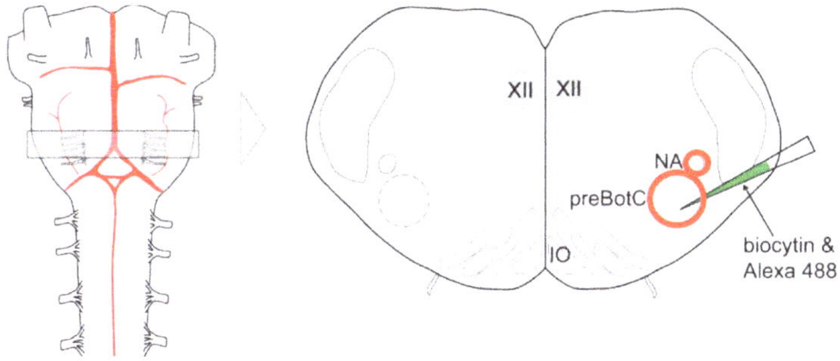

This is a transverse section of the brainstem in a rodent, showing the relative location of the breathing pattern generator to the Nucleus Ambiguus, one of the nucleii that coordinates the Connection System.

This is, among my colleagues, many of whom are quite learned in this area, something of a revelation. Yet this makes sense very deeply because the NA in the brainstem is the nucleii that assembles the ventral vagal complex, or social engagement system. This is the system that unites the neural regulation of the face, voice, tuning of the middle ear, larnyx, and pharynx, and turning of the head and neck with the heart and breath in the presence of an attuned caregiver.

Think about a baby sucking and swallowing. The sucking is a brake and release. Suck, swallow. Suck, suck, swallow. Every time we swallow, breathing pauses. This coordination of sucking and swallowing with breathing is, in fact, the first autonomic task of a newborn, after she crawls up the mother's belly and attaches to the breast.[2] It happens easily because of the proximity of these two Centeral Pattern Generators, which are right next to one another in the brainstem. This brake-and-release, happening in the pulsation of safety (bathed in oxytocin), coordinates the Connection System (new Vagus) with the digestive processes in the deep belly (old Vagus), uniting the

2 Although this happens extremely rarely in the modern world, human infants are hardwired to use a sort of lizard-crawl up the mothers's belly and attach to the breast. These are autonomic motor patterns available at birth.

Connection and Grounding systems.

The precision of this coordination, between breathing and sucking and swallowing, because the swallow is in fact an apnea–a temporary cessation of breathing–is a foundation metric in the Autonomic Nervous System. A foundation metronome, we might better say. There is a complex coordinated rhythmic sequence here that establishes the living rhythm of connection.

Because this living rhythm of sociability is neurally architected in the Nucleus Ambiguus, it makes sense that it would be adjacent to the rhythmic and movement pattern generators in the brainstem, including the breathing pattern generator, which is comprised primarily of inter-neurons, which by definition, are inter-coordinating different systems.

Safety is a pulse.

Sucking, swallowing, and digestion is a rhythm: the rhythmic braking and release, braking and release, of the ventral vagus. And here in the breath pattern, is a great mystery, again revealed in the physiology: breath communicates with all other aspects of your autonomic physiology.

Research by Feldman and colleagues (Feldman, 2011) has shown us that this rhythmic pattern generation has a complex autonomic function. Your Autonomic Nervous System is monitoring the body, and sending this information to the pattern generators that will adjust and micro-tune your body to the necessities of the present moment. The inhalation is the data flowing up to the pattern generators. It is the surveillance of the felt interior of the body that is corrected by the exhalation. The breathing rhythm is a continuous software update, modulating the parameters of the body to be most subtly responsive to the needs of the moment.

The breathing rhythm is a continuous autonomic software update for the body.

The breath is a subtle and refined mirror and expression of autonomic state. The rhythm of the breath encodes your autonomic state.

What this means, and yogis and mystics have known this since the dawn of time, is that your breath is encoding all of the rhythms of who you are. If, as we noted in Chapter 6, the nervous system is the altar of the spirit, the breath is the living manifestation of all the rhythms of who you are, meeting the present moment: the offering on the altar.

The breath is the offering on the altar of the spirit.

For a moment I want to dwell here on the degree to which here, at the intersection of breath, the domains of neuroscience and spirit are telling us the same thing. Anyone who has meditated seriously is aware of the degree to which meditative deepening– entering the flowstates of contemplation– is a phenomenon directly connected to breath. I have studied contemplative traditions from around the world, including a number of Buddhist lineages, animist contemplative traditions, shamanic lineages, and others, and every tradition I have seriously engaged centralizes the breath in its practice.

The breath is always understood to be a central doorway to spiritual experience: a primary governor of awareness. And what we have just been discussing, the foundational neural architecture, shows us how the breathing rhythm encodes all of the autonomic parameters of the body.

What this suggests, what it points to, is that if there was some way that we could see, with precision, the waves riding upon waves that comprise the breath, the complex and subtle interference patterns of rhythm, we would be looking at the governing patterns of the entire autonomic physiology, the precise nature of the tide, incoming and outgoing, of the flow of life. If you spoke mathematics with enough precision, the peaks and troughs of the wave of breath would teach you the entire topology of the body's rhythms. All of the information is held in the carrier wave of breath.

Many healing modalities have intuited this with varying degrees of refinement. Cranio-sacral therapy, for example, arose from the detection of a pulsative wave traveling the length of the spine (from the base of the cranium to the sacrum) by an osteopathic physician named John Upledger, who was at the time a Professor of Biomechanics at Michigan State University. The entire field of psycho-acoustics, as applied to healing, the intentional use of music and rhythm to modulate physiological and psychological state, is as old as humanity. Our ancestral trance dances, drum circles, chanting, and shamanic rituals have all worked with and through the rhythms of sound and the breath, coordinating inputs to modulate this fundamental rhythm.

Lovemaking involves the coordination of breath.

Singing itself, even speech, which involves the rhythmic coordination of breathing with biting, snapping, and the percussive modulations of the diaphragm that allow us to terminate the breath in a plosive– the sound of the letters p, t, k, b, d, and g all require us to coordinate stopping the breath– rely on our felt ability to modulate and control the breath: the fire that fans the flames.

PRACTICAL GUIDANCE FOR USING BREATH AS A PORTAL TO THE CONNECTION SYSTEM

FOUR PARTS

The breath has four parts: an inhalation, a pause (fullness), an exhalation, a pause (emptiness).

We tend to be naturally more aware of the inhalation and the exhalation, and less aware of the pauses.

INHALATION & MOVEMENT

The inhalation is mediated primarily by the Movement System, and on inhalation the heart accelerates. This is why, among other things, people preparing to fight or exert will breathe explosively in through the nose (breath of fire) or intensify inhalation (just before sprinting for example). Increasing attention to, or intensity of inhalation activates the fight/flight physiology.

EXHALATION & CONNECTION/GROUNDING

The exhalation is mediated primarily by the parasympathetic (Connection and Grounding) branches of the Autonomic Nervous System. This is why, among other reasons, people who are seeking to calm or settle the nervous system bring attention to or extend the out-breath.

4-7-8

Breathing patterns can help us bring structured cadence to these rhythms in ways that have intentional physiological effects. The 4-7-8 breath, for example, is a structured breathing pattern that can be used to help settle the nervous system.

People say it is calming anxiety, which is not false, but the mechanism is by enhancing parasympathetic tone (connection/grounding). In the 4-7-8 breath, you inhale quietly through the nose for 4 counts, hold the breath (fullness pause) for seven counts, and then exhale forcefully through the mouth, pursing the lips, and making a 'whoosh' sound for 8 seconds. Repeat the cycle 4 times. This controlled breath uses timing to exert pressure on the physiological drivers of breath, containing the sympathetic side and tuning the parasympathetic sides.

SURFING BREATH

In addition to structured breathing patterns, which are often 'doing' something to the breath, it can be very useful to simply bring attention to the natural rhythm of the breath, and more deeply experience how dynamically it changes on its own in response to internal and external conditions. Simply experiencing the breath is innately connective.

When people are beginning to use breath as an object of attention in meditation, often they find it difficult to notice the breath without changing it. When I'm teaching this, I like to use the analogy of surfing. What we want to do is ride the wave of the breath with our attention. We are not trying to shape the breath. We are not trying to influence the wave. We are simply riding it with our attention, letting the attention rest lightly upon the breath. The simple process of bringing attention to this wave is usually regulating (unless the body is severely dysregulated).

EARPLUGS

One of the simplest and most effective methods in the world for deepening connection to self, and finding a deeper state of settling and integration, is to put in earplugs and listen to the breath. With earplugs in, the acoustics of the breath are amplified, and it makes it easier for people who are novice in mindfulness to hold the breath in attention. You begin to listen to the breath with the same kind of attention that you might ordinarily use to listen to music. You can get lost here in the most marvelous way: completely merge into the indigenous/endogenous rhythm of the breath breathing itself.

NOTICE YAWNS

The body is naturally oxygen-seeking. We breathe, yawn, stretch– all of these things to help us draw breath deeper into

the body. See if you can notice when your body is doing these things and bring attention to them. Attention potentiates experience: it makes it deeper. The more conscious of the yawn you become, the more deeply you feel it from inside, the more effectively it does its work of delivering oxygen.

NOTICE THE PAUSES

Most of us don't naturally pay a great deal of attention to the pauses in the breath, but they are physiologically significant because the body is in equipoise. When the breath pauses, at the peak of the inhale, and the trough of the exhale, there is a temporary balance, and what is happening at these moments is, in a way, the revealing of the underlying baseline of the autonomic physiology. (When the tide is moving in and out, it is harder to see the seabed than when the water is still.) If you use a breath pattern that exaggerates the pauses, or brings deeper attention to them, this can be a doorway to working with deeper content beneath conscious awareness. It can be easier to access this content in the pauses.

With the breath, as with all the doorways to the Connection architecture, become intimate with it as a portal. Explore, play, dance, have fun. Intimacy with the portal is the route to deepening.

FOREST BREATHING

If you want to really open your mind to our nature as part of Nature, practice breathing meditation in a forest. Find a place to sit in the midst of trees. Before you begin, take a moment to acknowledge that the breathing of the trees is the inverse of your breathing. They are, to put it one way, your upsidedown twin. What they breathe in is carbon dioxide, and what they breathe out is oxygen. What you breathe in is oxygen, and what you breathe out is carbon dioxide. So through the medium of air, you and the trees are part of a single inhalation/exhalation system. Our symbiotic relationship with plants is profound.

Through the medium of air, we can consider ourselves part of a single larger organism, breathing.

This simple awareness is a good motivation for us to fill our homes with plants. They are nature's air producers and air purifiers.

CHAPTER 22

FACE & VOICE

If someone is particularly emotionally transparent, available, easy-to-read, we will often say that they wear their heart on their sleeve. But this is not true: we wear our hearts on our faces and in our voices.

Think about someone you know really well. Is it not true that if you phone this person, and spend just a moment with them, you can tell if something is wrong? You say, "Hey, how are you?" and even in a slight delay, or the micro-intonation of their voice, you can tell something is amiss. They say two words, and you say, "What's wrong?" They say, "Nothing." But you can hear it in their voice.

How is this so? The Connection System that we are learning about is a neural complex that wires together many of the cranial nerves involved in sociality, including your 10th cranial nerve, the vagus (its ventral branch), with the facial (IX) and trigeminal (V) nerves, the glossopharyngeal (VII), vestibulocochlear (VIII), as well as the accessory nerve (XI). The facial and trigeminal nerves innervate the face, the glossopharyngeal innervates the voice.

Let's return for a moment to the neuro-development of the fetus, and look at the direction of growth of this wiring. We start with the heart, which grows up to the brainstem. Here in the brainstem are the terminations of the vagus, which are adjacent to (slightly behind) the roots of the facial nerve, and the trigeminal ganglia. The heart flowers into the brainstem, the petals open into the face and voice. The signature of your moment-to-moment heart-rate variability is expressed in your face and your voice all the time. It cannot not be.

This is so literally true that there is technology developed by a psychophysiology laboratory that can extract the signal of your Heart Rate variability from your voice. So the reason that you can hear if something is wrong through someone's voice is that their moment-to-moment HRV is encoded in their voice. It is also visible in their face. When you know someone well, you can feel this. Sometimes you can feel it even in people you don't know well.

Now this doesn't mean that the face is necessarily expressive. Nor that the voice is. Because when we are in defensive states, the full expressivity of these systems is neurologically dampened. So often, if something is wrong, what you are hearing is not the expressivity of the voice, but the flattening of its melodic qualities. The voice becomes tight. Clipped. Distant, withdrawn.

Conversely, the better someone is feeling, the more their voice and face tend to express this. Again, paradoxically, a feature of dominator culture (sociological whiteness) is the suppression of full facial and vocal affect. I'm going to just name that, and then let it be, because it is outside of the immediate scope of this book, but I have written extensively about this elsewhere.[1] We are all, be we victims or oppressors, the unwitting children of empire.

Dominator culture is not interested in sovereign expression. It is interested in control. Nor is this a particularly modern phenomenon. The Roman empire forbade certain expression of mourning, because they recognized that there was tremendous energy moving through grief, and that people were easier to control when they could not successfully metabolize trauma.

Empire has long understood that people were easier to control when they could not successfully metabolize trauma.

Ancestral grief rituals happened in community and were healing rituals *for the community*, created in part to honor the departing of someone from the community (or their transition to the realm of ancestors) in order to restore vitality in the wake of the loss. Think for a moment about the unhindered expression of loss. People wail. They ululate. They beat their breasts, tear out their hair. They weep uncontrollably. They

---

1 See my book on empire in modernity, *So, About that Death Cult you Joined?*

faint, collapse, fall into one another's arms. All of these actions are, beyond being cathartic, beyond being ways of honoring the dead, ways of metabolizing tremendous amounts of energy that would otherwise congeal into grief and immobilization and get frozen in the guts.

The Romans understood well that people who were healed were more likely to be sovereign, and less docile subjects. These rituals were therefore outlawed. Keep people stuck in immobility and they are much more docile and passive imperial citizens. Cut the roots of relatedness off by freezing the deep belly.

Setting aside the cultural training of suppression of the face and voice, what I would like to tell you, since you came to this book to feel better, is that your wellbeing is intimately linked to your ability to express through your face and voice. The face, the voice, the eyes, the ears, the turning of the head and neck–each of these is a doorway to your Connection System. And because of the nature of the neural architecture of this system, if you are able to activate it through any doorway, you activate the whole thing.

I am going to say this again because it is so important, and so poorly understood. The house of your Connection System has many portals (we could say it has many doors and windows). Opening any of these doors and windows can change the climate in the entire house. What we want to do here is find the doors or windows that you find the easiest and most enjoyable to open. Once you are able to predictably and repeatedly open one of these doors and windows, you have way more agency in turning your Connection System on. Wellness is no more or less than spending more moments with this system online, organizing your Autonomics.

PRACTICES FOR OPENING FACE AND VOICE

MAKING FACES

You can practice making faces in a mirror, and simply exaggerating the range of expressivity beyond what you ordinarily express. Notice interoceptively, particularly in the belly, how it feels to make the face more expressive. It can be a powerful awareness practice to simply sit in front of a mirror and make various facial expressions, and take note of how they impact your visceral state (how it feels inside your body). This is particularly useful for people who have been socialized into suppressing facial expressions. If you have been told you have a poker face, or a resting (I'm not gonna say it) face, you have been neurologically trained to do this. This exercise is for you.

MIRRORING FACIAL EXPRESSIONS

You can practice mirroring facial expressions with a partner where you reflect back to them the face that they are making. As you go make the expressions progressively larger and more animated. Again, notice interoceptively.

You might try this sitting in front of a mirror while you are having a phone conversation with someone, and watch your face reacting.

Or you might record yourself in conversation with someone and watch your face. When does your face light up? When are you attuned and engaged? When does your face go blank or lose expressivity? With whom?[2]

All of this is vital health information. If you learn to see your face as the doorway to your heart, you'll begin to realize that watching faces (your own and other people's) is one of the best ways to develop a perceptual competence that is called *Nun-*

---

2 We will naturally exhibit greater facial expressivity (unless we have been neurologically trained not to) with people we feel safer with. So noticing who your face gets less expressive around can tell you who your body doesn't feel safe with. Sometimes your mind will not know this. Micro-aggressions emerge out of this disconnection.

chi in Korean. *Nunchi* is the ability to accurately recognize the emotional state of another person.

OPENING THE VOICE

We learn pretty early on to constrain the size of our voices. Most of us do not have much experience with vocalizing loudly. If you yell and scream in a city it tends to attract attention from neighbors and cops.

Can you find a place where you can make your voice as loud as you would like? Can you really let loose with screaming somewhere without getting the cops called on you? It can be really useful to have a place where you can make your voice enormous.

Particularly for folks whose voices have been silenced (how we might put it colloquially) or socialized to be quiet, experimenting with making big sounds can be both provocative and liberating.

IF YOU COULD HAVE SAID ANYTHING

One of the most elegant ways to help you come back to a baseline of safety in the wake of a confrontation is to get together with a trusted partner, and disinhibit around the voice.

If I have a meeting with my boss, power dynamics dictate that I may not be able to say everything I am feeling. If, afterwards, I get with a trusted friend in a place I can make noise, and they ask me, "If you could have said anything without fear of consequence, what would you have said?" it can give my voice permission to outflow excess energies. One of the most interesting things to notice here are the repertoire of gestures (motor movements) that accompany what you say. These are deeply connected to resetting the Autonomic Nervous System.

The next chapter explores another helpful vocal practice...

CHAPTER 23

SINGING IN THE SHOWER

Several years ago, I attended a volunteer day at the Occidental Arts and Ecology Center in idyllic Occidental, California, because I had been told that their biodynamic gardens were among the most beautiful in the world. This turned out to be true, and I spent the day outside, a good part of it sitting beside a vegetable bed weeding by hand as hummingbirds zipped through azure sky above me, the clip of their wings like feathered two-inch fighter jets. Sitting together about ten feet from me, and close enough that I could follow their conversation without straining, were two other volunteers, one of whom was a singing instructor.

At a certain point in their conversation, a propos of what I do not know, the singing instructor said to her friend, "Everybody can sing. Every single person I have ever met who thinks they cannot sing has been told by someone that they cannot sing." Her comment was unexpected, and struck me like an arrow, and I had to turn away because my eyes began to leak, and I was embarrassed. I sat quiet and motionless in the garden for quite some time, staring at an enormous oak tree and feeling myself inwardly. I was remembering the exact moment when I was told I couldn't sing.

When I was seven years old, I was uprooted from my place and village by my parents, because my father lost his job during a recession, and they decided to move back to St. Louis, where they were both from, to receive financial support from my mother's parents, who were affluent. This event turned out to be a defining fulcrum of my life, because it shifted my nervous system from a baseline of near total ventrality into nearly complete shutdown. I experienced the event for several decades as having been a drowning.

Shortly after this move, I took part in an elementary school musical production about American folk history. I didn't have words for it at the time, but in the wake of being uprooted, I could no longer feel myself from inside. My interoception was totally out of whack, I could not feel myself inwardly, and

I couldn't track rhythm. My body had moved into a strangle tangle of blankness and anxiety, which I couldn't articulate, didn't understand, and had no idea how to fix. I thought I was just nervous about being in a new school, or being around kids I didn't know. My school counselor seemed to notice something was going on with me, as it was not uncommon for him to put a hand on my shoulder, and when he did this I often felt like I would break down in tears, without knowing why, and I fought against this without realizing it would have helped me. I didn't know how to explain to him what had happened. The simplicity of naming my grief eluded me. I had no idea I was completely heartbroken.

Our school counselor was named Perrin Stifel, and he was interestingly a storyteller of significant renown. He narrated our musical in a voice filled with kindness and melody, and told folktales.

I was assigned to sing a folk song in the musical about John Henry, the steel-driver, reading the lyrics of which, forty years later, I grow emotional. The song is about his race against a stream drill. This beautiful black man with a great heart, and a great spirit, battling against a soulless machine. In the end, John Henry wins, but he also dies from exhaustion, and it seems that in truth he was imprisoned at the time, and had been born enslaved, though of course that's not what the song is about. And of course, despite his valor, it is the steam engine that triumphs ultimately. The song is wistful and opaque in the way folks songs are, recalling a moment of human heartfulness triumphing over mechanical greed, and condensing a nostalgia for a pure-hearted American past that never was. Never mind the death, the prison, or the enslavement part, the song seems to say: let's focus on the valor. Okay, whatever. I was eight years old: I didn't know any of that.

I remember sitting there, on the risers in the school gym, which filled with parents and all the other students at the school, both older and younger. I began to get very very nervous. I kept running over the lyrics in my own head, without realizing that my mind was starting to race. I was wearing a fringed leather cowboy vest, with riveted pockets, and a plaid shirt and jeans. I started messing with the tassels on the vest. Mrs. Norris, the music teacher, was sitting at an upright piano positioned in front of the stage. I found myself staring at the basketball hoops, which seemed incongruous above the heads of the audience. For some reason the gym was carpeted tan, the basketball lines woven in red.

The carpets went a few feet up the walls, and beyond that the walls were powder blue up to the ceiling, thirty feet above, which was punctuated by skylights. My solo was about halfway through the musical and I don't remember anything before I sang: I was elsewhere. I remember hearing my cue, perking up, and standing. The piano accompaniment began, and I started singing. The voice coming out of me didn't feel the way it had when I had been practicing, and this made me more nervous.

The more nervous I got, the more I wanted to be done. The more I wanted to be done, the faster I sang. The faster I sang the louder my heart beat grew, the more nervous my body, and the more the only thing I could hear was my own urgent voice.

John Henry was the man, I sang, *John Henry long and lean. John Henry was the man who could...beat the machine.* I got through the first verse. Then the refrain. Then the second, and the refrain, and finally the third. My heart was pounding in my chest. I hadn't forgotten the words!

And then I took a deep breath and came back to myself and looked out and Mrs. Norris was staring at me with eyes wide in surprise and still playing the piano. We stayed like this for a very long time, and at some point I realized that I had gotten so far ahead of her I had sung the song in half the time at twice the pace of the music. She had to finish playing the piano accompaniment, and it went on and on and on. My face began to burn. I didn't know where to look. I started swallowing and choking back tears. The music finally stopped and I sat down, and there was a long awkward space-filling pause.

And then my friend, Elizabeth Amber, started singing the next solo, about Annie Oakley gettin' her guns, in time with the actual rhythm, and the entire auditorium exhaled as one and the show carried on.

I don't think what someone said was, "You can't sing." I think what they said was, "It's ok you can't do everything." Of course, what I couldn't do was quite obvious.

All of this swum back to me in the garden. By this point I was in my early thirties, living a very different life, at a very different place in my healing journey.

Shortly after this encounter, which was probably in 2005, I attended an artist residency in Brazil, on an island off the coast of Bahia, and began learning Portuguese. Portuguese, of the variety spoken in Brazil, is one of the most melodic languages on earth. Brazilians talking to each other pretty much sounds like singing. This period of time in Brazil corresponded with a moment of significant heart opening for me, and also with falling in love with Brazilian music, particularly Tropicalia, and many of the Brazilian singers who had started out in the sixties and seventies, and whose music I had never heard before. Gilberto Gil, Caetano Veloso, Maria Bethânia, Gal Costa, Elis Regina and so on. Their music vibrated something open in my chest, and somehow it didn't occur to me that I couldn't sing in Portuguese. This was how I finally found my foothold.

Singing, someone once pointed out to me, is a powerful way to vibrate yourself on the inside. Vocalization *is* vibration. When we sing we awaken, hopefully, the interior resonance of the body. Singing is a portal to the connection architecture for sure.

CHAPTER 24

LANGUAGE AS DOORWAY

We are talking about the face, about singing, about vocalization, and so it is appropriate to talk about language itself. I would be remiss in this section on portals to connection, if I didn't address the deep and fundamental way in which language can either open or close the doorway to connection. Language is not in and of itself a physiological portal, though vocalization certainly is, and yet it is one of the defining characteristics of our species, which is not to say that many other animals don't have languages, but rather that the degree to which our experience of self is enmeshed in language is most probably uniquely human.

Birds have language– I have been a student of bird language for a number of years. Many animals have many different forms of language, but my sense is that humans get our sense of self more tangled up in language than any other animals. The language you are reading this book in, which is the language that I grew up speaking, is profoundly problematic in this regard. I first started to realize this when I was about 19 years old and began meditating, and consistently began having experiences that I could not translate into English. I then proceeded to have experiences in nature, and in ceremony, and with plant medicine that I could not translate into English at all. And then with land, and oceans, and an island, and trees, and animals. Very few of them yielded into English, though some of them yielded to other languages: most consistently Indigenous languages, which I have been told by many people, including speakers of the aforesaid Indigenous languages, rose up from the earth, the plants, and the animals themselves, in a particular place, so this makes sense.

About a month ago, I had the good fortune to receive a friend of mine from Bozeman, Montana who was traveling through California with a delegation of Indigenous women fundraising for shared projects. Two of them were Lakota, and two were Yokuts, a California tribe whose traditional territories are in the foothills of the Sierra Nevada, a couple hundred miles east of here. The valley in which our eco-reserve is located has been continuously inhabited for 8,000 years at minimum, and on arrival they met me at the ridgeline that forms the edge of the basket of the valley. This spine divides the county north/south, separating the rural western part, where the land we steward is, from the populated eastern part where the towns are.

When I arrived they were moving into ceremony on this spine of the mountain, arrayed in a small circle on a fire road, burning sage with a drum and Eagle fan out. I parked my car, walked up to them, and one of the women began singing, in Yokuts, her native language, as I watched– *it was visible, obvious*– the land around us remembering her language.

The hawks began to call and to circle above us, the animals perked up their ears, the trees themselves, the grass, the stones, even the wind began to listen to her song. The air itself brightened. The drum was passed, and the next Elder began singing in Lakota. Her song was beautiful– I know just a tiny bit of Lakota– and yet the land did not know the language in the same way.

Later, I spoke with the woman who had been singing in Yokuts, about what I had seen, what I had felt. She looked at me kindly and explained that her tribe had been passing through this valley singing, on their way to the coast, for thousands of years. The language was born near here. She was speaking the local dialect back to the earth. The earth understood what she was saying. It, and all the beings living on her, had missed her and her people. I had felt them reach out to hug this daughter who had returned, singing.

A language is a map, and perhaps you'll not be surprised to learn that the English language has entire continents missing from it. In linguistics, a place where a language could have a word, yet does not, is called a lexical gap. I wrote an entire book about the lexical gaps in English called *Keywords: A Field Guide to the Missing Words*, and I wrote it to help myself feel less alone. The English language has characteristic patterns of absence when it comes to describing interiority, and relationships. This is, in a way that's quite similar to the way that the view of allopathic medicine structurally prevents it from seeing certain things about the mind-body connection, a byproduct of the origin of the language etymologically (Latin, etc), and the domination lineages that have informed the worldview that built the English language.

Without straying too far into the esoterica of this, which you're welcome to do in the other book, the way that we talk to ourselves in our own private minds, and the way we talk about ourselves, and to others impacts directly upon our capacity for connection.

For many of us, if our internal dialogues were exposed, we'd be locked up. Many of us have a tendency to talk to ourselves with a level of harshness that we would find appalling if it was directed at someone else. It would be clear that it was a form of verbal abuse. I'll just say this plainly here. Your autonomic nervous system does not understand sarcasm. There is a place within us that has reservoirs of listening that are endlessly sincere, and that believe what we say about ourselves. Because of this, if our effort is in the direction of experiencing deeper connection to ourselves, it is imperative that we speak to ourselves more kindly in our own private minds.

I know a lot of people who call themselves stupid in their own internal dialogue, because sometimes this makes its way into the way they talk about themselves with others. I know a lot of people who consistently belittle themselves, mock themselves, speak poorly of themselves. I know a lot of people who make things impossible for themselves to learn, because they make declarations of permanent lack of ability. I find it useful for us to remember that when we speak to ourselves, we are speaking all the way into our deepest child places. I think it bears repeating that often when talking to ourselves, we have internalized the voices of others who have harmed us, betrayed us, or at least do not have our best interests at heart. It helps to think of your body as being undefended against the things you tell it, since you live inside your body.

It helps to think of your body as being undefended against the things you tell it, since you live inside your body.

What this means, in essence, is being nicer to yourself in your internal talk. This is an active form of self-care; of self-compassion.

One other important piece of linguistic technology is that if you tell your body it can't do some thing, it tends to believe you. There's this magical little word: *yet*. If I say to myself, *I can't do math*, there is a part of me that is listening in my body and actually believes this. If I can train myself instead to say, *I can't do math yet,* this leaves the door open for some future moment when I might be able to do it.

Since at a certain stage in personal development, a lot of what people end up working on are deprogramming unconscious and subconscious limiting belief systems that they have built, you can save yourself some time by not building them in the first place.

Another quick tip: take ownership of qualities that you want to bring more deeply into your experience, but do not take ownership of things and attributes with which you do not want to be associated. For example: take ownership of your brilliance. Take ownership of your power. When you talk to yourself, talk about *my* brilliance. Talk about *my* power. On the other hand, don't take ownership of anxiety. If you're experiencing anxiety, and you're talking about it in your own private mind, don't make it yours. Don't say, *my* anxiety. Say, *the* anxiety. How we deploy pronouns has the capacity to move things and relationships either closer to us or farther away. See if you can use this on purpose to more deeply affiliate yourself with the kinds of emotional and mental states you want to cultivate, and to distance yourself from those that are not useful.

If you really wanted to open the doorway to connection through language, I would encourage you to learn the old Lakota language. My friend, Tiokasin Ghosthorse, of the Cheyenne River Lakota, calls it a 'non-mathematical quantum mechanical language of intuition.'

Lakota describes the movement of energies. We would say, "I see the red ball bouncing back-and-forth." Lakota, which is verb-based, would say, "Bouncing back-and-forth." We would know it was you seeing it because you are talking. We can all see it is red.

Lakota describes the motion of energies.

Most of the words in English are nouns. This places us into a world of things. Most of the words in Lakota are verbs. This places us in a world of relationships. This is actually congruent with what modern science (quantum mechanics) tells us. We live in an impermanent world of vibration. Matter is congealed energy (congealed music). The world of connection is a world of relationships in flux, energies moving. How deeply relieving it must be to inhabit a language focused on describing this movement, rather than on permanizing a world of things that seem inert and dead.

I suppose if, in English, every time we said the word *Sun* we caused ourselves to feel his light on our faces– if we were more concerned with evocation than categorization– we would get close to what it means to speak Lakota, which seems truly a language of connection.

The word *kotodama*, in Japanese, is the word that I have come across– a word that fills a lexical gap in English– that acknowledges and attends to this interiority of the word. It means, literally, the spiritual interior of the word. You know how you can say, *You know I love you* to someone and mean it with your whole being? Or mean it as an admonishment? Or plead it? Or declare it resentfully? This is a change in the *kotodama* of the word.

We deepen into connection with another when we practice what is called in Hawaiian *kina'ole*. On the surface, *kina'ole* means 'without flaw'.[1] Yet the *kotodama* of *kina'ole– To say the right thing, to the right person, in the right way, at the right time, the first time.*

That is an expression of the quality of attention through language that opens it into a doorway to the heart. Although we can't explain that in English, for there is no adequate word in the language, you can learn to speak like that from your heart.

1 Kina (flaw) + Ole (without), which we leared earlier with Ha'ole (without breath).

CHAPTER 25

I SEE YOU

I'd like to have a cat named Eyeball, probably as an antidote to an earlier cat of ours, Arselan, who was blind (we found her that way, one eye missing, the other a socket that looked like the pinched-up end of an eggroll). I can imagine myself searching for the cat, walking around our house calling, "Eye Ball…Eye Ballll."

Cats have very expressive eyes. They are large, and remarkable in darkness, and convey nuance. When a cat looks at you, you often have a pretty good idea of what is going on in their mind beyond the eyes. This is why we say that the eyes are the windows to the soul. Notice here in conformity with our observations about dwellings and psyches, that the eyes are configured as a portal. A window is a portal to the house. We know that the eyes are a doorway. It's right there in the phrase.

One of the things that is most important about neural platform being fluid, about our neuroception shifting under us at times, sometimes unconsciously, is that it makes it so that our experience of a situation can change even if the situation doesn't. Eye gaze provides a pretty interesting window into this phenomenon, because we interpret eye gaze very differently depending on whether we are experiencing the water, steam, or ice versions of ourselves.

When we are in the connection state, eye gaze is generally experienced as connective. As long as it is not overlong, or overintense, most people find it deepens an experience of connection. As soon as we leave the realm of safety, however, slipping across the threshold into defense, our experience of the eye gaze of others changes. If we move into the steam state, and in the direction of the flight response, the same eye gaze from someone else that a moment ago felt connective is experienced as intrusive. Their gaze feels prying. If we shift into fight, their gaze feels confrontational. And if we shut down, their gaze is overwhelming.

This dynamic interplay happens in stark ways when we shift

strongly from one state to another, but each of these states also exists on a continuum. In the same vein that there are aggressions and micro-aggressions, and expressions and micro-expressions, our state shifts can be macro or micro.

If we are talking to someone we don't know well, there can be a moment when trust is established, and eye gaze becomes connective. But if they stare at us for too long– and too long is highly subjective– this can shift. If they then say something offensive, connection may evaporate, and the meaning of the same gaze may radically shift. Or, they may say something ambiguously offensive, and we find that suddenly the meaning of their eye gaze becomes ambiguous, and we are not sure how to relate to them. This kind of micro-moment happens in charged social encounters all the time: restaurants, bars, clubs.

Eye gaze that is too long or too short conveys charge. Is he flirting with me? Is she avoiding my gaze? The traffic between gaze and interpretation, innuendo and meaning is sometimes a dance, sometimes a streetfight. Wars have been started over errant glances.

We have a number of words for this. We can *glance* at someone, *gaze* at them, *ogle* them. All three might have the same emotional tone or intent, the difference may simply be in how nakedly this shows through.

If we operate from the awareness (which our culture both denies, and English has inadequate words to articulate) that we are not simply disembodied heads walking around on sticks, but embodied animals, we have to acknowledge that much of our interactions are energetic exchanges.

This is a significant part of why much of communication is non-verbal. Our bodies are talking for us.

The way we take up space, stand, move, gesture, use our faces, our voices– all of these convey. All of these communicate. I don't fully understand how or why, but I know that specific information passes through the eyes. A person's gaze can cut you. It can provide an invitation or a suggestion. It can exhibit dominance or submission. It can convey rage. It can plead. Or it can open a doorway to connection: a stream of vitality. The eyes can do all of this. I have experienced all of this through my eyes.

There is a famous study that was first formulated in 1997 by psychologist Dr. Arthur Aron, called *36 questions to fall in love*. The goal of the questions was to accelerate the creation of intimacy between two strangers. Two people would sit down across from each other, and both would answer a series of questions. There was no time constraint on this at all. The questions begin broadly: Question One – *If you could invite anyone in the world to dinner, who would it be?* Then, both people answer. After setting some broad context, question 8 begins to bring attention to the two specific people who are together: *Name three things you and your partner appear to have in common*. By the late twenties, the questions have become decidedly personal and intimate. Along the path they've shared their life stories. Answered questions like, *Is there something you've dreamed of doing for a long time? Why haven't you done it?* At the end of the 36 questions, the instructions are to stare into your partner's eyes for four minutes in silence.

Four minutes is a long time. You can listen to Ben E. King sing *Stand by Me* in two minutes and fifty seven seconds. You've just spent several hours talking to this person about things that are very real and very deep, learning to see one another.

The first partnered group of two strangers to do this together? They got married six months later. From my perspective, it is the eye gaze that seals the deal.

CHAPTER 26

THE ERRANT EYEBROW

Of all the reasons dogs have become our best friends, perhaps none is greater than the work of an errant eyebrow. Dogs are the descendants of wolves. Over the past twenty thousand years we have succeeded in pulling them closer to us, domesticating them into all the roles that dogs have played in human lives. We bred them to hunt: from elk (Elkhounds) to fox (foxhounds) to birds (bird dogs) to rats (the rat terrier). We bred them to guard livestock (Great Pyrenees) and shepherd (Australian Shepherd). Eventually we bred them for affection and companionship (lap dogs). But why dogs? Why not guard penguins? Or pigs? Or pigeons? Why was it dogs?

It is because of how they look at us. It is because to human eyes, the dog face registers a certain emotional intelligence, a certain expressivity, and the reason that their faces do this is because we are hard-wired to read faces through our autonomic physiology. The great success of dogs, and you can whip this out as an interesting little anecdote at parties, is due to the Levator Angle Oculi Medialis - or LAOM- a small eyebrow muscle, which is innervated by the facial and trigeminal nerves, the same nerves that innervate our human faces. This is the muscle that allows your dog to look pensive, his face resting on his paws, or to give you those puppy-dog eyes. It is what causes your dog to look like he is paying attention, or slightly embarrassed. It is why, when he cocks his head slightly to the side, and looks at you, you can hear him thinking, "Are you ready to go yet? Can we go now? Please?"

This facial expressivity, paired with the eyebrow, allows your dog to respond to you facially through reciprocal social engagement cues, and we can read these cues. We are granted, therefore, although I imagine that most of us are pretty anthropocentric about this, access to our dog's inner worlds. This is the way in which the face is a portal.

This is not the only reason dogs have become so successful in embedding themselves in our lives. They have learned over the past 20,000 years of history to become very focused on our needs. But without this portal of the face, the connection would not be as deep. In fact, research has been done on the differential rates of adoption between dogs who have this eyebrow muscle and those that do not. As you might imagine, dogs with expressive eyebrows are adopted at statistically higher rates. It is because we feel more connected to them. And we feel more connected to them because we can watch their faces responding to our social engagement cues.

This quality of attunement to the nervous systems of their humans makes dogs what my friend Jimbo calls 'remote interoceptors'. What a beautiful phrase. Sometimes the reflection we see in our dog's faces tells us more about our own internal state than we know ourselves in the moment. I have more than once been experiencing a sadness of which I was dimly aware, only to have my dog come plunk his face in my lap and stare into the depths of my eyes as if to say *I'm all the way with you in this grief, whatever it is.*

One of my favorite things to do with our dog is make faces. I recorded a whole video with him, in my best made-up German, doing psycho-analysis with him on the couch. His head was in his paws, and I started analysing him, using made-up words and intonation. And he would give me big eyebrow shift. I would unleash a stream of response, and he would give me another eyebrow shift. We went on like this for some time. Sometimes he would puff a giant exhale. Sometimes he would look worried or disconsolate, and his eyes would shift from side to side, the brows crawling up and down. I don't think he understood a thing I was saying, because *I* didn't understand a thing I was saying, but he could feel the rise and fall of the intonation and melodic (or lack of melodic) quality of my voice, and he responds to that immediately.

Sometimes I will just sit and play with him by making faces. We can just sit there watching each other for ten or fifteen minutes. I'll begin to move my lips, quiver them, caterpillar them slightly, like they are worms on my face. He'll mirror this back at me. I will play-snarl at him and almost reveal a canine tooth. Fanging. He will begin to growl in the back of his throat. I will flash my teeth at him and he will flash his teeth at me, his hackles dutifully beginning to rise. I will then withdraw slightly, pull back energetically. He will sneeze as the pressure releases. I will cock my head to one side, and he will cock his head to one side.

This is a game that we are playing, to be sure, but there is also something serious and important going on here. He is my pal, my dog-friend, and also my responsibility, and I think he sees me in similar ways. Shortly after I got him, I had a dream in which I fell asleep in a place that was unsafe, and he came and woke me up. Let me tell you a bit more about this.

We adopted our dog, Brix Valentine, from the Humane Society on Valentine's Day four years ago. It was winter, and he had been on the street for several weeks, and had scars on his ears and the top of his head. The first couple of weeks we had him at our house, I slept on the couch with him. For the first week he had nightmares every night. I would awaken to find him growling or whimpering or twitching against me, fast asleep. He was terrified of water, which was not that surprising because he had been unsheltered in winter, and there was a lot of rain that year. When he flailed at night that first week, I just held him. Often he would wake up from one of these dreams exhausted, and realize I was there, and then he would just sort of soften against me and all the tension would drain out of him. After about a week the nightmares stopped, and then he just melted into my nervous system. I hadn't had a dog since I was a kid, and the dogs we had when I was a kid were not attached to me. They were my parent's dogs, really.

My dream about our dog happened after this period of several weeks. I think I realized, someplace deeper in awareness, that I had a new animal companion. I love Brix. I really do. He is also, in certain ways, not the sharpest tool in the shed. And he knows the following words. "Brix, sit, walk, drop, stop." He also knows the word *shampoo*, because that word is his nemesis, and comes just before his least favorite thing. This means that what I can communicate to him verbally is pretty constrained. "Brix, walk?" is probably his favorite (short) sentence in the world. It's also about as sophisticated as we get verbally.

But there is a lot of communication that happens between us, and clearly it is not verbal. The game of faces we play is one way that we deepen our connection and play together and talk. Playing with a big dog is also interesting because you can feel and see that within the envelope of connection they calibrate the physical intensity of their responses to you very precisely. Brix has only broken my skin with his mouth once, and that was an accident. We were playing with a big stick I was holding, and he lunged at it, and my finger was in the place he bit down. This was actually quite enlightening, because it made me realize how profoundly he is softening his bite when we play. (When he wasn't softening, he just tore my finger open.) Our relating happens within the domain of sociability, and even when there is very high energy play and he is exerting intensely, like when we are wrestling with a rope, he knows exactly how much pressure he can use with his mouth.

Beyond our play, there are health benefits for me, and likely him, in this game of faces. Because our face is a portal to our Connection System, wired through the Facial and Trigeminal nerves right into the brainstem, and from there into the heart, the expansion of facial expressivity opens the ventrality of the heart. Expanding facial expressivity expands our connective range, and activates the connection physiology.

Making faces makes you healthier. This seems like a good thing for people to know, and like a good thing for white people in particular to know. Because part of being socialized into whiteness is socialization into dampening the affect. In whiteness you are not supposed to be too expressive. You aren't supposed to gesture greatly, or get too passionate, or animated, or facially or vocally expressive. The history of whiteness, for example the movement to extend the mantle of white identity to the Italians, who in the early 1900s in the United States were not considered to be white, came straight up against this prejudicial attitude toward expressivity. Italian culture is notoriously expressive. It is passionate. People talk with their hands. This was not considered to be 'white'.

So here's a physiological connection between socialization into whiteness and ill health. I recently heard a massively successful black entrepreneur presenting at a technology conference to a white audience. He walked on the stage, beamed at the audience, who had offered a very luke-warm reception, and said something like, "Wake up. We're gonna have a good time. We're gonna get together. We're gonna learn something. Let me hear a little bit from you." The clapping got a bit louder, still tepid. "You're happy to be here," he said. "This is gonna be good. *Tell your face.*"

Tell your face. I loved this moment. A successful black entrepreneur sneaky-teaching neurophysiology to a white tech audience. Because if you tell your face– if you allow yourself to allow the 'happiness of being here' to register on your face– guess what, you'll feel happier to be here. Attention potentiates experience. And when we flow attention through the doorways of the connection system, guess what...it turns on. And then, you feel better. That's how this whole thing works. So, *tell your face.*

CHAPTER 27

The smallest muscle in the human body, instrumental to your wellbeing, and that you've probably never heard of, is the stapedius. The stapedius is a muscle in the middle ear that is about half a centimeter long. Tiny. Think of a grain of sushi rice. It is situated in the middle ear, in the tympanic cavity that receives sounds through the ear drum, and transduces them mechanically through an articulated chain of bones into the cochlea. If you want to wax rhapsodic about the genius of the body, study the ear. Its strangeness is so deeply otherwordly as to simply engender awe.

Sounds from the world-at-large enter the outer ear and are funneled down the ear canal to your eardrum. The outer ear is the part you can touch, the part that makes contact with the outer world, the part that your parents are forever warning you not to stick things into, for fear that you will perforate your eardrum. The outer ear terminates in the eardrum (the tympanic membrane), which, as you might guess from the inclusion of the word *drum* in its name, is a stretched membrane that is tensioned to receive sounds incoming from the outside world.

The middle ear begins on the backside of this membrane, where a chain of bones, the auditory ossicles, translate the

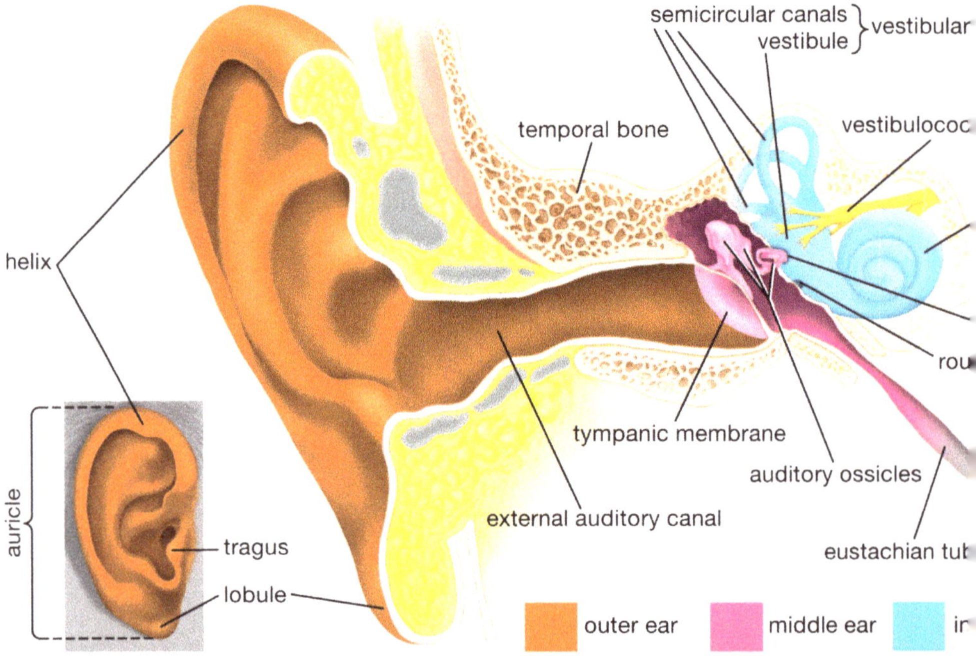

vibrations from the eardrum to the cochlea. There are three bones in the chain. The first, shaped a bit like a hammer, is called the malleus (Latin for hammer). This sockets into a somewhat ball-shaped joint, the incus (or anvil), which in turn abuts the stapes (stirrups). True to its etymology, the whole chain appears to have been fabricated by God's tiniest black-smith, and catches soundwaves (vibration) from the backside of the eardrum and channels them to the cochlea, in the inner ear, which is filled with fluid.

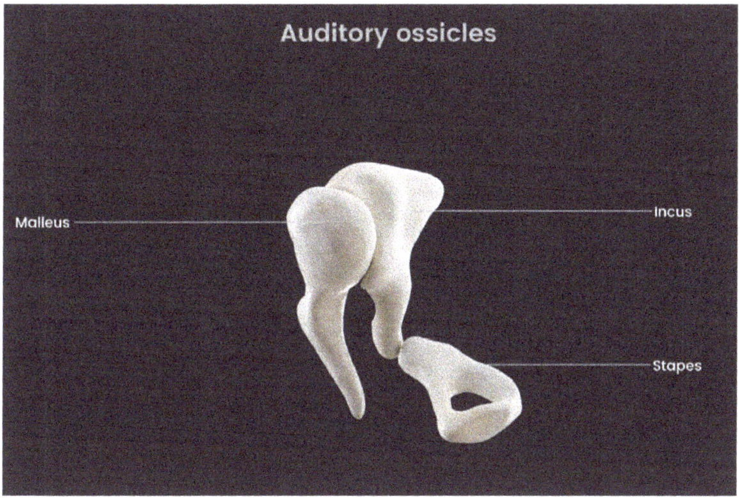

The middle ear, though not accessible from the outside of the body is however a cavity filled with air, the presence of which deep in your head is accomplished by inclusion of a tiny fleshly tube that runs from the back of the throat into this cavity. Called the Eustacian tube, it functions to normalize the pressure in the middle ear– if you've ever flown in an airplane, and chewed gum to try to get the pressure in your ears to equalize– you've felt what happens when your Eustacian tube is not opening enough to match the pressure in your middle ear to the pressure from outside. This tube, incidentally, changes angle as you grow, and is often the cause of earaches in infants and young children, when it is unable to drain fluid that accumulates in the middle ear. You can think of the tympanic cavity, the space of the middle ear, as a sort of cave behind the eardrum that translates vibrations from its backside across an inward spatial gap and into the fluid medium of the inner ear, where through an even stranger mechanism they will be processed into the neural inputs we interpret as hearing.

Within the middle ear is the stapedius muscle, which is connected to the stapes (bone). Anatomically, the muscle stabilizes the auditory ossicles, attaching to the posterior end of that bone, where it abuts the incus. The stapedius develops from the second brachial arch, which is interesting in terms of evolutionary anatomy (a portion of the Polyvagal Theory is premised on the arising of a ventral regulatory nucleus in the brainstem that coordinates breathing, sucking, swallowing, heart, and breath, and seems to have arisen to coordinate the gill arches (brachial arches) of ancient bony fish), and is served by a vertical branch of the facial nerve, which, at the point where they touch, is of a similar diameter to the muscle itself. There is a degree of confusion, in the literature, about the size of the muscle. In my literature review for this section, there were confident assertions that the stapedius was one millimeter, six millimeters, and nine to eleven millimeters long. The only literature I found where the muscle was actually shown, in a journal article on micro-surgical anatomy, where the authors removed the muscle from 32 cadavers, and photographed it, contends that it is much larger (9-11 millimeters) than anyone

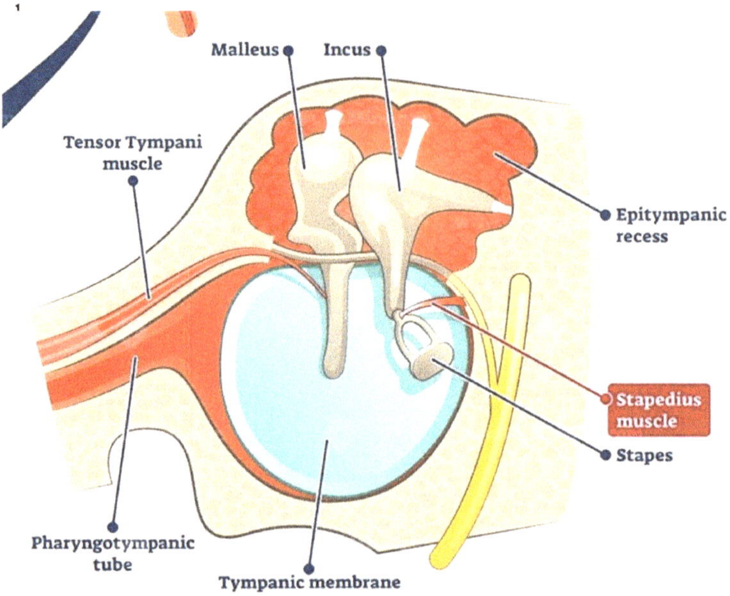

Lest we neglect the mechanism entire, sounds that make their way down the auditory ossicles are transduced, at the far end, through the flat base of the stapes, which rests, like the bottom of a stirrup, against the cochlea. The cochlea is a spiralling fluid-filled alien snail that flew inside your head when God blew breath into your body, and translates vibration into neural inputs.

Okay, I'm just kidding. It is not an alien snail that flew in there, though it looks like one. Yet it might as well be because it is actually weirder than that. The cochlea has two primary parts, both of which make use of stereocilia, a sort of isolated hair cell. The hearing function is accomplished in the cochlea, which is fluid-filled, by the vibrations of fluid across a set of about 12,000 of these hair cells in an area called the organ of Corti, where they are translated into electrical impulses that your brain interprets as sound. In another section of the cochlea, there are a set of semicircular canals that are the home of your vestibular system, responsible for balance and spatial orientation, where the stereocilia are arranged three dimensionally and arrayed along different axes, and where tiny, as in quasi-microscopic, rocks (otoliths) made from calcium carbonate are suspended in fluid, knock against them, and create your sense of balance and spatial orientation. Calcium carbonate is a primary component of chalk, eggshells, and gastropod shells. Maybe it *is* an alien snail.

Hearing happens because sound waves, which are vibrations, register on a tiny drum, are translated into the movement of a series of tiny bones, which then push into a fluid-filled bulb where they shake tiny hairs that translate them into neural impulses.

Cool, right?

The stapedius, it turns out, has a very interesting and crucial role in all this. Remember that the stapedius is governed by a branch of the facial nerve, which is wired together in the brainstem with the entire social engagement complex, which governs the neural regulation of the face, the voice, the eyes, the tuning of the middle ear, the turning of the head and neck, and the heartrate and breathing. It is an integrated complex. And this complex, in sophisticated ways, governs the way that the stapedius is shaping sound. Technically, the stapedius is tightening and loosening (it is a muscle) to dampen or undampen certain types of vibration. Functionally, it is deciding, based on whether or not we feel safe, to focus our hearing on vibrations in the vocal frequency of the human voice (by dampening low frequency vibrations) or to pick up lower frequency vibrations by dampening sounds in the human vocal range.

This determination is directly connected to our present moment felt sense of safety. When we feel safe enough, we can focus on relating, and when we are relating the sound of other people's voices is crucial incoming information. So in situations where we are bodily experiencing safety, the stapedius shapes our hearing to focalize human voices. By contrast, when we are experiencing danger, because ancestrally most predator signals were lower frequency, when we experience threat the stapedius shapes our hearing to focus on lower frequency bands, and we lose the ability to track (and also process, though this happens in different parts of the brain) human vocal signals.

You know this already, even if you don't realize that you do. Have you ever been out at night, in the city, hanging out with friends when something happened that made you nervous? Let's say you are walking down a street with two close friends, at night, and you suddenly detect someone lurching unsteadily toward you. Have you ever noticed that in the moment of that threat detection, as your body becomes aware of it, you lose track of the voices of your friends? In my early twenties I practiced martial arts, and I learned to box. My boxing coach was a guy named David who talked in a sort of breathy half-whisper, and held up pads for me to punch into. "Strike it clean, Gabriel," he would say. "Strike it clean." When people are learning to box, when they first step into a ring for a fight, and it gets real, they often stop being able to hear their coaches. This is because when you are learning to box, your actual honest-to-god fight response comes online. "Everyone has a plan: until they get punched in the face," former heavyweight champion Mike Tyson reminds us. When you get hit hard, your threat response activates. And when this happens, the part of your nervous system that experiences threat re-tunes your hearing. Suddenly you can't hear your coaches talking to you. Mouths move, but we don't receive the sound.

So there are frequency domains associated with feeling safe, and there are frequency domains associated with threat. Friendly sounds are made by instruments like violins, flutes, and piccolos. You've probably heard Prokofiev's *Peter and the Wolf*, where the friendly forest creatures are voiced by instruments in the vocal frequency of the human voice. Peter himself is voiced by strings, often a violin. The violin is the instrument that most closely aligns with frequencies of human vocalization.[1] The bird, a friendly, is a flute. The cat is a clarinet, which begins to move in the direction of the bass: possible overtones of danger. The wolf, French Horns. The hunters? Tympany drums. Real deep threat. The body, innately, and physiologically, responds to frequency domain, both of safety and of danger.

Think about how movies are scored. Why was *Jaws* so terrifying? Can you remember the music? My wife saw the trailer, just the trailer, at 5-years old, and will still barely go in the ocean. It is the frequency that does it.

Another example: think about how we talk to babies. Watch the way the cadence and prosody of the voice ticks up. A big burly dude will kneel down to talk to a baby, and his voice will go up several octaves. If it doesn't, the baby will burst into tears. Or think about how we talk to dogs. We know, instinctively, about the cadences of safety and menace. Consider how we

1 The cello is the instrument most aligned with the human body.

make our voices melodic when praising them, how we flatten our voices when scolding. *Good boy* is juicy and mellifluous, punctuated with praise. The dog wags at us, comes closer, seeks to engage. *Bad dog* is flat and clipped, a sharp admonition. The dog cowers. Remember, they don't speak English. They don't understand what you are saying. They are responding to the prosody of relating, or its withdrawal. You could say *Rufus Mashed Potato* to them, or *Machine Engine Bungle*, or *Sesquecentarian debacle*, and if you say it with the cadence of *good boy* you'll get a wag, with the cadence of *bad dog* you'll get a cower. We know this stuff already.

Fortified with this awareness, and the automaticity of the transformations in hearing, we can consider how we shape our voices so that they evoke signals of safety in others. Safety is prosodic. Safety is a pulse. A rhythm, a music. It has a particular psycho-acoustic profile.

When we feel safe, our voices express this. They have to, because our hearts are expressed through our faces and voices. When we feel dangerous, this too is expressed. A friend of mine recently spoke about noticing that at times in his life his voice was laced with 'serrations of menace.' What a phrase. Think about the cutting edge of a knife. Several months ago, a mentor of mine shared with me a video of an Elder. I was in my car, driving home from a tennis match when I heard it. And something deep within the voice itself arrested me. Unbidden, from some deep place in my own awareness came the words, "That man has a knife in his voice."

Both safety and threat have acoustic signatures. These experiences are right here, close to us. As a portal to the connection architecture, the ears therefore conduct us in the direction of safety and connection when what we are hearing has specific acoustic and prosodic features. This means that how we nourish ourselves acoustically– from the voices of other people, to the music we hear, to the sounds in our environment– all of this matters to our daily experience of wellbeing. It impacts upon our felt sense of safety.

I didn't realize this or reflect deeply upon it until writing this chapter, but I grew up with a set of voices in the living room of my home. On Saturdays, we would listen to Garrison Keillor's *Prairie Home Companion* on the radio. It is impossible for me to remember this without feeling safe. It was the tangible visceral felt quality of that voice that told me I could exhale. I grew up reclining on Garrison Keillor's voice the way you just surrender your weight into a hammock. I also remember listening to Ray Suarez, the Host of NPR's *Talk of the Nation*. There was, for me, a whole world in Suarez' voice. Think back to the voices that made you feel safe. What was it about them in the felt quality? Those voices were providing you a form of nutrition. Whose voices were you in love with? How did they make you feel? Where in your body did you feel this?

Take a moment to think about music. With no criticism or endorsement of any particular genre, how does the acoustic signature of your music support or undermine your sense of safety? Do you intentionally utilize music to help evoke states of safety and connection? Could you? We make playlists all the time. Could you make a playlist whose express purpose was to increase your felt sense of safety? Whose music would be on it? Would it be vocal or instrumental? Who do you want to hear singing you home to yourself?

And what are the acoustic features of your daily environments?

Your home? Where you work? I've thought a lot over the past decade about the acoustical features of hospitals, because they so directly undermine the purported aims of healing. Hospitals are filled with right angles, hard walls, easy-to-sterilize surfaces that also bounce and amplify harsh sounds. Gurney wheels scrape on linoleum. The low humming sounds of ventilation and machinery basically nest in the frequency ranges of predation. The weird knocking, electronic whirring, and strange chirping of the gradient coils in an MRI machine sit in the acoustic signature of danger. Ambulances arriving at the ER come in with sirens ablaze. Whatever your conscious mind is telling you, stepping into the hospital is to enter a zone of acoustic menace. And then they hand you a garment that is open in the back, so that if you had the audicity to rise from your bed, everyone would have a full view of your ass crack, and make you sign a bunch of liability releases. *Relax, we'll take good care of you*, they say. But your body? The feeling animal? It says, *Yeah right. Get me the fuck out of here.*

It does not have to be this way. We could design healing spaces considering the acoustical features of safety. We could soften those hard surfaces, sound-wise. We could mitigate against these bass frequencies in the HVAC systems, in the machinery. But to do this would require that architects and designers of these spaces spoke the language of the felt, and understood the neurobiology of connection.

Over a lifetime of hearing a barrage of sounds, over a lifetime of stressful experiences, the actual shape of the stapedius muscle changes for many people. The grain of sushi rice, which should be wide around the middle, tightens and narrows. And when this happens, the ability of the muscle to exercise the full range of modulation diminishes. Stephen Porges, who developed the Polyvagal Theory, recognized this and developed a sequence of phased acoustical interventions that use mathematically-modulated music to stretch, exercise, and retrain the stapedius itself.[2]

The intervention is like doing reps at the tiniest gym in the world. Your stapedius, all half a centimeter of it, learns to curl, learns to lift weights. I was astonished, the first time I experienced this intervention, by how fatigued I felt afterwards. I was somewhat amused by this. I think about lifting weights, maybe bench-pressing a hundred or a hundred and fifty pounds. And then I visualize my tiny stapedius, the size of a grain of rice, working out. In the image, I dress them in gym shorts. Yet the

2 https://patents.google.com/patent/US10029068B2/en

effort required by this neural training was real.

At the present time, this intervention (the Safe and Sound Protocol) needs an autononic diagnostic tool on the front-end to calibrate the dosage for individual nervous systems of varying sensitivity so that people do not overdose on it. But the intervention is beautiful, and future versions and variations of it will be even more powerful and refined. Who would have imagined that our entire experience of safety and danger could be modulated by a portal into the human connection system governed by a muscle the size of a grain of sushi rice?

CHAPTER 28

HANDS

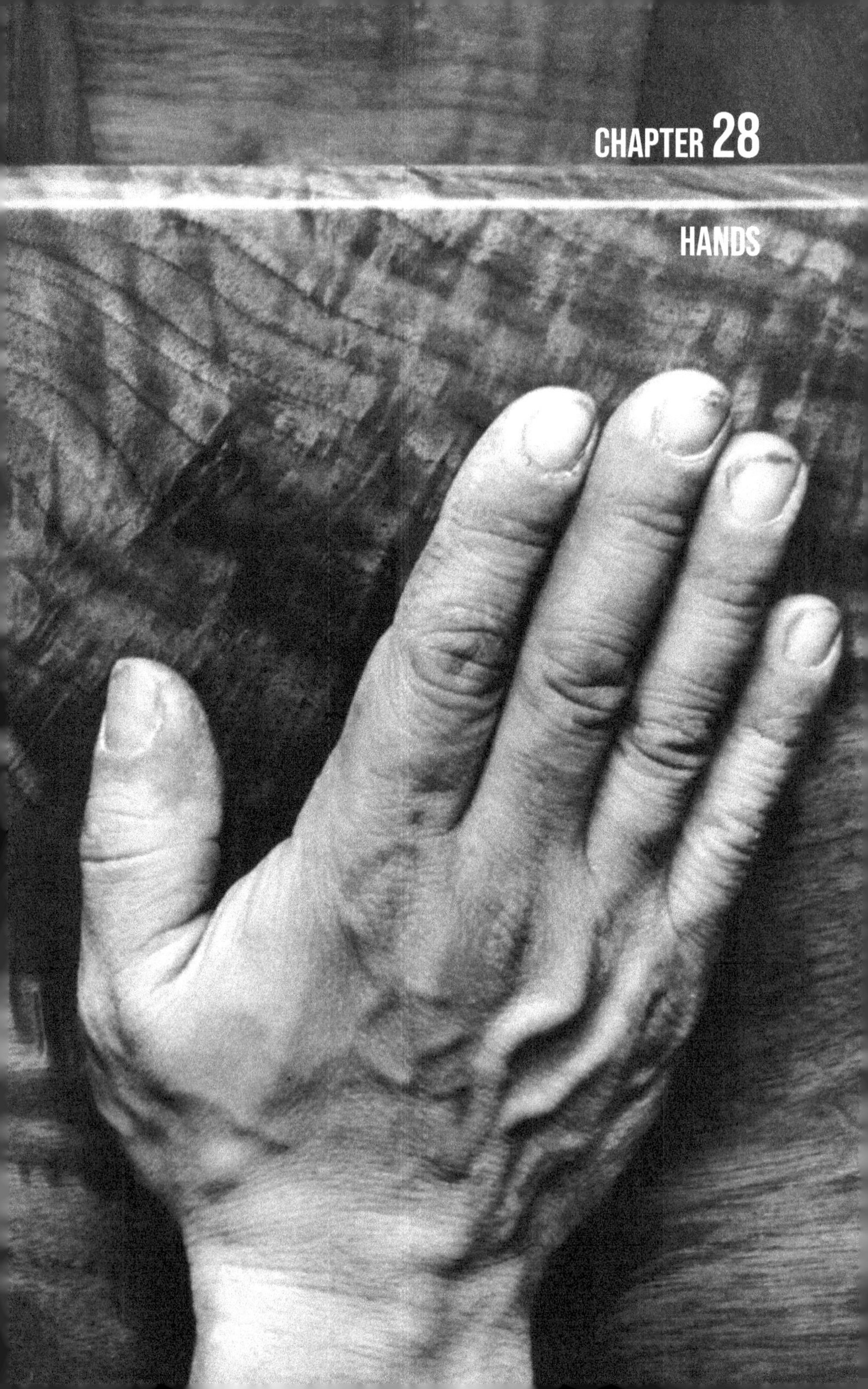

We have moved now through the face/voice/eyes/ears social engagement complex that Stephen Porges identified: the neural architecture of connection when a baby is born. We now move into the neural architecture of the Connection System that establishes itself in the womb-with-a-view period: the first 18 months of life, before the closing of the fontanelles. I will not necessarily do this in the order in which the circuits myelinate, but rather in a gestalt. From the core architecture (social engagement system) that unites the first of the infant's connection portals to birth attachment, we now move to the portals that complete the embodiment of connection.

If you meditate on the origins of the human species, deep in the mists of ancestral time, and you reflect on what distinguished us from other animals, there is the movement to walking upright on two feet (8 million years ago), which rotates the hips in such a way that we are now looking out, looking forward, looking toward the future rather than toward the ground– and there is the intricate use of the hands.[1]

Our hands are *magnificent*. Take a look at them. Able to grip and grasp with great force, they are also capable of learning to make movements that are almost unimaginably refined. Our opposable thumbs give us tremendous dexterity. The hands can strike, and they can caress. Guide a jackhammer or fold oragami. Wield a sword or a paintbrush. Wield a sword like a paintbrush and a paintbrush like a sword. Just, *wow*.

An experienced surgeon or master painter or elite tennis player is able to control the movement of their hands with breathtaking precision. To use a scalpel in such a way that it cuts through exactly the tissue required and no more, or to paint the finest line that is perfectly straight, or to take the pace off

1 There is also language, about which we've already spoken.

a blistering 100 mile-per-hour forehand and arc a dropshot crisply inches over the net–in each of these cases the intelligence of the hands is extended through a tool that merges with and transforms them into instruments of something else. In this example, one tool is a form of knife, one a form of brush, and one a form of racquet–yet each merges with and amplifies the capabilities of the hands– a modern tool that extends our abilities in a way that harks back to our most ancestral roots.

Carlos Alcaraz, the Spanish tennis phenomenon, on a clay court in Madrid gets ready to undercut a dropshot, taking all the pace off. He will take a shot struck at a hundred miles an hour, scoop it from beneath with a deft roll of the wrist, creating intense backspin, causing it to dramatically change trajectory, arcing just over the net, and surprising his opponent, who is over eighty feet away behind the court's opposite baseline. To do this, he will use not simply the hand that holds the racquet, striking the ball, but the other hand, floating in space, the twin wing, which will provide balance and ballast for this ultra-precise kinesthetic ballet.

The potential sensitivity of human hands is so enormous– our ability to detect nuance with them so extraordinary– that in 2013 Swedish scientists showed that people can detect nano-scale wrinkles while running their fingers across a seemingly smooth surface. Mark Rutland, Professor of Surface Chemistry, says that the human finger can differentiate between surfaces patterned with ridges as small as 13 nanometers in amplitude, and non-patterned surfaces. "This means that, if your finger was the size of the Earth, you could feel the difference between houses and cars," Rutland says. "We discovered that a human being can feel a bump corresponding to the size of a very large molecule."[2]

This intelligence of the hands, which is an intelligence wired directly into the heart, for the hands indeed bud out of the heart during our embryonic development, is a portal to trans-

2 https://www.sciencedaily.com/releases/2013/09/130916110853.htm

formation. What a pity that in this modern moment hands have been reduced to pressing and typing and swiping when they have the power to open the door to such profound transformations in our wellbeing.

It was Maria Montessori, the Italian educator and founder of the movement that would become Montessori schools, who pointed out first in the modern era how deeply the development of the hands relates to the development of the intelligence of the child, yet this is something that was deeply understood ancestrally, if not spoken aloud much. In Montessori's beautiful classrooms, children spend a great deal of time, particularly at the earliest ages, doing hand work, and developing hand-skills, as they should. From learning to tie buttons and laces–if you can tie your shoes you can master pretty much any hand skill– to teaching mathematics and spatial volume with colored wooden blocks, Montessori created the most beautiful and comprehensive hand-based teaching tools I have encountered.

What Montessori did, I believe, was translate an ancestral zeitgeist, an ancestral knowing, into a palatably modern form. In Italy at the time people didn't want to send their kids to the farm to be educated, and so Montessori took what was intrinsic to the ancestral environment – the farmyard, the town square, the shops of artisans, the market– the felt tactile living environment, the way that the hands help us know things through our system of touch, and brought it into a classroom context. She refined and imported the essence of the thing, packaged it in more modern form. She is, in a way, my primary mentor as a software designer. The software I have built (The Autonomic Treatment Atlas) is an homage to her approach: an ancestral artifact rendered in a digital medium.

Montessori classrooms create an invitation to touch. The desks and chairs are made of wood, which is a deeply human and ancestral material, and they are smooth and inviting to caress. The tactile pleasure that we derive from touching their smoothly sanded surface, or moving the bentwood chair around, helps us enjoy being in a body. Moving outside into the garden children are encouraged to get their hands in the soil, to touch and hold and examine seeds, and leaves, and the bark of trees: to learn to wield trowels, and spades, and rakes. The teaching of cursive writing, which many argue is not necessary in this era of keyboards and voice commands, is rather a love letter to the hands. Children in a Montessori classroom learn to write first with chalk on a chalkboard, then begin practicing with a pencil, and spend several years learning to control the looping forms of capital and lowercase letters on pink and blue colored lined paper, sized particularly for the task.

One of my favorite photographs in the world is a picture of my daughter at three years old, dressed in a blue and turqoise dress gleefully posed before a chalkboard green as deep ocean, her name written in loopy cursive letters that rise up the board behind her, the cursive T as big as her head. A snapshot of the moment she succeeded in writing her name for the first time. A triumph of three-year old hands.

Mike Merzenich, one of the world's leading neuroplasticity researchers, performed an experiment in which monkeys were taught to use a small stick to entice ants out of a hole. Imagine an ant farm with a small hole in the side of it, just large enough for an enterprising monkey to insert a slender stick. Ants, it turns out, are wonderful monkey snacks. Just crunchy delicious little morsels: very motivating. The monkeys loved to engage in this behavior because they liked the delicious result. They learned to manipulate the stick with ever-increasing dexterity to reach further and further into the hole and extract ants from deeper and deeper in the cavern. As this learning progressed, Merzenich imaged the neural connections in the areas of the brain associated with these refined movements, and discovered *massive new neural growth*. Many new trees in the inward forest. As the dance of the hands became more refined, the brain areas, both motor (movement) and sensory (sensation) affiliated with the hands grew more and more dense with connections, until they no longer resembled the brain areas in their simian cousins who had not learned these movements. This is self-directed neuroplasticity.

Our neurobiology is an expression of the flow of our attention. What we put our attention on, over and over, wires into our nervous systems as we grow a forest of new neurons to support these connections. As this process happens, and as the neurons myelinate, the behavior gets faster, smoother, more refined. How we use our time thus sculpts our nervous systems.

One of the many reasons that modern people are so unhappy is that most of our daily lives, for many modern people, have become profoundly abstracted from the body, and our nervous systems are not getting the right inputs to grant us flourishing. Our modern digital sensory environment has compressed many varied tools that previously engaged our hands– maps, calculators, cameras, newspapers, alarm clocks, compasses, flashlights, rulers, pen and pad–into abstract digital bricks.

Your smartphone is a featureless digital brick.

From a tactile perspective it is about as interesting as a rectal suppository. It is bland and smooth. Its form factor has no relationship to its function. The compass doesn't work like a compass used to. There is no satisfying bubble of water, no careful reading of the dial required. The camera shutter doesn't really click. You can't point the flashlight as you used to do with a flashlight that was, in actual fact, a flashlight.

Just for fun, let's take an inventory of the analog objects that the smartphone replaces, in their analog form factors. Take a moment to consider how many hand movements have been consolidated or replaced by the change in form factor. From the pleasure of handling vinyl discs, to attempting to fold maps back up, to the rotary dialing motion? Our 'smartphones' are making our hands dumber.

In the absence of adequate tactile input, the primary entry point of which was, ancestrally, the hands, we spend more and more time in abstract domains, thinking.

Attention is like a river. When we point attention at thoughts, it strengthens thinking. The flow of words, of talk, of internal talk gets stronger and stronger. We look at tiny screens and we think, and our thoughts run faster and faster without tactile anchors to slow us down. And then people are surprised when they have a hard time turning off their thoughts. Yet if you've been stoking that fire all day long, why is it a surprise that it burns brightly? Like any passable fire, it burns for a long time after you've stopped adding wood.

Ancestrally, because of the requirements of actual survival–no grocery store, no refrigerators, no bank accounts–much of our daily living time was focused on the necessity to obtain enough calories to survive, which is not an abstract undertaking. If we are going to obtain food we need to find it or grow it, and the long arc of our hunter gatherer history was concerned with this active finding. What is common to both hunting and gathering is that we are using our hands all the time. Our ancestors engaged in an incredible array of daily living tasks with their hands. Let's list some of them here to help us remember.

On the hunting side, we would have been tracking, which involves using the hands as radar. We might have used spears, or bows and arrows, which were propelled using the hands. These we could say, are a form of hand tool. If we were successful in the hunt, we would have used our hands for skinning, then butchering, hopefully with a knife: another hand tool. Let me pause to note that we would have had to learn to wield these tools skillfully. This wouldn't be automatic. We also, through most of our lineage history, would have had to make these tools.

Knives probably got made first, because to create a spear we need to sharpen a stick, and to do this is easier with a knife. To make a bow we needed to string it, and this was probably done with gut or sinew, again hard to separate from other tissue without a knife. The first knives, and those for many millenia thereafter were knapped from flint or obsidian or some other form of conchoidal fracturing stone. There are obsidian fields in Canada that archeologists have recently determined were the site of pilgrimages for tens of thousands of years, as people came from near and far to gather obsidian, which is a glasslike volcanic rock formed from the rapid solidification of lava without crystallization. Glossy and jetblack, obsidian fractures like glass and would have been used to make razor-sharp points of all kinds.

Here is a point fashioned from obsidian and attached to a bone shaft. Note the slight translucence near the edges where the stone is thinner. Each facet, each angle, was fractured by a distinct precise strike of the stone.

Traditionally, flint or obsidian or other similar stones would have been knapped into points using a series of progressively finer tools. Again, if we think about this activity through the lens of neuroplasticity, and the refining of the brain areas associated with the movement and sensory processes of the hands, learning how to do this requires finer and finer motor control. First we would have used another hard stone or possibly an antler to flake off a suitable section of flint to make a knife or a spearpoint or an arrowhead. This is difficult and bloody work, because you are holding the flint in one hand, and bashing it with the other. The angle at which you strike the flint determines the way it breaks, and what we are seeking is a controlled shatter that flakes off a piece of the approximate size and shape we want, such that it can be further refined. These pieces that flake off are often razor-sharp. (This is more or less like shattering pieces of glass. You can imagine that both the piece you are holding, and the pieces that flake off could cut you pretty seriously.) This danger is a good incentive for us to become precise so that we are not seriously wounding ourselves on the regular.

Once we have a primary piece, there is further refinement required for the cutting edge, which generally requires a fine point. This work could have been done with the tip of an antler, or a sharpened piece of hardwood, which would have required a previous knife to sharpen. Once we have a passable point, it has to be affixed to a handle, or we have to make sure that the tail end of the tool doesn't flay our hands open when we pick it up. To affix it, we'll need some kind of cord or rope or other substance to attach it to a shaft. The shaft in turn, likely of wood or bone, needs to be shaped to receive the point, likely grooved to receive the cordage. This kind of technical knowledge in small hunter gatherer bands, about how to make tools, is prized knowledge. Millenia of trial and error, of learnings and refinements, are carried by the band. I encourage all of us to attempt to replicate this kind of early tool-making, because it gives us an appreciation for just how hard it is to do so, and it helps us take our profoundly built environment less for granted.

As modern people, most of us are one-click away from nearly any tool we can imagine, which was a situation inconceivable for 99.9% of human history. During most of our lineage history, if we wanted a tool: be it a knife or a paintbrush or a pen, we had to make it ourselves, or find an artisan to do so.

These are the ornate spiralling tips of Venetian Glass pens, made on the island of Murano, one of the places that developed the most artistic blown glass in the world. You dip the pen in a bottle of ink, and the ink remains in the spiraling lines. The tip has a star-shaped end: you use one vein of ink then turn the pen to the next star point. These have been made by glassblowers in Murano since the early 1600s. No Bic pens here.

CHAPTER 29

CRAFT IS THE LANGAUGE OF THE HANDS

Craft is the language of the intelligence of the hands. The intelligence of *making*. From learning how to make cord, rope, to thread, to carding wool, to knitting, to weaving, to dyeing fabric, to cobbling shoes, tailoring clothes, gardening, to pruning, to milling, to brewing, to fermenting, to working with clay, to painting, to writing–what all have in common is the intelligence of the hands transforming our environment in ways that serve. If you begin to think about natural materials– wood, stone, and clay for starters– we can notice that there are entire disciplines of craft that have arisen around them, to transform the raw or elemental materials into forms we can use. The crafts of carpentry, masonry, and ceramics are all responses, by way of the hands, to these respective materials. And the way that a culture views itself, and views the world, ends up implicitly expressed through the craft.

One of my favorite venues for exploring the way culture expresses through craft is the distinction between traditional Japanese carpentry and European carpentry. Both are clearly disciplines of craft that emerged to translate wood, through the use of the hands, into dwellings, and the implements with which to furnish them. In order to do so, both traditions developed a common vocabulary of tools. Tools to cut wood: various saws. Tools to shape wood: various chisels. Tools to measure wood, tools to mark straight lines on curved surfaces, etc.

Yet because of what wood represented in these two different cultures: because of their different cosmologies and epistemologies, the tools are *profoundly* different. Without placing value on this, I point this out because tools implicitly express the culture from which they come.

My favorite example of this is the difference between Japanese and Western saws. If we distill the difference between the two types down to its essence, it would be the following: Japanese saws cut on the pull stroke, European saws cut on the push stroke. Yet in this seemingly minor difference, reflected in the angle of the biting teeth on the tool, there is in fact an entire worldview.

In the Japanese Shinto religion, which is animist, wood is viewed as being alive. Trees are clearly alive. I don't think Europeans would argue with that either, but the wood itself is understood, in Japanese culture, to possess a spirit. Because of this spirit, it is something that we want to have a relationship with, and carpentry therefore, as it evolved in Japan, yearns for intimacy with wood. I have known many woodworkers in my life and its not the case that woodworkers trained in the European tradition are not intimate with wood. It is not the case they they do not experience love, and even at times reverence, for the material.

It is rather that, a thousand years ago, the *felt vocabulary of this reverence* for the spirit of the material in Japanese culture was a worldview that led to the creation of a saw that drew the wood toward you. If wood is a being, and we want to have a relationship with that being, we increase our intimacy with it. This worldview led to the development of tools that drew the wood toward you. In Japanese carpentry, the carpenter who is cutting a piece of wood often uses his or her own body to brace the wood, and then uses the saw to draw the wood toward them. If the cutting teeth on the saw cut on the pullstroke, this works. Japanese carpentry has clamps as well, but they are used differently than European clamps. Japanese carpenters wear soft shoes (something I was taught as a child was dangerous, in fact a bias), shoes with soft flexible soles that can conform to the wood, and they embrace the wood and draw it closer. This cosmology that comes from and through the heart into the hands then informs the tools that are built. These days they are built from more-or-less the same kinds of steel as European saws. But because of this foundational intimacy, Japanese carpentry focused on joining wood with wood in ways that drew it closer together with itself. This view gives birth to the extraordinary joinery that characterizes the Japanese carpentry tradition.

It is significantly harder to create joinery like this with European tools. European saws cut on the push stroke, or bidirectionally, and the European technique is to use clamps to hold the wood and cut by pushing away from you, or fixing the wood and then cutting both directions. European carpenters viewed wood as a beautiful *yet inert* commodity. They did not believe the wood they were cutting had a spirit, an interior life of its own, or its own inclinations. They did not believe it was imbued with *yearning*. They viewed it rather as a blank slate, an empty canvas, a material characterized by its type and hardness, but void of interiority. A thing, not a being. This view extended to Nature herself, and to Others who were different. Things we do not bother drawing toward us. Things we push away. Things we dominate.

Beings we draw toward us, things we push away.

European carpenters have made, and continue to build extraordinary structures and furnishings. Their techniques are extremely refined. Yet they do not share the intimacy, the relatedness, the neurobiology of connection, that characterizes Japanese technique. In Japanese carpentry, emerging from an animist worldview, wood is a being, a verb, and we draw it closer into relationship. In European carpentry wood is a thing, a noun, and we push it into the shape we want.

What I'm hoping to show you here is how cosmology and worldview, which are expressions of the heart (*kokoro*), make their way into tools, which are expressions of the hands. From the heart, to the hands.

And what I'm hoping that you feel, as you read this chapter, is a yearning to use your hands, and to MAKE. When I put on my clinical hat, and I look at the broad field of who I know, and how livelihood relates to temperament, something that I notice is that people who work with their hands are often more even-tempered while they are working. Have you ever noticed this? Why is this? Most craftspeople I know, when they are

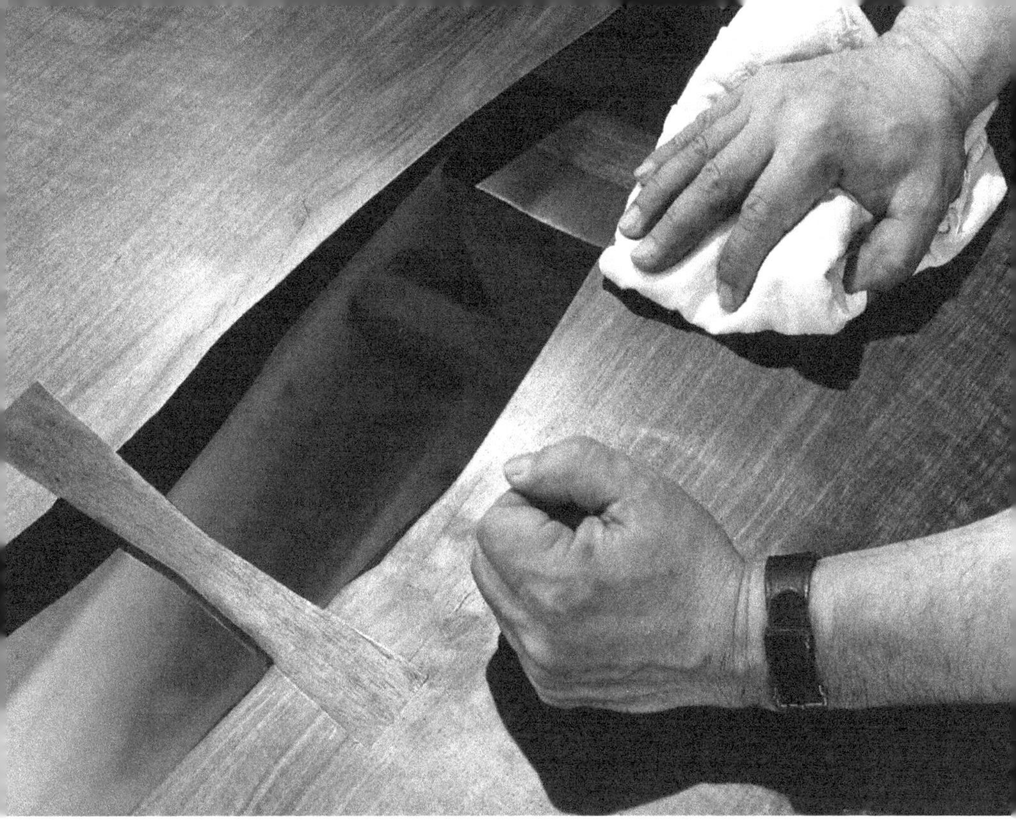

Japanese-American furniture designer, George Nakashima, wiping down a table he built from live-edged slabs, butterflied together.

crafting, enter a very balanced state.

Over the past fifteen years, I architected a neuro-developmental model of wellbeing based on interviewing more than 5,000 wellness professionals about how they take care of themselves. These interviews led to the creation of a database of more than 350 restorative practices, clustered into three broad areas: things people did by themselves (self-care), things people did with others (relationships), and things people did in nature (nature). Within each of these broad categories, there were sub-categories. So within the realm of self-care, there were movement practices, embodiment practices, meditation practices, and so on. One of the largest categories in self-care? *Using Your Hands.*

There are an extraordinarily large number of people who use practices that involve working with the hands to restore their wellbeing. Carpentry, wood-working, furniture-making, whittling, sculpting, metalwork, masonry, ceramics, painting, drawing, calligraphy, glassblowing, jewelry-making, beading, carding, knitting, weaving, quilting, sewing, gardening, pruning, cooking, baking, fishing, hunting, butchering, fermenting, canning, soap-making, candle-making, fixing things, tinkering, mechanical work– all of these are handwork– all of these are things people told us they did to restore themselves, and all of them, I should point out to you, are the kinds of ancestral work that people have been doing with their hands since we became human.

So what are the common elements of activities that involve the hands and are regulating to your nervous system? Well, clearly they involve using your hands. They also involve using your attention. If you look at the list in the previous paragraph, these are all activities that require you to run attention through your hands. This means that, as opposed to thinking or ruminating about other things while you do something with your hands, which is something I do a lot when I'm folding the laundry, the activities listed above invite and even require your attention to be in your hands.

I will even go so far as to say, after having done this work for thirty years, and having worked with tens of thousands of clients, that simply putting your attention in your hands is regulating. Even if you are doing something that is not an ancestral activity, even if you are simply driving or typing, or cradling your face in them as you read this, intentionally putting your attention in the hands will generally increase your experience of bodily regulation.

When I say - *put your attention in your hands*, what do I mean by this? I mean *interocept*. Feel your hands from the inside. If you are typing, bring your attention to the sensations of contact between your fingertips and the keys. Become aware of the temperature of the air around your hands. Notice the muscular exertion in your fingers. Notice how they spring back into place.

Notice what if feels like to just rest your fingertips lightly on the keys between strokes. And notice that you can do all of this while you are typing, without losing focus. Notice that you can balance your attention in such a way that it includes more of this information coming in from the hands. Same if you are driving. If you are driving, notice the sensations of your hands on the wheel. Notice the way it feels to squeeze the steering wheel. The tactile quality of the wheel itself. Notice the pressure when you turn. Notice the sensations that make their way up the arms.

I'm talking about sensing the body from inside. This is *the felt sense*. Most people, ancestrally, through the bulk of our lineage history were resting in this felt sense in every moment. There were no moments when they were not in contact with it. This interoceptive field of the felt sense of the body becomes the ground of intuition. This field of sensing, emotion, and memory is a way of knowing: the fountain of myriad ways of knowing. It's part of what I'm trying to hook you back into through this book. It is a universal ancestral inheritance that has been de-natured by trauma and domination. It is yours to reclaim. To do so, we have to re-inhabit the body, and change the way that we are flowing energy through it. We are not designed to flow energy primarily through thinking. This happens as a byproduct of trauma. We are designed to flow energy through the body entire, up from the earth, out through the hands. The ventral surface of the hands. The ventral surfaces of the hands are part of your Connection System.

They form a triad with your face and your heart.

RECOMMENDATIONS FOR USING THE HANDS TO SUPPORT WELLBEING

Like all connection practices, we are working first with attention. Attention is a derivative of autonomic state. To use the hands to support wellbeing, we want to flow attention through the hands and we have to have enough safety online to generate enough free attention to do this.

1. Consider what you enjoy doing. Consider for a moment whether you generally relax alone or with people. Consider whether you find it more relaxing generally to be inside or outside. Do you feel most at ease alone and inside or with people and outside or some other combination thereof? Think of this as your preferred *mode*.
2. Once you've identified the mode: e.g.., alone or with others, inside or outside, think about *categories* of tasks. We might group these in terms of craft, nature awareness, sociability, daily living, etc. Think of this as a *domain*.
3. Now that you have a sense of your preferred mode (alone/together), and domain, what are you drawn to? In the domain of craft, what kind of materials speak to you? Stone, wood, wool, paint, ink, canvas, paper, thread? In the domain of nature awareness, what environments call to you? Forest, mountain, beach? What elements call you? Water, fire, wind, earth? In the domain of daily living, what tasks do you enjoy? Cooking? Mending? Cleaning? Sweeping?
4. There are thousands and thousands of practices that involve your hands. The wonderful opportunity is to find practices that flow attention through your hands, and are repetitive yet varied enough that you can find a rhythm with them without being consumed and without getting bored
5. they have enough complexity that you enjoy them and enough ease you don't find them frustrating
6. they are meaningful to you
7. If, for example, you like being outside, with people, and are drawn to water, join a sailing club. Lots of handwork

on a sailboat. Or if you like being inside by yourself and are interested in tactile craft, learn to knit or crochet or tie flyfishing ties. Make sense?

Just for fun, I'm going to list the things I like doing with my hands these days, acknowledging this is only my personal list.

1. Typing very early in the morning, before it gets light, at a wooden table in my kitchen while I drink Yerba Maté from a clay mug whose surface that is rough enough I can draw my fingertips across it with satisfaction.
2. Tending a forest, right now with a chainsaw, because it is pruning season. I try to merge with the saw, to feel through it.
3. Making wooden signs with handtools including chisels, a grinder, and various saws and sanders, then painting them.
4. Sanding wood (furniture, sculpture, etc.) with the kind of sandpaper that is used to polish glass (up to 10,000 grit) until when I run my hands along them with my eyes closed they are slippery. Since we can feel something the thickness of a large molecule with our fingers, closing the eyes and really noticing if the surface is uniformly smooth is a major source of neural inputs. Very satisfying.
5. Preparing simple nourishing food for my family from available ingredients, especially if I have a connection to at least one ingredient or sourced it myself, e.g., lemons from our backyard, herbs from the garden, mushrooms from the forest
6. Foraging in general
7. Flipping through large format books on design, particularly when they have beautiful pictures, layout, and typography. The way a page flexes under the pressure of your fingers as you turn it.
8. Hard ground tracking with other people in Abbott's lagoon and elsewhere
9. Stopping by the side of the road to harvest a section of broken-off cactus that I can replant (I generally keep a kit in my car with a knife and pruning tools for such eventu-

alities)

10. This winter I planted about 80 Buckeye trees in tennis ball cans and more generally I am pleased by thrusting my hands into soil, planting things, gardening and watching them grow
11. the use of most well-made tools, which includes sharpening my knives by hand with whetstones of progressively finer grit. There is a profound tactile pleasure in sharpening knives, and in sanding surfaces, I will note here. When you sharpen a knife with a whetstone, you move from progressively rougher stones to smoother. The smoothest stones you dunk in water to make them slippery. When you are sharpening a blade you have to keep it at a specific angle relative to the stone. So there is concentration involved. It is also very satisfying, at the end, to have an extremely sharp knife. I find something generally to be very meditative about taking care of hand tools in this fashion.
12. Anachronistic handtools generally, like a pair of very old hand planes I bought in an antique store for twelve bucks apiece, and then using them. I am fascinated by old hand tools generally, again because of the degree to which they encode culture and worldview. They are intricate and beautiful.
13. Playing tennis
14. When I drink coffee grinding it by hand if possible. This adds an element of tactile satisfaction to the process, as does pour-over coffee and all other made-by-hand elements of crafting a cup of coffee
15. making fires in a woodburning stove or hearth or campfires generally, and all the hatchet work, gathering of kindling, splitting of wood, making of the fire itself, and blowing on it associated with that

It seems to me that we are happier with dirt under our fingernails, the occasional splinter, hands that are healing from a thornstick. I like to try to not be too precious with mine. I like to pick up things that are sticky or gooey or slippery or oozing or slimey, things my wife thinks are disgusting, and to really feel the sensations associated with these things. Not to be squeamish about it, but to revel in the fact that I can feel the slipperiness, the oozing, that it has a particular tactile signature. Things with weird textures, a banana slug here and there, the occasional dog turd when necessary in order to chuck it, because I don't like plastic bags. As an aside, part of the reason I like to wear jeans is 'cuz you can wipe stuff on them and they still look ok. This is important if you pick up something kinda foul. When I look at my hands these days I'm satisfied to see I can't get all the dirt off, even after a good washing. I'm satisfied to see they look a bit feral. We are animals, people, animals.

Mammals with intricate housing and many complex potions to obscure our body odor. Our hands are fucking glorious. Let's use 'em like it. You won't regret it.

CHAPTER 30

YOUR HANDS ARE RADAR

At a certain point many years ago, as we turned our attention to the intersection between neurophysiology and ecology, we began seeking out mentors who were really deeply nature connected. We did this in several ways, scouring our networks for people that we knew were deeply in love with Nature, deeply comfortable and at home in the Living World, deeply attuned to the rhythms and cycles of life. And what happened, strangely, as we met more and more of these people, was that at a certain point it would turn out that they were trackers. We did not plan for this to happen. It surprised me a great deal, in actual fact.

At the time I knew almost nothing about tracking, except that learning how to follow animals through the forest by finding and following their pawprints seemed about as useful to me as learning typewriter repair. *We have Whole Foods, you know*, I would say, if pressed on this point. The only tracking skills I need are to be able to make my way to the meat counter, where they keep slices of the animals you are out looking for. But I couldn't get around the fact that, again and again, the people I met who had the kind of relationship with the Living World that I wanted, that I yearned for myself, had these skills. A certain percentage of them were, in fact, somewhat obsessed with tracking.

The first time that I saw a San tracker was on film. The San people of the Kalahari are one of the earth's source cultures with a continuous oral history that goes back several hundred thousand years. Yes, you have read that correctly. And the San are acknowledged, by the people who know these kinds of things, to be among the finest trackers in the world. They can do things, consistently, that other humans cannot do. As an aside, all of the source cultures of the earth I know about, which include also the Australian aboriginal peoples, and the Lakota peoples, and possibly many others, are extraordinary trackers. Simply extraordinary.

So I found myself watching a film called *The Great Dance*, which was made by Craig and Damon Foster, South African filmmakers who were responsible, more recently, for the stunning film *My Octopus Teacher*. *The Great Dance*, which you can find online for free, follows a group of San trackers in the Kalahari, with much of the film watching them as they track down and pursue, running hours upon hours zigzagging across the desert, a large kudu: a giant gazelle. What stunned me was the way they use their hands. They listen through their hands and they hear through their hands. When the San track, one of their hands is held out in front of them like a dowsing wand, and they are kind of jiggling it, I would say, kind of shivering it in a trance-like fashion.

Those hands, I thought. *What is the deal with those hands?* What *IS* the deal with those hands? This was before I understood, thoroughly, that the palms of the hands are one of the body's ventral surfaces. If you want to do a fun experiment to understand the density of sensory neurons in the body, and this is something that you can do at home, in a DIY fashion, take two pushpins, of the kind you use for sewing, wash them off, and cut a strip of paper into a rectangle about two inches long and half an inch wide. The density of sensory neurons in the body will correlate to the fineness of the spatial resolution with which you can distinguish unique sensory inputs. So, push the pins through the paper around half an inch apart, so that the points are sticking through the same distance, and then take this contraption and set it lightly against your palm. Can you feel the two distinct points? Yes? Good. Now set it on the back of your arm. Can you still feel them? Place them against your back. How about now?

The topography of your skin is woven with sensory and pressure receptors, as well as a kind of autonomic fiber called a c-tactile fiber, which contains nociceptors (generally understood to be pain receptors). Our biology is designed in such a manner that certain parts of our bodies– our faces, our tongues, our genitals, for example– are disproportionately dense with sensory neurons. What this means is that the sensory inputs from these parts of the body are more robust than from other parts. This mapping is consistently so skewed, favoring particular body parts, that the brain areas that map sensations in the body hold a body representation that is quite distorted. This representation is called a somato-sensory homunculus, and it is a version of you with outsize lips, tongue, genitals, and, you guessed it, hands and feet.

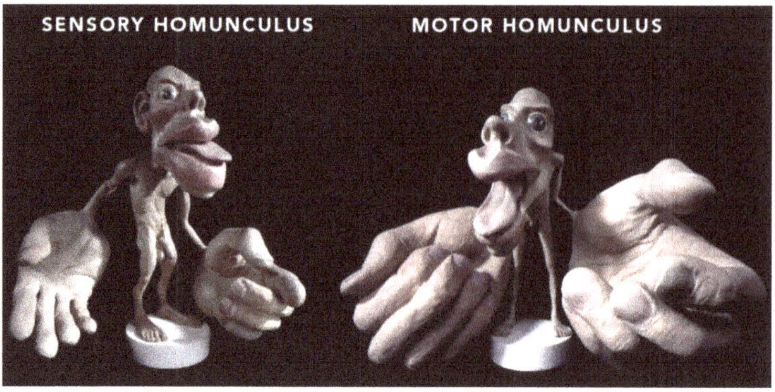

Move the pins closer, to a quarter of an inch. Can you still differentiate the sensations on your palms? What about the back of your hands? How about your arms? My guess is that you can no longer differentiate the points on your back. Move them to an eighth of an inch. Where can you differentiate them now? A sixteenth?

What I'm hoping you are discovering, what you are witnessing through this exercise is *intimacy with the neural fabric enveloping the body that you are wearing*. And the point I'm making is that your hands are very high resolution, sensorily speaking. Do you understand that there is a way to use your hands as radar? That there is a way to listen with your hands: to give yourself over to their sovereign listening intelligence? To let them begin to feel for things that your mind is not aware of? This is what the San are doing. Their hands have become wired into their tracking.

LEARNING TO USE YOUR HANDS AS RADAR

When people are training this embodied faculty for the first time, we'll start just by moving our hands around in front of us, over objects. You can try it now. If you are sitting on the couch reading, hold your hand out in front of you over the floor and close your eyes and just sense through your hands. Imagine your palm as radar. Then move your hand over the coffee table. See if you can sense, through the hand, a difference. It might be very hard to verbalize how you know this. But can you feel it? The feeling may be subtle, but can you feel that there is something closer to your hand than the floor? Open and close your eyes. Put your hands above your glasses, or a book, or a plant, and see if you can feel the object through your hands. See if you can *sense* it.

This process is a form of *sensory calibration*. It is the fine-tuning of the senses, in this case a sense you didn't know you had, an ancestral sense that we moderns haven't been taught about. You are using another sense, vision, to help calibrate the radar of the hands.

The next step? Walk through your house with your hand held out in front of you, eyes open. Let your hand sense. Walk down the hall, letting your hand sense both the openness of the hall, and the proximity of the walls on either side. Good, now do the same thing with your eyes closed, letting your hand guide you. See if you can feel the walls as you breathe.

The next time you are walking about in the house at night with the lights out, next time you wake up to pee, just close your eyes. Slow down. Find yourself in space through feeling. Stop straining your vision, which doesn't work in the dark, and open up the sensing of your hands, and your body, which can activate irrespective of the lighting conditions. Awaken your body in space, awaken your hands. We moderns are completely over-reliant on vision. Let vision fall away. Find you way through the house with your hands.

Next step? Now point your palm toward something alive. A creature. Can you begin to notice it with your hand? What does it feel like to begin to point your hand at something alive and open to it? To know that creature through your hand?

Go outside and ask a tree for permission to touch it. I know that sounds strange, but asking first is better manners. And it changes something in the communication. You probably wouldn't just walk up to a stranger and touch them without asking. That would be rude. Asking permission is a way of acknowledging sovereignty. If the tree says *yes*, then touch her with this quality of awareness. Let the mind fall quiet. See what the tree communicates to you through your hand. Now move to a different tree, ask permission, and touch this. Notice if the communication you receive from the tree is the same.

When I say communication, I'm not suggesting that the tree

is going to talk to you, but don't rule this out. I'm suggesting that the tree is going to *communicate* with you. This whole book you've been learning a language that is outside of speaking. Listen with your hand to what the tree is communicating. Listen in your whole body.

Now that you know these two trees, see if you can point your hand at the tree and receive this communication. Point your hand at the tree as a listening device, and see if you can pick up on the communication of the tree remotely. Do you have to touch it to find it? Or can you touch it remotely? Can you, by listening with enough awareness for the felt signal of the tree, find it without your hand resting on it?

Good, next step. Can you distinguish with eyes closed, between a pet and a person? Between your cat and your partner? Ask permission to touch your cat– again, this sounds strange, and you don't have to ask permission in words. Communicate your request in a language the cat can understand. Then sense the way it feels, eyes closed, to be in communication with the animal. Feel its heartbeat, its breathing, the aliveness of it. The *particularness* of this animal. Then back off. Now without touching the cat, see if you can point your hand in his direction and pick up communication from him. Try this with your dog. Now try the same thing with your partner, including asking permission to touch.

At this point, I believe you have some sense of what the San are doing out there, in the bush, with those hands. If you have been refining this awareness your entire life, how subtle a signal might you be able to follow, as a beacon, to establish communication with the animal? In a way, the deepest art of tracking is not finding animals. The deepest art of tracking is connection phenomenology. It is finding *relatedness*. It is building the ropes of connection with All That Is. Once you have built those ropes of relatedness, tracking is simply the refinement of picking up the signal that leads you back to where you and the animal are in the same geographic place. Relatedness means that from the

CHAPTER 31

TOUCH THE GROUND

Did you know that the proper way to greet an infant is to grasp their feet as though you were grasping their hands in a warm embrace? The feet are the first hands, the back hands, and in a baby they are the doorway to enter if you want to be received. They are the portal that is our access to the ground: the gate through which our terrestrial mother speaks to us.

If we are going to re-nature the human body, and we must if we are to thrive, we must address the severing of the root of connection with the Living Earth.

This hacking away at the root of what it means to be human is one of the principal undertakings of Empire. This severing has been approached with methodical violence by empires since before Rome, although Rome was perhaps unsurpassed in its annihilatory glee.

A central part of the Roman undertaking was the absolute destruction of earth-based peoples. Rome sought not only to kill the people, but to destroy their very gods. It cut the heads off their statues to deprive them of breath. Aware that they were dealing with animists, and that spirit could come even into the stone, they sought to destroy everything that was connected to the good Mother Earth. A signature Roman maneuver was to destroy the temples to the original gods of a place, which were often constructed at features of geomantic significance, places where the beneficent energies of the earth are most palpable, and to erect Roman temples on top of them. Christianity later adopted this specific tactic. It is quite common to find Christian churches (consecrated to analogues in the Christian pantheon of saints rather than their Indigenous counterparts) sitting atop the decimated ruins of Indigenous sacred sites and temples. A shrine to the Virgin thus sits atop a temple originally consecrated to Athena, for example. People are then, in this age of forgetfulness, astounded by the curative powers of the church, while in actuality they are experiencing the curative powers of the Our mother, The Earth.

This severing of our root of relatedness to the earth is apparent in the origin stories of western civilization, as I have noted. It is also experienced, viscerally, in the fact that most modern people's feet are not touching the ground. I grew up in a family where I'm not sure anyone's feet had actually touched the ground for several generations.

When I say no one's feet in my family touched the ground for generations, I'm being slightly provocative. On the literal level, I mean the statement to be true in the sense that my family, like every other modern family I know, grew up wearing shoes. In addition to protecting your feet, which shoes undoubtedly do, they prevent your feet from ever actually touching the good earth.

Rubber is a very good insulator. It was the first insulating medium used when electricity was discovered. Your nervous system is an electro-magnetic system. The earth, which is the source context, has a core made of molten metal, and expresses itself electromagnetically through geological formations and geomantic lines of force. The ancient Chinese discipline of *feng shui*, (wind water) is a refined codification of the art and science of working with these energies in the Living World. In English, this is known as *geomancy*, and in Arabic *Ilm al-raml*. While most of our technology is too crude to measure these, a recent study with advanced satellite optics confirmed that the songlines of the aboriginal Australians actually map to lines of geomagnetic force. Since the Aboriginal Australians possessed no 'advanced' technologies, but the ancestral neurotechnologies of non-cognitive ways of knowing, it seems safe to observe that they could FEEL these lines of force. So can you if you get connected, again, to the Source.

There are some places on earth that carry the particular signature of well-being of a place. There are also places on earth that hold the memory of tremendous harms. Both of these energies can be transmitted from the earth up into human bodies.

While we don't have a word for this in English, the Lakota peoples, who have a 200,000-year-old culture, one of the original human lineages, call this *anamagoptanye*. This word has been translated to me as "the earth is constantly uploading information to you." It is one of numerous Lakota words that convey information that is passed upward, either from the earth or from ancestors.

Because the earth is doing this, all the time, it is useful for us to be able to receive the information she is sending. The earth is, {newsflash}, *alive*. Forgive me if I am telling you something you know already. She is sentient. She is our Mother, capital M. She knows you are here. My indigenous mentors tell me she misses us.

Your birthday suit, *aka the body you are wearing*, has been uniquely and sacredly engineered to receive the information she is sending you. Unfortunately, because she communicates partly in electromagnetic signals, having rubber between you and her – I'm going to resist the temptation to follow this analogy further – may interfere with the transmission.

If you want to be able to hear what Mother Earth is saying, it's a good idea to be able to feel her with your feet. One of the specific ways that our research extends autonomic physiology is the awareness that the vagal system is architected through the ventral surface (soul-side) of your feet. Your feet are part of your connection architecture, and your grounding architecture. They wire into it as you begin to walk, perhaps when you begin to crawl, perhaps before.

And so the first thing I'm saying about my family, is that it had been many generations since people had direct contact with the earth in this way, through the Connection System of the feet.

Non-cognitive ways of knowing speak the language of energy. Your body cannot help but being an electromagnetic transmitter and receiver. The electromagnetic oscillations produced by your heart are detectable by a magnetometer from eight feet away. What that means in practical fact is that the radius of space that you take up vibrationally extends from your heart eight feet in all directions.

One of the interesting features of COVID-19, which globally mandated six feet of distance between humans for two-and-a-half years, is for that period of time many people, without realizing it, experienced the actual amount of energetic space they take up for the first time. I don't know if you had this experience or not, but after being sensitized to having six feet of space around me for two years, I could feel it if someone was closer to me than that. I could feel it whether or not I could see the person. If your vibrational field extends eight feet in all directions, you should be able to feel anything that happens within that space. Most of us hadn't been afforded that much space before the COVID pandemic, and when we are accustomed to having our boundaries transgressed, it is hard to know where we end and other people begin. Although interoception is generally co-terminous with it, you don't end with your skin, or the edge of your body. You end at the edge of your energy body, and even then it isn't quite so clear.

I have worked with clients to find the edges of their energy perimeters, which is fairly easy to do. You need a safe open space–it could be park, or a field, or a gymnasium without anybody besides you and the person helping you. Safe enough that you can close your eyes without feeling a sense of alert or danger. If you are using a public space, you might do this very early in the morning. The basic method is to stand in the center of the field, slow down, get grounded, and then close your eyes. You will have your assistant approach you from the front, back, and both sides, starting somewhere between thirty and a hundred feet away. They will walk slowly toward you, and you will notice when you can feel them. As soon as you can feel them they will stop, and you will mark how far away from you they are in that direction.

This detection will be done with your eyes closed.

One of the things that astonished me when I started doing this with clients was to realize that some of them could detect people very far away. I worked with someone who could reliably feel a person about eighty feet away from her. Her personal energy perimeter was about a hundred and fifty feet across. We don't have a word for this in English that is part of mainstream language. There are some New Age words for this, but there should be a neurological word, because it is a neurological, not a mystical phenomenon. (It is mysterious, but *neurologically* mysterious.)

We can't properly use the word interoception, because interoception refers to feeling the interior of the body. But if you can feel something with your eyes closed that is five or ten or eighty feet away from you, you are feeling something because it has crossed the perimeter of some form of boundary. One of the other interesting things about this I've noticed is that often this energetic perimeter is not circular. It may be asymmetrical. Larger on one side or the other. It may extend farther forward than backward, or backward than forward. Having a sense of your energy perimeter is a good idea. Because if you can feel things that are thirty feet away from you, your neuroception (moment-to-moment detection of safety or danger) may include things happening in rooms you are not standing in, places you cannot see, etc. Some aspect of your awareness is monitoring a space much larger than your physical body. This changes how you conceive of yourself moving through the world.

A number of the clients who I've worked with who have energy fields that are this large do not know how to make them smaller. I am out on a limb here, but I believe that this phenomenon is in part a neurobiological sequelae of unresolved shutdown.

Shutdown, as a neurological phenomenon, disconnects the deep belly from the ground, and it dissolves our sense of clear boundaries. The deep belly, when we feel safe, connects us to the earth. This happens through energy flow up from and through the heels, and into the hip sockets, then up into the lower dantien. This grounded energy circuitry helps us contain our energy, and have a clear energetic edge.

Part of what happens when a person has a lot of unmetabolized shutdown is that this grounding connection between the deep belly and the earth is lost. When we cannot ground our energy fields, when we cannot drop our roots into the earth, our energy perimeter becomes diffuse and permeable: cloud-like around us, not anchored down. Awareness floats, diffuse.

Since we always create an electro-magnetic field, it is of crucial importance that that field be grounded, so that you don't float off into space. So the other, and slightly more abstract way in which I mean that my family's feet hadn't touched the ground for generations, is that like many modern people, their energy fields were not grounded, and so they were permeable to ambient energies that crossed their boundaries without their awareness regularly. They also didn't inhabit the lower parts of their bodies somatically.

If you had to draw a map of the bodies of my family, they would look kind of like the avatars in Meta. Head, torso. No legs. No feet. I may catch some shit for saying this, but it is possible that the avatars in Meta are how Mark Zuckerberg experiences his own body, but which I mean I'm not sure he noticed they didn't have legs. I don't think that guy's feet are touching the ground either.

I am pointing at something rather pedestrian here, rather obvious, but also profound. When your root of relatedness to the earth, your earth-based nature, is annihilated, it is impossible for you to feel your feet on the ground. To be deprived of place, within the communal body, is to be deprived of rootedness. The Romans made damn sure that no one felt their feet. The Romans cut everyone off at the roots.

When the Romans adopted Nicene Christianity as the empire's official faith, with surgical precision they excised all of the earth-based elements of the religion. An empire doesn't want you to feel your feet on the ground. Remember - if you can feel your feet on the ground, you might become sovereign again. You might remember that you belong to the Land, and that the Romans are colonizers. You will certainly become much less compliant.

People uprooted from place have had their roots cut. Our roots pass through the soles of our feet.

The Empire needs you to believe that there is a spiritual hierarchy, and that you are not worthy of feeling sovereign. Sovereignty is reserved for the elite. There are spiritual intermediaries you have to pay or placate to gain access to the Source. They are the gatekeepers to the providence of Heaven. This is called organized religion, and it is *always* political. The price you pay for admission is the code of ethics promulgated by empire that keeps you in line. Religious cults that don't get crushed figure out how to align themselves with power.

Feel their feet on the ground, and their bodies from inside, and people run the risk of becoming sovereign. Sovereign people don't enjoy very well tolerate being enslaved, and they don't tend to follow the rules of others that don't feel right to them. If you can disconnect people from the ground, and teach them that feeling the energy moving through the inside of their bodies is immoral, that's a pretty successful empire business. And make no mistake that empire is a business. An extraction business, to be precise. It extracts life.

In order to touch the ground, one of the things you can do is remove your shoes. You can't remove them everywhere, because many of us live in cities that have all type of dangerous garbage on the ground. In these cases, you can wear barefoot shoes. Barefoot shoes happen to be the most effective antidote to anxiety I've ever encountered. Essentially, anxiety is what happens when you have a flight response and lose contact with the ground. Barefoot shoes make it much harder to lose contact with the ground, since you're getting constant proprioceptive inputs from her. With barefoot shoes on, you can feel her through your soles.

CHAPTER 32

YOUR FEET BEGIN IN YOUR BACK

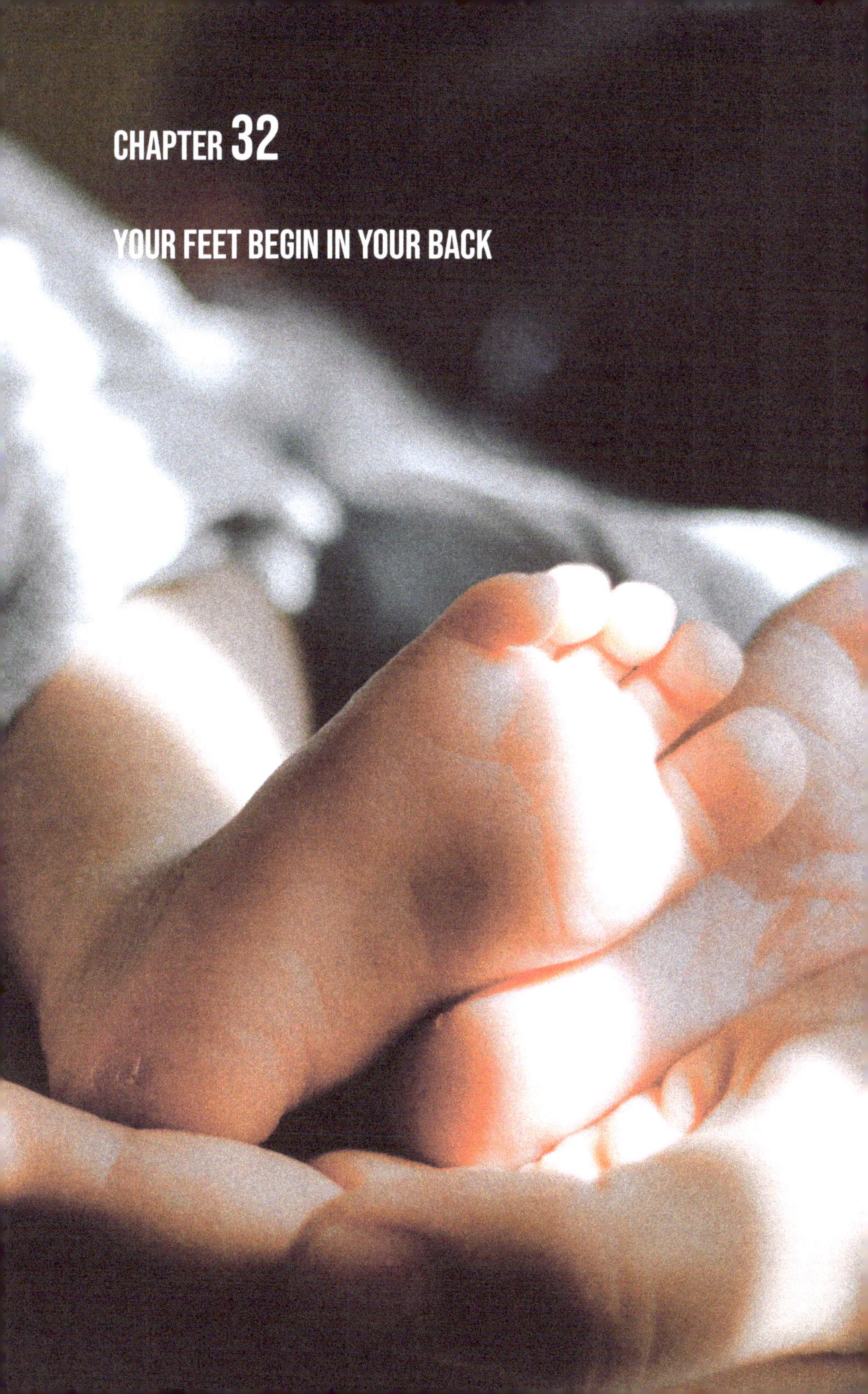

Something that feels important to remember here is that while the portal into the body is through the soles of the feet, the feet begin in the attachment of the psoas muscle to your spine (and the anterior branches of lumbar spinal nerves L1-L3). This is where they emerge from neurologically. So the energetic conduit that opens up the feet passes through the opening of the hips. This is of particular importance, because one of the very specific effects of traumatic shock– any neuroception of threat, in fact– is that it tightens the psoas muscles, which are a set of long spindle-shaped muscles running from the transverse processes (the bony flange that projects off the side of each vertebrae) and vertebral bodies of T12 - L5 (your lower spine) down to the lesser trochanter of the femur. Basically, the psoas connects the vertebral column with the inside of your hip joint. When you contract into defense, it tightens, locking in tension.

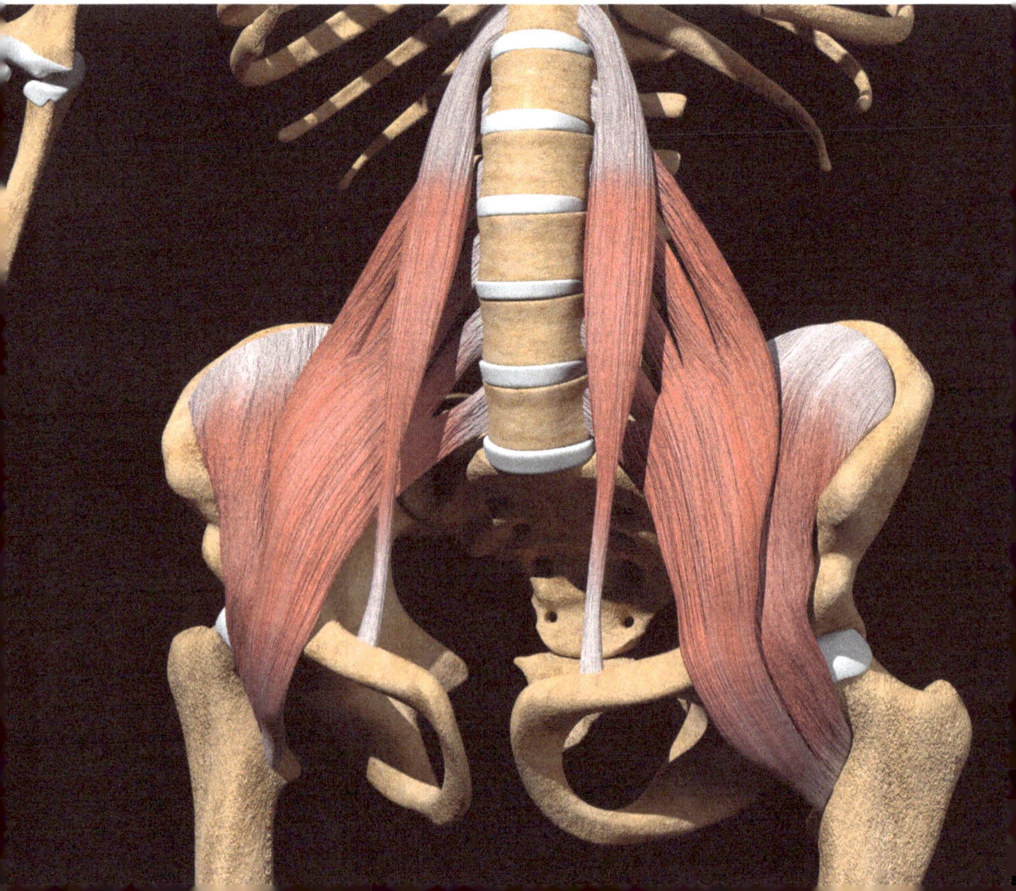

Take a very good look at the diagram, because you need to be able to feel the structure of your psoas clearly from inside. A hallmark of modern people, we who are the inheritors of multi-generational trauma, is that we are carrying unmetabolized allostatic load, undigested trauma, in the psoas. Do you see where the muscle passes outside of the pelvic cradle and then to the inside of your femur? If you set the palms of your hand adjacent to your genitals, in the inner seam of the leg, you are touching the inward surface of the psoas.

Anatomically, because shock gets archived in the psoas, this is part of why trauma can disconnect us from the ground. If the threat is great enough to activate our lifethreat response– which is to say that if we feel that we might die– this awareness will lodge in the psoas.

The psoas is the place where our torso roots into our legs. And our legs are our roots. If we are fearful enough that the living tree of us is going to get cut down, this will show up in the psoas. Terror and lifethreat get archived in the psoas.

Tightening of the psoas is where our embodied roots get cut off from the ground.

We feel our feet through the conduit running from the spine through the hips, down through the knees, to our feet. If the psoas tightens enough, we don't feel our feet in contact with the ground. So in a way, a preliminary to opening the contact with the ground is to invite opening of the hip joint: to restore mobility here, to help the muscles release.

People who are multi-generationally traumatized draw pictures of themselves, in virtual reality, with no legs and feet because they cannot feel their legs and feet on the ground because the psoas has locked in trauma and shut down the cord of energy running down to the ground.

Absent this groundedness, people get lost in their thoughts. People with this configuration who get 'spiritual', particularly Eastern spiritual (Buddhist, etc.) end up becoming expert in bypassing. If you cannot feel your feet on the ground, and your psoas has locked in trauma, you don't mind sitting on a meditation cushion for hours. You don't mind that you cannot feel your feet. It doesn't bother you because you cannot feel your feet anyway. And that is how we end up with a bunch of dissociated Jewish Buddhists. I'm gonna get myself in a lot of trouble for saying this, except that I know, from personal experience, what the fuck I am talking about, because I was one of them.

Most modern people should not be practicing seated meditation, because modern people are already dissociated, and seated meditation makes it worse. Modern people who want to become spiritual should be doing embodiment practices. We should be learning to move from our cores. We should be transforming and resolving the infant motor movement sequences that got inhibited.

There are a number of ways to stretch this part of the anatomy out, and a number of trauma modalities that have evolved to work specifically with releasing allostatic load (stress) accumulated in this muscle grouping, because it is so important to our ability to fully embody.[1]

If we are stressed out this area is going to be highly tensioned, and restoring mobility to this joint will feel really good. This restoration of mobility in the hips is one reason why some people (often men) habitually turn to sex for stress relief (in addition to the oxytocin release). Yet the restoration of this mobility doesn't have to be sexual. It can simply involve mov-

1 See David Bercelli, PhD: Trauma Releasing Exercises (TRE).

stretching out this muscle.

The hip opening exercises in most standard yoga classes will accomplish this. Bound angle poses, both seated and lying down will bring your awareness to the psoas. A cautionary note here is that if the psoas contracted initially due to traumas that were sexual in nature, the pose itself may carry bodily memories, and feel *extraordinarily* vulnerable. DO NOT force this muscle to open. Forcing any muscle does not work, but if your psoas is tight, it is not simply mechanical. Armoring here is emotional in nature: an attempt to get safe.

Experiences of violation are not abstract. They don't live in cognition, but are held in implicit memory localized in the body. Because some part of us knows this, we often make unconscious postural and behavior modifications to avoid provoking this awareness in the body. The earlier in our lives such adaptations occurred, the deeper they are in the neurological substrate of habit. So don't be cavalier about doing hip-opening exercises, and if you have experienced sexual violation approach them both with trusted support from someone who is a trauma-informed practitioner, and with very nuanced attention to the levels of activation that accompany the exercises. You will not be stretching out a muscle, you will be renegotiating trauma.

If you begin to do a hip-opening exercise, and you are not moving, and your heartrate shoots up, it is a physiological indicator that you have unprocessed information held in sensation. If this happens, slow way down, and instead of focusing on the muscular tension, attend to what is happening emotionally. Do this with supportive therapeutic accompaniment. Unwinding the psoas can take a very long time (like years or decades).

As the hips begin to unlock, as you begin to restore mobility in this joint, bring your attention as you stretch the hips, with your feet on the ground, to the flow of sensation moving from the soles of the feet all the way up the legs and into the body. Notice how opening the hip opens the conduit of sensation in the feet. See if you can feel the way that opening the hip translates into deeper and broader awareness of the sensations coming from the entire sole of the foot into the body, particularly the heel.

Although the psoas is the primary hip flexor that stores allostatic load, it bears mentioning that there are five hip flexors in total, including the psoas. The full range of motion in the hips depends on coordinated movements of all five muscles in each hip. It is worth spending time and attention opening up the hips and working to become interoceptively familiar with, and stretch out, all five of these muscle groups, because all of them relate to being able to feel your feet on the ground. After you have stretched out your hip flexors, notice the degree to which you are able to feel the ground through the soles of your feet with deepened awareness. Most of us moderns, particularly since we sit so much of the time, are well served by spending time opening the hips, and really practicing rooting into the ground.

TOUCH THE GROUND

Find a place where the earth feels benevolent to you. A place in nature. And then take your shoes off. And bring attention to the sensations of contact between you and the earth. Settle into the flow of sensation contact between your feet and Mother Earth. This can be a kind of homecoming. See if you can allow yourself, like a plant, to drop roots into the earth. Allow yourself to visualize this.

Do this each day.

Visualize in granular detail and feel the contact with the ground extending through the soles of your feet porously and into the earth.

In your visualization, explore how you see these roots. Are they like giant taproots that drop straight down? Are they diffuse? And can you visualize, in detail, the tiny hairs on these roots that are the actual mechanism of conduction of nutrients from the earth into these roots? Allow your visualization of roots to become precise and particular.

Now take a look down at your feet. Most modern people have been taught to walk with their feet in parallel. But I want you to imagine that you're going to have to squat down right now and take a big dump on the forest floor. I want you to imagine that you're gonna have to squat in an ancestral position with your bum as close to the ground as possible.

And I want you to allow your feet to adjust their position so that this is possible. Take a moment and actually do this. Set yourself up so that you can squat all the way down to the ground.

In fact, go ahead and squat all the way down to the ground if you are physically able to do this. Now take a look at your feet. Are they parallel anymore? No, they are not. The position your feet are in now is the correct orientation for allowing the energetic circuitry of the earth to flow through your body.

Your heels will be slightly closer together than your toes. Your heels are a primary energetic grounding into the earth, because this is the stalk of the body that bears the most weight.

In the front of your feet there are two other grounding poles. One of them is under the ball of your foot. One of them is under the outside of the front of your foot just at the base of the smallest toe.

The point of your heel and these two points in the front of the foot come to together to form a triangle.

When you set your feet on the earth, take a moment to feel the grounding of each of these points. Start with the heel of your foot. Press it in to the earth, standing. Then move your attention to the ball of the foot and lean your weight onto the ball of the foot. Move your attention to the outside of the foot and

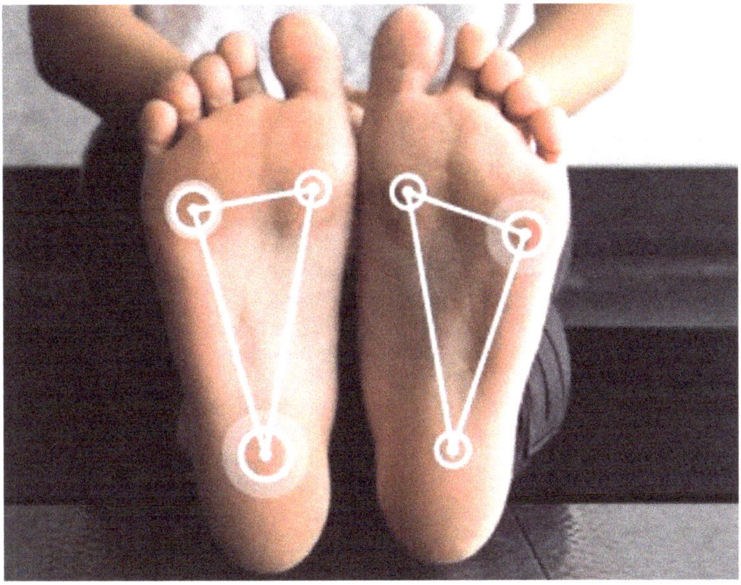

Each time you practice grounding into the earth, take a moment first to rotate your feet into a position so that you can squat. And then establish these three grounding points with the earth. Find the specific balance that your body wants in this moment.

Allow yourself to simply settle into the sensations of contact with the earth. As you do this, setting thought down, notice if the contact is pleasant. If it's not, there are a couple of possibilities. One of them is that the place that you've chosen to root into the earth has energies that are not that beneficial. Another is the possibility that you're not yet relating to the earth from a place of deep respect.

THE FEAR

As you actually begin to come back into contact with the Mother, it is likely that you will pass through fear. One way to understand this is that most of us are living lives that harm Mother Earth as a daily matter of course. It feels important to understand that you have been doing this because you have been de-natured by millenia of Empire. Empire would invisibilize this fact, because Empire doesn't want you to know that the Earth is alive. As you come back into contact with the Mother, your knowing–your recognition of your complicity in these harms– will be yours to repair. I cannot tell you how to do this. I recommend reparations. I recommend, personally, that you surrender to her completely. This will involve grief.

We can know, and we can say that we are children of the Universe, but until we FEEL this it is only words. We are made in the image of the Divine, children of the Mother Earth, but we are not acting like this, and until we act like it, it is hollow. We must find our way back to life.

We are told in the Essene Gospel of Peace. *And Jesus answered: Seek not the law in your scriptures, for the law is life, whereas the scripture is dead. I tell you truly, Moses received not his law from God in writing, but through the living word. The law is living word of living God to living prophets for living [humans]. In everything that is life is the law written. You find it in the grass, in the tree, in the river, in the mountain, in the birds of heaven, in the fishes of the sea; but seek it chiefly in yourselves. For I tell you truly, all living things are nearer to God than the scripture which is without life...God wrote not the laws in the pages of books, but in your heart and in your spirit. They are in your breath, your blood, your bone; in your flesh, your bowels, your eyes, your ears, and in every little part of your body. They are present in the air, in the water, in the earth, in the plants, in the sunbeams, in the depths and in the heights. They all speak to you that you may understand the tongue and the will of the Living God."*

The fear is what you must dissolve to come to know this reality in your body. I recommend that you lay yourself down upon the Earth Mother and dissolve completely. On the other side of the fear is re-union with the divine feminine expressed through our Mother, the Earth. So let yourself pass through the fear and across it to the far shore. You will claim your inheritance

CHAPTER 33

MOVEMENT

Your brain is in fact your entire body. Embryonically your skin develops from the same surface as your brain and spinal cord, and the density of neural receptors throughout your body (the oak in summer) means that your entire body is a sensory organ. The entire body is a FEELING apparatus, from the cellular level up. People refer to interoception as the sixth sense, but what is more deeply true is that the entire body is a FEELING sensory organ of immense sophistication. This feeling organ undergirded most ancestral ways of knowing. Much of learning ancestrally was to train and calibrate the intelligence of this knowing-through-feeling: modes of awareness that the modern world is categorically not attending to, partly because we don't know (and sometimes refuse to know) that they exist.

One of the most unfortunate byproducts of our modern brain-centric bias is that we have misplaced the division between the brain and the spinal cord. Seduced by a brain-centric view and anatomical appearance, rather than functional, evolutionary, and developmental understanding, we observed the junction of the 'brain' and spinal cord and mis-characterized it.

This separation is so deeply entrenched in our thinking about neurology that if you have studied this before–if you are a physician or a psychologist or a body worker–you may find what I am about to say ontologically painful. It will, if you let it sink in, deeply interrupt your understanding of how we are organized.

Modern neurology says that our brain consists of a forebrain, midbrain, and hindbrain. That our brains develop from back to front, from bottom to top, and that the brain itself consists of these three parts that junction with your spinal cord to form your Central Nervous System. But we've put the division in the wrong place. The primary division is not between your brain and spinal cord. The primary division is between your fore/mid-brain, and your brainstem/spinal cord. The brainstem and spinal cord are an integrated functional unit that houses the Autonomic Nervous System, and contains the Central Pattern

Generators of the body. They communicate with your midbrain and forebrain.

The cutting-edges of brain-based trauma therapy research is mis-directed right now because it doesn't understand this. We have as a field, grossly neglected the Autonomic Nervous System in neurology because we have categorically failed to understand that the brainstem and the spinal cord are a single functional unit with its own particular developmental trajectories that come online in sets of movement development sequences.

Our measurement tools, including fMRI, EEG, PET scans– most of our neural mapping technologies– have gotten fixated on peering into the brain and we have forgotten that cognition is riding around on top of an animal organized through movement. We don't understand the autonomic neurology of that animal and we don't have methods for investigating it neurologically in vivo (HRV excepted). It is only recently, as evo-devo (evolutionary developmental biology) methods have helped us understand deep homology, that the evolutionary (genetic) underpinnings of this division have been more clearly demonstrated. And yet, unfortunately, this awareness has not filtered back through and updated the neurological foundation models.

When we put the division in the correct place, and start to understand brainstem and spinal cord as a single integrated system, a coordinator of rhythmic movement, a number of things that were difficult to explain start to re-organize and move into relationship and synchrony.

As we peer back into the deepest mists of ancestral time– into the origin of the lineages that would evolve to eventually become human, we gaze back at the moment when body plans of organisms shifted from being radially symmetrical, like the starfish, to organizing along a verterbral column (phylum Chordata).

Our dear starfish ancestor has a mouth at the center (a belly mouth) and five radially symmetrical limbs. It does not have a spinally-organized Movement System.

When the body plan of chordates[1] emerged - this orga-

1 In biology, the phylum Chordata (corded, as in *spinal* corded) contains three sub-phylum: vertebrates (of which we, as well as all mammals, birds, fish, amphibians, and reptiles are included), lancelets, and tunicates. There are about 30 known species of lancelets. Lancelets are small, eel-like organisms that live in the ocean. They are filter feeders and use cilia to filter food out of the water. They anchor their tails in the sand and let the water wash over their mouths. They have a nerve cord, but no brain or vertebrae. Tunicates are also known as sea squirts. Adult tunicates are simple organisms. They are basically a barrel-shaped sack with two openings or siphons that water passes through. They draw water into their body through one siphon, filter out food like plankton, and expel the remaining water out of the other siphon.

nization of a spine with paired sets of limbs allowed for the development of entirely novel forms of locomotion. In plain language, the starfish cannot walk, or move away with speed. When the chordate body plan emerged, with pairs of symmetrical limbs (wings, fins, legs) a whole new range of movement possibilities emerged, as did an entirely new set of neural architectures that we are calling the spinal Movement System rather than the sympathetic nervous system to denote its coordination of not simply fight-flight responses, but all centrally-generated patterned movement.

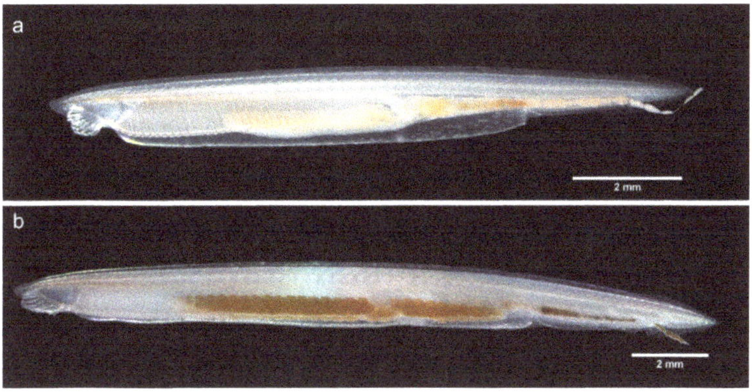

The illustration above: Amphioxus Branchiostoma californiense: a) immature organism b) mature organism (female). This is the basic body plan of chordates.

Notice, as compared to the radial symmetry of starfish, that we have a notochord, a nerve cord (spinal cord), a dorsal hollow, pharyngeal slits (these are early gills on fish, will move into the brachial arches undergirding our social engagement system in the voice and face), and a post-anal tail. Undergirding these new biological structures (notochord, nerve cord, pharyngeal slits, post- anal tail) that give rise to new movement

Tunicates spend their adult life attached to a fixed object like a rock. In the larval stage, tunicates look like little tadpoles. They can swim and have all of the characteristics of chordates - a notochord, a dorsal nerve cord, pharyngeal slits, and a post-anal tail. As tunicates mature, a sticky substance forms and they attach themselves to a rock or other fixed surface. Once they are attached to a surface, their body absorbs all of the structures in their tails. It only takes a few hours for a tunicate larvae to mature. Its main purpose is to find a place to settle so the adult tunicate can form. Some tunicate species live alone, others form colonies of many tunicates.

plans were a set of novel neural architectures: a spinal movement system containing central rhythm pattern generators. We have already spoken about central pattern generators in our chapter on breathing, but these pattern generators undergird a whole range of behaviors that we don't often consider rhythmic. When we think about rhythm, we think of a beat or pulse: heartbeat or the breathing. But there are rhythmic pattern generators undergirding swallowing, vocalization, speaking, vomiting, digesting, peeing (micturation in science parlance), pooping, and orgasms. Rhythmic pattern generators also undergird all forms of locomotion in vertebrates: walking, running, swimming, flying. And rhythmic pattern generators undergird our emotional range.

Central pattern generators govern rhythms both internal and visible from the outside.

Most of us are so accustomed to moving that we have completely lost touch with how neurologically complex it is. Take a moment to swallow. Easy, right? You can do that without thinking. See if you can slow the swallow down enough that you can feel the progressive wave of muscular contraction required to make this movement. Swallowing requires the time-phased coordination of twenty-six pairs of muscles that fire in rhythmic sequence. It is one of the first rhythmic patterns you have to master when you are born, just after sucking. It is so automatic now that you don't have to think about it–don't think about it most of the time.

Yet if you slow it down enough, you can feel the complexity of organization required. Learning to walk takes many months of daily practice. We have forgotten the extraordinary neurology undergirding these movement patterns we take for granted. The time required to bring our legs online neurologically: to coordinate the ability to sit, to hold ourselves upright, to be able to turn over, to crawl, and eventually to learn to walk? The learning of this rhythmic movement coordination takes most of us nearly half a year of constant daily practice.

I have had the good fortune to study birds pretty closely for a long time. Several years ago, in the condo complex where we live, my daughter discovered the nest of a pair of Great Horned Owls, about eighty feet up in a Eucalyptus tree. When the chicks fledged I spent several months nightly watching them learn to fly. Mostly this happened very late- between midnight and dawn. Watching baby Great Horned owls learn to fly is a great reminder of just how difficult it is to learn to organize these movement sequences. The pattern coordination required to maneuver in the air is quite refined.

The three owl chicks learned to glide long before they could coordinate the rhythms required to raise or lower themselves or turn effectively. Practically, what this meant was that for several months I watched them crash into things at velocity. They would take off from a limb about thirty feet high- at a certain developmental point you could see that they were anxious to be able to fly. They wanted to figure this out!

And then they would sail several hundred feet horizontally, more or less straight, until something interrupted their flight path. Sometimes it was a tree, and they would ploof into it in a shudder of wings, claws scrabbling to grab something as they smashed through the foliage. Sometimes it was a building. It was like watching a blooper reel– night after night. Some of the crashes were appalling. One night I watched a fledgling crash into a second-story bedroom window, slide down the glass, and drop into the person's backyard. I almost threw up.

We, in neurology broadly, categorically fail to understand the importance of movement, and the completion of motor movement patterns, to the state of the Autonomic Nervous System. As we have learned previously in the chapter on breathing, a central pattern generator is characterized by being able to create a rhythm on its own in the absence of neurological signals from the brain. The generator, as a relay loop, can function autonomously. It can also be influenced by information coming down from the brain, coming up from the body (senses), or by neurochemistry.

Central pattern generators come pre-programmed with a whole host of evolutionarily determined movement patterns waiting to, and needing to, express. Many of these are sequences linked to survival behaviors, both predator and prey sequences. We come hardwired with patterns of predation around biting into food with our fangs and snapping it off. (See the bite/snap fighting rhythm of Bonnie Bainbridge Cohen). We come hardwired with symmetrical victory gestures (see the next chapter on Gesture). We also come hardwired with escape sequences. We are replete with movement plans accumulated from four billion years of evolution. These plans live in the Autonomic Nervous System, up and down the spine, not in the brain.

At the bodily level, movement functions continuously on the inside. Our insides are alit with autonomic pulsations, even when the physical body is still. Our hearts beat, our lungs expand and contract, our digestion pulses. A slower cranio-sacral pulse moves up and down the spine. Even when our skeletal muscles are at rest, the interior of us is moving rhythmically. Movement is connected, primally, to our deepest instincts to approach and avoid. Approach and avoid plays out through the neuroceptive continuum between safety and danger. What feels safe and inviting we want to move closer to. What feels dangerous for any of a variety of reasons, we want to get away from. We might get away from something by attacking it, by creating a boundary, or by running away. The bodily decision-making of all of this does not happen in the brain– it happens in the central pattern generators. Here is the nuanced translation of approach and avoid into the patterns of movement, from the most subtle to the most pronounced, that govern our interactions with the external and internal environment.

In the wake of a recent coaching session with a practitioner focused on early infant motor patterns, we marveled at the motor movement sequences of early infants as they discover that the arms in the body they find themselves wearing belong to them. I remember vividly watching this happen when my daughter was born. Prior to her arrival, prior to my own becoming a parent, I really had no idea how unwieldy the body was for an infant. I remember lying on our bed in the afternoons with my daughter, she on her back at two or three months of age, and her arms floating above her head moving gently as seaweed pulsing in a deep ocean pool, her gaze trained on these tiny stalks, clearly entranced by their motion and as yet unaware that she could govern it. Every once in a while one of the floating arms would bop her in the face, much to her surprise. I remember the moments when she began to understand that she could steer them, the way that the pulsing of them slowly gelled down into movement with intent, which happened over a period of several weeks.

After my session on the arms, I awaken in the middle of the night, thinking that dawn has come already, conscious of the roots of the arms in my own body, their sockets aflame, neurologically. I am sleeping without a shirt, and when I cross my hands over my heart, and lay the palms against the opposite shoulders, the ventral junctions of my pectorals into the shoulders are cool to the touch, almost icy, as though devoid of blood flow.

As though the arms themselves have not turned on, have not myelinated yet, have not found themselves, but are the arms of a babe in the first couple of months after birth. Arms that are appendages yet unwieldy, not yet under the sway of conscious persuasion.

These are the months, at first, when the infant is swaddled, bound, the arms are held in at the sides. Arms before they have grown neural affect. A tadpole before it has sprouted arms. This is the sensation: a sort of fire in the roots of the arms, as though they are limbs that have not yet grown in. And is it not so neurologically?

I awake swimming in the depth of a profound knowing, an embodied awareness with its own innate certainty, that the arms are tools of emoting first. That long before they turn into hands, turn into palms, long before we become aware of our fingers with all the fine and detailed motor movements those digits are capable of, the arms are emotional figures. Their sweep and arc, their turning in gestures laterally symmetric, movements coordinated from side-to-side, are the limbs of the heart lateralized, extended through the chest, coordinating the rhythmic pulse of emotion.

The arms are the wings of the heart.

Awake to the innate knowing that the arms are the conductors of emotion- conductors in the symphonic sense, and conductors in the electric sense. That our emotion requests and requires the arms in which to play out, to make itself known. That the early movement and motor patterns of the arms are the very development and instantiation of emotional range.

The early movement and motor patterns of the arms are the devel-

opment and instantiation of emotional range.

And this is, of course, why we talk with our hands. Our arms are the movement vocabulary of emotion.

That what we are learning as we harness them, learning as we reach out to grab hold of the thumbs of our caregivers, learning as we push against, learning as we reach up towards them, are the extension of our hearts, the extension of our yearning, the metric, depth, and grasp of our emotional range.

That the range of motion of the arms, in movements autonomic in origin, beginning in the roots of the arms in the chest, the place where they bud out from our torso, is the awakening of the full range of our emotions. Before the arms are fingered, before their play is of hands and digits, the arms are these emotional wings. I watched this happening with my daughter, without fully understanding it then.

I awake lit with a vocabulary of gesture, felt, a vocabulary of knowing the innate rhythmic coordination of the arms as stalks of emotion. Feeling the ways that awakening neurological wiring ties the autonomics into the emotional centers of the heart, and surely the midbrain. Conscious again of the degree to which the earliest developmental pattering in this womb-with-a-view period, the period from zero to 18 months of age, is absolutely crucial for the neurological development of all that follows in our lives.

If during this period of reaching out, learning our arms, learning our hearts through them, there is no response from our caregivers– if no one reaches back to us when we reach out, reach up– then how are we to develop the emotional vocabulary of relatedness? How are we to learn the call signs of love? How are we to become who we are if our caregivers are too busy scrolling on their phones to respond to these primal communications? If they grew up in households where their caregivers were too busy to attend to their primal communications?

And thus trauma is passed, multi-generationally, through the failure of call-and-response in the movement and gesture, the outreach of the infant learning her emotional range.

Sometimes, for a variety of reasons, we do not make the movement patterns that the Central Pattern Generators activate. The body does not express them. If we reached out for the parent, reached out for the caregivers, and they continuously failed to respond? If we reached out for them and were shamed, yelled at, told to stop? Told they were too busy…these movements get inhibited. We may also have been socialized into the conformity of stillness. Socialization– in the sense of acculturating to a society and its norms often involves the suppression of movement. Children are to be seen, not heard…is this not an acculturating norm? Have you ever found yourself in a meeting where your entire body wanted to leave and you made yourself stay? You know what I'm talking about.

When this happens, we accumulate localized allostatic load. It may be that deeper evolutionary structures– older evolutionary structures– starfish structures– inhibit them and so we do not feel safe to move. It may be that downcoming cortical signals (thinking signals) inhibit them.

When we are trained–and this is a neurological training, to be sure, although one that we tend to give our children accidentally– not to move... When we learn to ignore, over-ride, and lose contact with these endogenously generated patterns...

From early infancy, the sovereignty of our movement is overridden by well-meaning adults who want to make us eat our spinach. They don't like it when we turn away. *Stop doing that*, they exclaim. *Eat your spinach!*

An infant gets told this enough times and the movement patterns shut down. We move into immobility, an older evolutionary state: the defensive state of the starfish. The entire reason that we need disciplines like Authentic Movement is because modernity has trained us not to move from our centers, not to move from our innate impulses. We have been neurologically trained to suppress our innate movements, our innate responses.

When these movements are suppressed, what is constrained is not simply movement, but our emotional development, sense of autonomy, and sense of agency. Our ability to move our arms in particular undergirds finding our emotional range. Our ability to move our legs undergirds our sense of ourselves as active agents in our lives, our authorship of our sovereignty. There are movement patterns of expansion that we must make and feel ourselves making. And then there are movement patterns of defense that if interrupted prevent us from returning to a felt sense of safety, as we explored in the chapter in the second movement of the book entitled 'We've Lost the Housekeys.'

When we neurocept danger, for example, central pattern generators throw up a movement sequence. This sequence is a motor program, a kind of autonomous sequence of movement that the body can execute. This is not done by the conscious mind, not done by our ordinary sense of self. We don't 'choose' the movement sequence with our autobiographical selves. It isn't executed by our 'Default Mode Network'. The body itself, endogenously, selects the movement pattern that flows through the pattern generator.

The execution of the movement program, which might involve a physical movement of, for example, boundary setting with the left hand in a forceful gesture that says, *Back up!*, is what restores the signal of safety moving through the Central Nervous System. Since safety is a pulse, the completion of motor movement programs of defense is what moves us back into that pulsative rhythm. We express the movement, the person who is too close backs up, and we are able to return to a sense of safety.

But let's say that I do not complete the motor movement sequence. Let's say that I don't set a boundary with the left hand because I have learned that will make things worse for me (it elicits more extreme violence, for example, from the source of the distress). In the absence of completing the motor pattern the body elicited, what happens next is that the body is unable to return to the pulsation of safety, because it has a cued-up incomplete survival-level motor program stuck in the central pattern generator.

This is not a thought, not a cognition. It is not an emotion. It is something else. A movement program, autonomic in nature, and localized at a certain level of the spine. Remember: because the central pattern generators are able to autonomously execute movements, it did not need permission from your brain to do this. This is *the body itself* responding to the environment. It is held endogenously in your left arm, and the muscles that would have permitted you to make this movement. Yet now, with the movement incomplete, we are removed from the pulsation of safety, because the pulsation of safety cannot flow when there are incomplete motor programs waiting to be discharged. Sometimes with enough nuanced awareness, a person might feel this in their arm– but most modern people don't.

When a central pattern generator prepares to fire, and cannot do this, allostatic load accumulates and it locks in as muscular tension in the body. When this begins to release at a neurological level we experience what is sometimes called a *neurogenic tremor*. This shivering or shaking movement is connected to the movement motor program that did not get to express.

Sometimes a movement pattern that is held, waiting in a central pattern generator, will express at the gateway of sleep. These we call *hypnogogic jerks*. They are waiting for the disinhibition of the movement in order to come online and execute.

Sometimes this happens on the threshold of sleep, as our waking cognition passes the baton to sleep. Have you ever jerked yourself awake like this? Sort of spasmed in your bed and thought, *what was that?*
If it doesn't get to complete, this program is latent in the body part that wanted to express it. And interestingly, your body will start executing that program, trying to complete it. You'll start throwing that left-handed boundary-setting gesture up as you talk, as you move through the world interacting with your daily life. Most modern people are fairly constantly throwing up (the image of vomiting is not wrong here) the host of incomplete motor responses their bodies are attempting to complete. We cannot help but do this. A good part of our gestural repertoire, when we talk with our hands, and as we move through the world, can be incomplete survival level autonomic motor programs seeking completion. Often programs from very early infant developmental stages.

Interestingly, we can train ourselves to make contact with the layer of awareness in the body where these motor programs are retained. They are often just beneath the threshold of conscious awareness. There is a particular kind of mindfulness of the body that can be used to bring our attention to these incomplete motor programs. It can be evoked partially by attending to patterns of chronic muscular tension in the body, particularly in muscles that are connected to the vertebral column, because these are the muscles that will be holding incomplete movement patterns from the central pattern generators in the arms and legs. (Think about how much tension most people carry in the psoas muscle, which connects the spine to the femur, or in the shoulder girdle, which connects the arms to the spinal column.)

Most of the gestures that we have not completed are residing here, in the arms and the legs, or along the length of the spine, waiting for us to have the right quality of awareness to evoke them. Lest you think that these movements are simply defensive, they are not. What is also waiting here are the hugs that your body wanted to give and you did not. The expressions of joy you failed to give yourself permission to make. As well as the things you wanted to (and needed to) say but did not. Those live in the muscles of the pharynx and larynx. Again, they are not thoughts, though with language there may be concepts involved. They are the animal vocalizations– of distress, boundary setting, or love– that we have not given ourselves permission to make. The full range of guttural noises, moans, whimpers, distress calls, weepings, wails, keenings… all the sounds our animal bodies know how to make that our socialization suppresses. They can also be the sentences we have uttered inwardly and not brought to voice.

Something that I hope you'll notice about these gestural categories– for they are a specific kind of gesture– is that they have emotional tone attached to them. The hug you wanted to give someone but were afraid to because you didn't want to be that vulnerable, or you were afraid they would misinterpret it– has an emotional tenor. It is in fact an emotion. The feeling animal wants to move and its movements toward and away have emotional tone. The boundary-setting gesture carries irritation. If un-acted upon, or ignored, this irritation turns to anger. Motion births e-motion.

Where is the line between movement and emotion? Is there one? Etymologically, the word emotion has the word motion within it. The Latinate derivation of the word is *ex* (out) - *movere* (move). An outward movement. Yet when we think of emotion we do not think of motor programs generally. Somehow (is it our brain-centric view?) we have detached them from the body, we think of them as being abstract. But have you watched someone feeling an emotion recently? You can see it in their face, in their entire body. Even the slightest registration of pleasure, emotionally, cannot help but be expressed through movement. Watch someone crying: that is motion. It is possible to imagine a world un-alienated from the body where an alternate neurology organizes in an entirely different manner, from the simple premise *I move, and therefore I am*. The

URE AS WAY

Now that our face/voice complex is awake, our hands are open, our feet are back on the ground, and we understand something about the foundational role of movement, let's have a look at how movement flows into gesture: autonomic gesture specifically.

In the wake of a challenge to our autonomic equilibrium, it is not the return to safety that shifts our ANS back into a connection state, but rather the completion of the motor movements of defense.

When we are pushed out of safety and connection into a fight or flight response, the mechanism whereby the ANS resets is to complete a defensive movement protocol. Think for a moment about two cats getting ready to fight. There are a whole host of specific movement patterns that accompany this interaction. Their hair stands on end. They snarl at each other. They move in an almost-dance, tiptoeing it seems, circling one another. Yet this tiptoeing has no quality of lightness, but rigidity. It is tensioned tiptoeing. The snarls and whines are guttural. Eye gaze is electric. Ears flatten. And all of this happens before the first strike. Often, dominance and submission is established without the animals touching– but energy is, assuredly, being combatively exchanged. Someone turns tail and slinks away. But sometimes not. Hackles rise. Someone gets swatted. Someone wails and moans. And then there is a full-on attack.

These movement patterns are not ordinary, volitional movements. Rather they are connected to this deeper layer of our system: expressed by the Autonomic Nervous System. They are connected to a particular level of activation: a certain increased heartrate, muscle tension cued to survival.

Successful completion of these motor movements of defense: both fight responses, and escape responses, depending on what happened and how our Autonomic Nervous System experienced it, can re-calibrate our systems. They are more-or-less the reset button on the Autonomic Nervous System. The movement pattern will be based on what happened. They are not generic movements, but motor sequences of particular defense: of fight, or of escape.

Striking, kicking, scratching, biting, leaping away, running, etc. Ancestral cultures were good at creating contexts where these movement patterns could be evoked. Trance states, ceremony, and shamanic intervention often create contexts for this to happen spontaneously. Part of what makes someone a gifted healer in these contexts is not necessarily what they are doing, but rather the context they create that allows for the disinhibition of what the body already wants to (needs to) do to come back whole. Often what is required is to get the mind, the ordinary sense of self, out of the way. In these cultures, people are often more in touch with their animal selves, so this is easier. The animal and the animist haven't been equated with the base, the evil, and so people are less reluctant to allow themselves access to modes of primal knowing and expression. They give themselves permission to kick like a chicken, to swat like a cat, to bite, to tear, to eviscerate– even if the mind doesn't understand this.

When people's egos are less solid, and they are more open to spontaneous expression, these patterns tend to emerge spontaneously. They are just beneath the surface, seeking the opportunity to arise. I've had some very interesting experiences with modern people (myself included) sitting in ceremony. In a teepee in the middle of the night, or in a journey on plant medicine, in non-ordinary states of consciousness, the concrete ordinary self can get out of the way and things can move through us. I've seen people who were otherwise proper and composed move like animals, shiver and shake and fluff themselves like chickens, stalk like jaguars, and move in ways that they would not allow themselves, in ordinary waking consciousness, to move. Things that, were they looking in a mirror, they would not permit because they would become too self-conscious. Something deeper, something more primal, takes over. This kind of work can also happen in other liminal states. When someone is feverish or very ill. When they are on the threshold of dreaming. Hypnogogic jerks– those sudden startling movements we sometimes make just as we are falling asleep– are one way that the body will complete autonomic motor programs.

Yet these completions don't have to happen so far away from our ordinary lives or our ordinary consciousness. We can learn to balance our attention in the body in a way that gives it permission to move as it wants to, not as we (our ordinarily constricted sense of self) want it to. If we bring the right combination of attention to these movements, as they move through us, they will help us shift our autonomic baseline.

I recently worked with a client who is in the process of liberating some of these responses. In her mid 60s, the part of herself that was enacting this gestural repertoire was somewhere between three and six years old. As she really got into contact with it, the necessity emerged as a very clear set of movement directives. She needed to strangle something. She needed to deliver it a death blow on the back of the neck. And then she needed to kick it into the fire. The specificity of the movements that her body wanted to make alarmed her. This ordinarily mild-mannered person needed to engage in a very complete set of killing gestures. Yet when she was able to move the ordinary sense of self out of the way, and the inhibitions on these responses, the amount of energy that this completion liberated in her system absolutely floored her. The object of this outrage? Some monster from childhood. Maybe a parent. Maybe an uncle? Possibly an amalgamation of a number of characters in her life. Maybe the monster under the bed? Who she needed to strangle was less important than that the body needed to actually do it. These survival responses needed to complete motorically. Her grown-up body was preoccupied with something she had needed to do since she was a small child.

One of the beautiful origins stories of Somatic Experiencing®, the naturalistic approach to the resolution of trauma developed by Dr. Peter Levine, is that it came in part out of his observation of animals in the wild responding to threat. He noticed that animals in the wild would spontaneously recover mobility by shaking after they had moved into a shutdown or immobilization state. I've seen extraordinary videos of this happening in the wild. Prey animals (gazelle, impalas, etc.) that have gone into a full shutdown state, so deeply that they seem dead, not breathing at all for minutes, spontaneously reviving when the threat has passed, moving into chaotic full-body shaking that suddenly coheres and organizes, then suddenly getting up, and galloping off. This response was endogenous to the animals in the wild. Animals in captivity, however, stopped displaying this resilient response. The observation that Dr. Levine made is that modern people respond to threat like animals in captivity.

Modern people respond to threat like animals in captivity.

This has a lot to do with the context in which we are socialized, yet what Levine noticed was that if he could shepherd people's attention in a very particular way to aspects of their experience, he could coach them through this spontaneous expression. He could *facilitate* it.

And thus the name of the modality. In a way, you could say that Somatic Experiencing is about creating a proper context for the body to do what it ancestrally did on its own, or in the context of community, ceremony, and ritual. For the undigested *experience* to completely metabolize *somatically*.

What makes a gesture autonomic rather than volitional? If we recognize that gestures that arise at the depths of the Autonomic Nervous System have a different kind of potency, how do we identify them, and how do we use them to help engage our Connection Systems? Since 2011, when I was studying Somatic Experiencing, and my daughter was about two years old, I have been fascinated by what I think of as *archetypal motor gestures.* I found her developmental trajectory quite extraordinary, as any first-time parent who is paying attention to their child does. I hadn't realized that it would take her a month to learn how to use her arms. I was astonished, as she was learning to walk, to come into her bedroom at night and find her facedown in her crib sleepwalking. These patterns of movement are deep within us. They are rhythmic in nature, connected to central pattern generators in the brainstem and sympathetic ganglia. The movements we see expressed through major muscles are only the ones we can see from the outside, but the interior of the body is essentially a rhythmic oscillator. From the drumbeat of the heart, to swallowing, to the cadence of the in-breath, and out-breath, to the peristaltic rhythms of rest-and-digest, to our circadian rhythms, our bodies are moving through rhythmic pulsing.

When my daughter was about two, and fully mobile, there was a set of gestures that she began to express that caught my attention. At the time she was about two and a half feet tall. What I noticed was that when she was distressed, she would joggle her hands and arms, as if wanting to shake them off. There was something about the movement that seemed unusual to me. As though it wasn't her ordinary sense of self that was doing it. Like the movement was coming through her, for lack of a better way of describing this. She would move like this, shaking her hands in this way– this kind of remote look in her eyes, her body like–*Yuck, this feeling*–and then turn to us, and raise her arms up, symmetrically at her sides. This second gesture also caught my attention. She obviously wanted us to pick her up. She had just a few words at this point. But the gesture spoke volumes. It also seemed to me archetypal. She would wring her hands, shaking off distress (the yuck), and then raise her arms and her eyes to us for comfort to scoop her up (where she would melt into us briefly).

Both of these gestures seemed to come *through* her. They seemed deep and essential. One of them was a gesture of distress. The other a gesture of connection. One got rid of something. The other restored relatedness.

I didn't get the sense that she had learned these gestures from somewhere, but rather that they were *innate*. They were both movement patterns, both a set of motor programs, but they were also indistinguishable from emotional and relational activity. She wasn't moving her hands so that she could move her hands–the goal wasn't movement–she was shaking off an icky emotional state. How did she know to do this? She didn't reach up to us to exercise her arms. She reached up to us to come back to the relational baseline of feeling love.

I became intrigued, at this point, by gestures like these. Gestures that seemed to come from someplace deeper, and with their own function. Gestures whose very role was to shift autonomic state.

And so then I began to hunt for them, and to find them everywhere. One of the places I've looked for them, and catalogued them fairly extensively, is in sport. I've looked at these phenomena in a variety of sports, including running, soccer, and tennis, partly because those are sports I play, and so they are familiar to me. There are entire categories of gesture that players use to give themselves integrative feedback, those that are designed to increase energy, those that are corrective. Most players express these gestures without much awareness, which means that they are not consciously harvesting the nutrition they provide. Some of them, often truly elite players, use the gestures as tools. They harvest the gestures, feed on them consciously. Many of the people at elite levels of sport who have learned to do this are trained in mindfulness, but most of them are not trained in autonomic physiology, so they are paying attention to the gestures and believing it is the awareness that makes them important, when it is in fact the awareness flowing through the movement components of the gesture that gives them power.

Attention is downstream of autonomic state. Attention is derivative of autonomic state, because autonomic state has survival priority and therefore governs how attention works. We have to be in certain states (enough safety) to have access to the kind of free attention that people think of when we talk about mindful awareness. Absent this safety, attention is recruited by survival-level processes, like identifying and mitigating threat.

When we look at gesture through this lens of attention, particularly autonomic gesture, we note that a movement can be made, a gestural pattern can be invoked, but if there is not sufficient focal attention on it, it will not reset the Autonomic Nervous System. One of the things we started tracking when I was first studying Somatic Experiencing, was the role of unconscious expression of these classes of gestures in our clients. People are making autonomically-charged gestures all the time without realizing it.

We throw our hands up in the air, or turn away in disgust. We set a boundary with someone when we can't tolerate their words anymore. We bear our fangs at an opponent when we are competing. These gestures seem to me autonomic in nature when they are pure communication. When they make words irrelevant. And each of these gestures seems tied to some deeper aspect of defense or invitation.

The gestures accomplish their regulating objective when there is sufficient awareness on them. In Somatic Experiencing, we learned to gently bring our client's attention to the gestural repertoire they were already expressing. Once a gesture was repeated a number of times, I might make the movement myself and show it to the client (this moves it into a mirror neuron process). The next time they moved in that way, I might invite attention to it. I might pause them and say, *I'm noticing that gesture again. I'm wondering what would happen if you take a moment to allow yourself to feel it?*

Often, this invitation came as a surprise. It would happen typically during the re-telling of a story, and sometimes the client would feel interrupted. I would apologize and insist. In that moment, the client's attention was usually on the narrative. What they were usually tracking were the words. But there was the body, holding the deeper energetic signature of the moment, beneath awareness. *I move, therefore I am.*

And so I would say, *What if you noticed that? What does that feel like to you?* And often the client would realize that the gesture was where the juice was. That the gesture was holding the energy of the conversation, the ember yet burning. The aliveness. The spark of what needed to be felt and transformed. The story has been rehearsed, whereas the gestures are authentic.

I haven't seen one yet, but I imagine somewhere there is a complete taxonomy of archetypal motor gestures. In such a map, we would require a category for gestures of defense. These are gestural patterns tied to the fight-or-flight response.

It seems important to note that there really isn't a repertoire of motor gesture for shutdown responses, because they are immobilization responses. Shutdown is associated with a *loss* of gestural repertoire. We could say that the gestures of collapse are affiliated with shutdown, and that would be true. Spine losing curvature, head falling forward, abdominal core losing muscle tone. But as for the movements of the arms and legs, which is where I tend to do most work on gesture, those are really not online when we get shifted into shutdown.

The taxonomy of defense would include gestures of fight, gestures of flight, and boundary-setting gestures.

Many of the gestures of defense involve the arms and hands. When I was working with incarcerated youth, one of our roles as facilitators was to make sure that there was never a fight in our groups. This was a non-trivial task, because sometimes we were facilitating groups with rival gang members. One of the sure-fire warnings something was wrong was that long before a punch was ever thrown, a kid would begin to ball their hands into fists. In tandem with this, you could often see them tensing their jaws. I've come to think of this as a sort of pre-fight sequence, like the cats circling. If I'm going to hit someone, I'm gonna do it with my hands.

There are a set of preparations that my body needs to make to organize that movement pattern. If I'm really going to strike someone, with force, the movement begins in the core of my body. It organizes in the pelvic floor, on the abdominal side from the transverse abdominus, on the back side through the erector spinae muscles. I mean, you can feel this right now yourself. If you are really gonna hit someone, and knock them down, you cannot do that with a flaccid core. And as you start to get angry, if that option of striking someone is really on the menu, the movement organizes deep down in your body long before your arm. In tandem with this, you have to organize the breath. If you begin to organize the core, the breath will start to change on its own. Now that core is moving differently, you are sitting differently in your chair. Feel that. You can't strike someone if your core is loose and you are slouching or leaning back. By the time the preparatory movement gets to your shoulder girdle, and then the muscles of your arm, the jaw is flickering. Nobody hits someone with a slack jaw. Only then does the preparatory signal get out into the muscles of the arms. Yet very early on in this process, there is an increase in blood flow to the fingers and hands. The body responds to this by clenching the fists. This typically happens before the volition to fight has come into conscious awareness or cognition. So here are like ten adjustments, pre-fight adjustments, that have already happened long before the body throws a punch. With ability to track this, we can see the punch coming sometimes before the person knows they are preparing to throw it.

I've been married for a very long time. My wife and I know each other pretty well. There are moments when she'll say to me–*you're getting irritated.* I'll say, *I'm not.* But she has already seen it in my lip. There's a particular thing my lip does, I am told, when I'm starting to get pissed off. When we know someone really well, and we're paying attention to these kind of things, we start to pick up on the subtle repertoire of gesture.

We were on vacation in Maui recently, and attempted to stay for several days at a friend's house. They were out of town, and had offered us their place, which was quite lovely. As we were unpacking our bags, I noticed an insect trap on the kitchen counter. As we were putting our food into the refrigerator, I noticed a small cockroach. My wife saw it. She turned away, and there was just the very slightest flicker at the corner of her lip. Nearly undetectable. A tiny flaring. It was an autonomic gesture of disgust.

The next thing I did? It was not to continue to put food in the refrigerator, I'll tell you that. I sat down with my laptop. My wife says, *What are you doing?* I didn't say anything because I was tired, but I was finding us a new place to stay. She didn't have to tell me that she wouldn't be able to sleep that night. I knew it already. The signal of disgust, though tiny, was actually a *very* clear signal.

Just generally speaking, flight activates the legs and feet. Flight is an interesting one, because there's a lot of social inhibition on us fleeing in this culture. Because of this inhibition, a lot of flight energy gets suppressed, and when it gets suppressed, people often lose contact with the ground, and shift into anxiety, or deeper into shutdown.

The flight response is very connected to our motoric relationship with the feet, and the ground. A pre-gesture of flight, which you can see everywhere, is people tapping their feet. Like the clenching and unclenching of the fist, the tapping of the feet is a way of working, metabolically, with increased blood flow to the feet, because someone wants to get away. Before this, adjustment to the hips and pelvis. If we are socially conditioned not to be able to escape– I see a lot of this gesture

in children in school classrooms– we still have to metabolize the energy somehow.

A third primal category on the defense side is boundary-setting gesture. Boundary-setting gestures are often related to the fight response, but rather than purely confrontational, are about defining one's space. Let's start with the classical, or archetypal boundary setting gesture, the one that begins this chapter, which I like to call *Talk to the Hand.* I love this photo, because we do not even have permission to see the face. We don't even get that much access. We have been stopped long before that.

Here's a picture of two variants on *Talk to the Hand.* One of them, over the shoulder, says 'Flip off.' The other one, the hand forward, says, "There's an edge here." Everyone has seen these, does these. We do variations of this a lot when we argue. We also know, most of us, like our dogs who can bite us without breaking the skin, how much force to put behind these gestures. Let us not pretend they are not a language.

I'm showing you this repertoire between two women, because if I show it with two men, or a man and a woman, it can start to feel dangerous. Yet let's look at this, together. Here's an artful one with a man and a woman. I like this photo because of the balance of power. The man, who is white (two centered social locations) is with his girlfriend, who is black (two less-centered locations) but he is back on his heels, and awkward, and though he is pointing, the gesture is close to his chest, contained. His girlfriend, however, has matched the finger gesture, but with more freedom. She's telling him to fuck off. Her gesture has power in it. There is contra-lateral balance from her hips to her heels and in her torso. Her finger gesture is tossed over the top of her head, complex motionally, yet in rhythm with the balance in her body. The other hand is relaxed, no unnecessary tension. She is sure of herself. At a visceral level, it is satisfying. She's like, *See ya later. I'm out.* Movement and emotion are coordinated.

If this is a scene in a movie we are watching, I think the next frame is him regretting what happened, but I'm doubting she does. Her movement is too sure. I think she's done. She is moving on.

I'm starting out here talking about defense, because one of the ways to get back to connection is to feel the energy of the defensive motor responses. If you find your body doing one of these things–and the common denominator here is that they are not done by your ordinary sense of self–bring a little bit more of your focal attention to feel it. Trust the impulses that it conveys to you. If you're in a meeting, and your feet start tapping, find a way to give yourself permission to step out of the meeting. If you can't do that, find a way to ground yourself, more fully and to let the energy build in you, because it's building anyway, just outside of your awareness. This acknowledgment of the intelligence of the body is a way of moving into deeper alignment with the survival-level signals that our autonomic nervous system is conveying to us. This deeper alignment brings the mind and body into closer harmony. This is good for you. Often people get sick because the mind doesn't understand what the body knows, or doesn't have permission to integrate what the body feels. We are healthiest when what we say, and what the body does, are in alignment.

Everything that we've been talking about so far, in the behavioral gestural repertoire of defense, relates to gestures that don't feel particularly good. Yet the other broad category here are gestures of invitation.

These would include gestures of integration, connection, and the category I think of as victory gestures. I've had people say to me, *Well, Gabriel , if that's the case, then, why don't I just make that gesture voluntarily?* And my response is that if you make it voluntarily its not an autonomic gesture. That is your ordinary

A gestural repertoire of prayer: receiving through the hands.

sense of self *doing* something. What characterizes an autonomic response is that it's flowing through you. It's not your ordinary sense of identity that's making the movement.

Gestures of connection include gratitude gestures, and the kind of reaching out, or reaching up gesture that my daughter was making that started out this chapter (the prayer photo above has some elements of this). When we reach out to shake someone's hand, or when we open up to give someone a hug, these are gestures of connection. Again, the import is when these desires move through us. We have to want to open into them. No one ever got more connected by forcing themselves to give someone a hug. It has to arise spontaneously through you. Yet again, when we're able to allow these gestures to arise through us, connection gestures grant us the deepest experience of being connected.

If humans freak you out, and there is good reason for a lot of us to be freaked out by humans– we are fairly well surrounded by shitheads and fools–look for opportunities to explore gestures of connection with pets, or with living Nature. You can give a tree a hug, and it will hug you back. A tree also will never stalk you, weaponize something you did out of kindness, or leave creepy comments in your Facebook feed.

I first became interested in victory gestures watching Rafael Nadal win tennis matches. When Nadal burst onto the international tennis scene, he was a 16-year-old from Spain with a penchant for wearing clamdiggers and sleeveless shirts, with the most ferocious attacking forehand I have ever seen. He stunned the tennis world in 2008 by unthroning the reigning Wimbledon champion, Roger Federer, who is probably the most kinesthetically beautiful male player to ever play the sport. At the time, Federer was so dominant in the men's game that he appeared to be functioning in an entirely different dimension than his opponents.

When Nadal won a big point, he would lean back on his heels, cantilever his hips forward, and pump both his forearms. It's a gesture you've seen before, because it's an archetypal victory gesture. At the time, I started looking at these gestural repertoires in tennis players, and comparing them to the repertoires of other athletes in moments of victory.

Perhaps the most classic victory gesture is the one where a person raises both hands above the head. Watch anyone win a race, or score a goal, and you'll see what I'm talking about. Again, this is a movement that comes *through* people. It is universal. You don't have to teach people to do this. It is a spontaneous movement expression. Sometimes the fists are clenched, and the arms pump. Sometimes they extend fully and the palms are open. Each of these has a different energy. Victory gestures feed us energy. You can watch a tennis match pivot around a victory gesture. Even at the highest level, tennis is a sport that tends to move in waves. One of the players will become dominant and win six or eight or eleven points in a row. Momentum in a match is very connected to autonomic state. The more we keep winning points, the more confident and relaxed we get, and our performance improves. The more we lose, the more likely we are to get frustrated with ourselves, and shift into a defensive autonomic state that tightens us up.

If someone wins a point after losing ten of them, you'll often see them do a single armed fist pump. This is what I think of as a minor victory gesture. This gesture is a kind of food for the autonomic nervous system. Victory gestures, in my experience, are still somewhat effective if you do them voluntarily. So you can try this. Here's a picture of the gesture. It feeds vitality back into the body. Often this is enough to help someone win the next point. If they win the next point, they'll often give a stronger victory gesture. If they win the next point, they often give a stronger hand pump. Often this is enough to shift the momentum in the match. You might then see them win ten points.

I'd like to invite you to spend some time exploring, in particular, these three categories of invitational autonomic gesture. See if you can begin to notice moments when your body is surfacing integration gestures. When it happens, lean into it. Feel them more deeply. Notice the moments when you open into gestures of connection. Even if it is simply a warm handshake. Again, really allow yourself to feel the connection nutrition that comes to you through these gestures. Finally, explore the realm of victory gesture. Let yourself get pumped up. Make gestures of stoke.

Our bodies are designed to move. Gesture is designed to flow through us. Much of our communication is not verbal, and when we disinhibit around gesture, when we let our arms and hands speak from within us, it grows our hearts. I am not speaking metaphorically. This is factually and neurologically true. Your arms are the wings of your heart. So again, my invitation to you is to release the inhibition around being expressive. Talk with your hands. Talk with your face, with your voice, with your eyes. You will feel this feeding back to you as increased vitality, and well-being. You are now learning how to harness the power of your neurobiology of connection.

CHAPTER 35

POSTURE AS DOORWAY

As we transition from exploring gesture to posture, we are shifting from the realm of phasic muscles (actions, movements, gestures) to the realm of postural muscles (muscles of core stability that support posture).

This is an over-simplification, and not exact, but you can think about gesture as happening in the limbs, and posture as happening in the core. When we look at gesture, we are often looking at movements in the arms and hands, the feet and legs. When we look at posture, we are looking at the stability and organization of movement, as it relates to sitting and standing, walking and turning.

Postural muscles include the deep muscles in your abdomen, pelvis, and back. These are the muscles that hold you upright when you are sitting. These are the muscles people are talking about when they complement your posture, or admonish you not to slouch. All of them are centrally located along the midline of the body, and connect to your spinal column.

MUSCLES BALANCE FOR DIFFERENT TYPES OF POSTURE

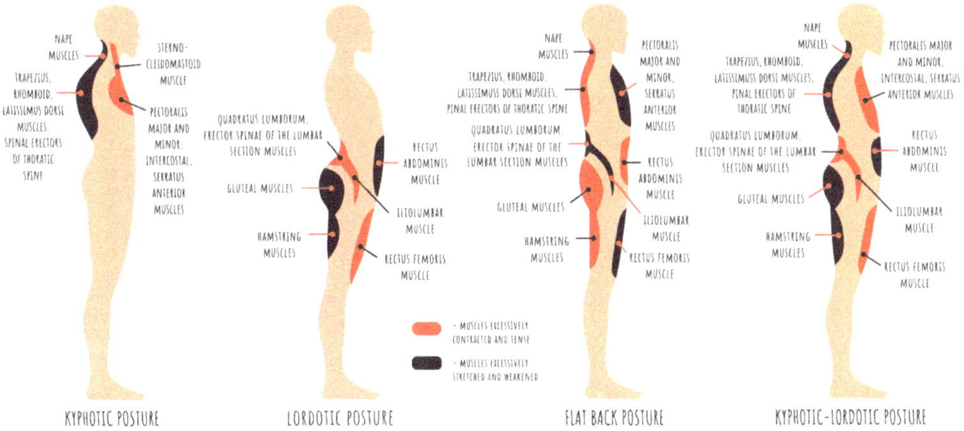

Structurally, your spinal column scaffolds the entire body. At the north end is the juncture with your head, the cranium, which is very heavy. The skeletal and muscular organization required to balance your head effortlessly on top of your neck is formidable. At the southern, or sacral, end of the spine it unites with the pelvic girdle, to scaffold the structure of your legs. And at a line essentially lateral to the heart, it ties in to the organization of the rib cage, to scaffold the arms. This bodily organization is so successful that it has given rise to the entire phylogenetic class of vertebrates. Vertebrates include mammals, birds, reptiles, amphibians, and fish. All of us share this common organization. All vertebrates have a backbone, a skull, an endoskeleton, bilateral symmetry, and two pairs of appendages.

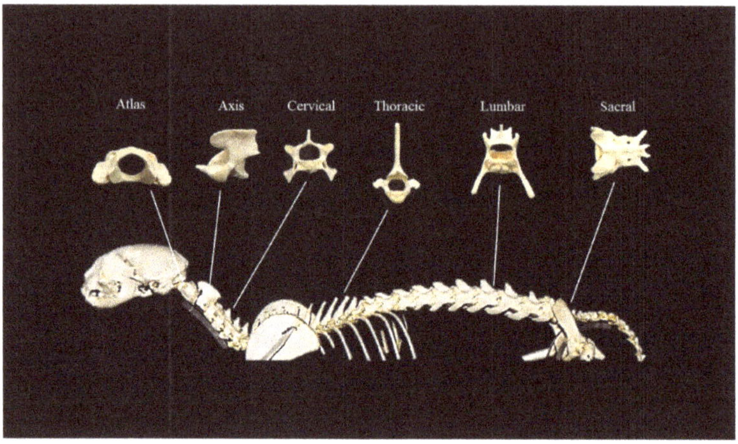

The vertebrate body plan scaffolds both a cat (above) and a bird (below).

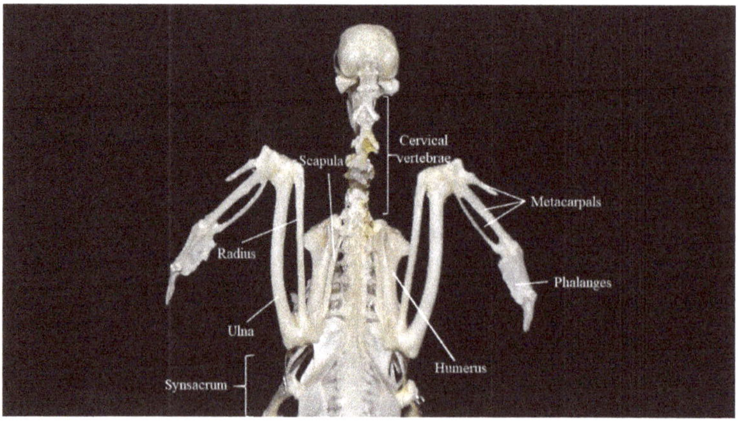

In terms of structural anatomy, we are not that different in our organization from a bird. Both of us are organized on a spinal column topped with a skull. Each of us has an internal skeleton, and we both have two pairs of appendages. The bird's happen to be wings and claws, whereas ours are feet and hands, but you can see that the design is shared. We are both structurally symmetrical from side to side.

The Autonomic Nervous System innervates postural muscle. Although you may not know this cognitively, you know this experientially, because when you're feeling relaxed, you sit and stand very differently than when you're feeling under threat. All of us know this at a bodily level. It's the reason that so much time and attention in yoga is dedicated to opening the hips, opening the shoulders, gently twisting the spine. It's the reason that you have neck pain and tight shoulders when you're stressed out. It's the reason that your back collapses when you fall into grief. Core Autonomic state maps directly into posture.

The Autonomic Nervous System innervates postural muscle.

When we feel safe enough to open into connection, our deep bellies soften, our chests open, our hips flex, our spines elongate, we hold our head high.

As we move into defense, our deep bellies contract, our hips gird themselves and armor up, our chests tighten.

You can feel the difference just by mentally adjusting your mood. When we are open to connection, the body is receptive. This receptivity expresses itself on the belly side. In the same way that the ultimate gesture of invitation for a dog is to go belly up, when we feel truly open and receptive, we open up on the front side. We can feel our bellies release, we can feel our hearts and chest open invitationally. If we get down to it, this is the posture of lovemaking. But we can feel it without taking off our clothes. When we are adequately met, and matched, although this posture is structurally vulnerable, we don't feel exposed so much as connected.

We feel vulnerable when we start to detect threat, and are aware that we are expressing that degree of openness. Just thinking about this, you can probably detect tension in the body. The way that it wants to retract itself.

That retraction is the movement into defense. Take a moment to notice how it presents in the primary postural muscles.

Notice how it changes the way the hips orient. Notice how it tightens the deep belly. Notice how the chest adjusts. The back. There are incremental changes in each of these, when we move into a sympathetic, or fight/flight response.

The changes are even more extreme as we move across the threshold into shutdown. As we move into shutdown, the spine collapses. We can no longer hold our heads upright. It is important to recognize that this is not a question of will. When we shift into a shutdown state, the neural re-prioritization changes the way our muscles operate. In order to organize the head on the neck in a way that is balanced, we have to have access to a certain neural tone. In shutdown states, this is not available, because endogenous opioids blockade the Movement System. The spine loses its curvature. The bounding box of the viscera made of the deep intrinsic muscles of the abdominal cavity loses its tone and the core collapses. The hips lock up. None of this is voluntary.

If we just isolate these three sets of muscles: the hip flexors (psoas in particular), the deep abdomen, and the back, we can see that they have a distinct set of patterns in the neuroception of safety, danger, and lifethreat.

If we start to view this through the lens of emotion, how autonomic state correlates with emotional range, we can see three distinct bodysets that represent these core neuroceptions. Contraction of postural muscles in the wake of trauma is so repeatable a phenomena that there are trauma healing modalities that have evolved to directly address the contraction of the psoas muscle. The psoas tends to lock-in traumatic experiences, holding the body memory of violation. By mechanically inducing flexion, and release of the psoas, people can often move through traumatic content. The movement will often awaken memories of the violation, hopefully helping them to re-organize. We are in the domain here of the title of Bessel van der Kolk's book- why he says that *the body keeps the score*.

How can we use this awareness of posture to help us activate the neurobiology of connection? One of the simple ways is to understand the bodyset of safety and connection. We don't really *have* a body. We *wear* a body. And we wear that body differently, and more pleasurably, when we experience safety and connection.

It's important for us to get clear how that feels. With respect to the postural muscles, we want to understand how it feels in the hips, in the deep abdomen, in the back and the chest. We want that to become our embodied physical baseline. When we feel this deeply and understand it, it gives us a greater awareness about when we are deviated from it. In order to fully inhabit it, I need to know what it feels like when my body feels good. Feeling good, at a postural level happens when my connection neurobiology is online.

In the hips, this translates to easy rotational motion. Without being lurid about this, think of the full rotational motion of the hips that is required in order for you to enjoy having sex. There is a full mechanical motional range that's required for both the male and the female anatomy to engage in the physical act of coitus pleasurably.

If you set down the erotic aspect of this for a moment, and just consider that sex is the act of greatest physical vulnerability, the moment in which we are least defended, most open at the root– at a biological level in order to move freely in this activity, we have to feel safe. For this activity to be enjoyable, the hips have to be able to open and be able to rotate. They have to have freedom of motion. That freedom of motion is available to you all the time, not just when you're having sex. Yet most people don't stretch, or even recognize that having that freedom of motion in the hips is part of what makes it feel comfortable to be in your body.

Range of motion in the hips is part of what makes it feel comfortable to be in your body.

Moving on to the belly, most of us, much of the time, are told to armor up our bellies. We are told to suck them in because it makes us look better. We're also often holding micro-tension in the belly because we move through a world filled with danger. Yet one of the bodily instructions most common to ancestral mindful awareness lineages is to soften the belly. Try it right now. Just let the belly release. And notice how different that feels from the way that you usually hold it.

I'm not talking about zero tone. I'm just talking about letting it be soft enough that you can feel through the *hara*. Through the deep belly. That you're treating the belly as a portal to awareness. For that is what it actually is.

Moving on to the chest, we talk about the chest being open. We practice this in yoga, by putting a bolster underneath the shoulders, and arching the back over it. But the mechanical aspect of it is only part of opening the heart. The postural aspect of this is, yes, to have the chest open. But the emotional or energetics aspect of this is as important, which is that we are opening to experience.

In order to open to experience, we have to WANT to invite experience in.

In order to open to experience, we have to want to invite experience in. And this requires having a life that we want to open

The physical posture of chest opening is important, but its essence is the opening up to experience. If that availability arises, this moves us toward the neurobiology of connection. If not, we are just stretching. Stretching is wonderful, but it is not transformational until underlying neural setpoints shift.

into. At some level, opening the chest is about finding what you love. It is about designing a life where you are spending your time involved with people and doing things that you feel passionate about. When you love your life, the chest *naturally* opens. When you love your life, *life* opens your chest. The universe conspires to aid you in this process.

We can practice these three bodysets of connection. We can physically stretch, massage, and open the postural muscles to give us more access to the feeling of the postural embodiment of safety and connection.

Retracting into defense is natural and necessary, because we live in a (potentially) hostile world. But if you practice opening the body, and you practice getting familiar with the bodyset of safety and connection, you will notice more quickly when you get pulled away from it, and you can correct. If the world-at-large is hostile, part of our opportunity is to string together moments of safety and people of safety and places of safety into islands of safety into archipelagos of safety upon which we can travel. Knowing, at an embodied level, what safety feels like is necessary for us if we are going to find and grow it in our lives.

The practice of yoga seems indispensable here. Any basic yoga class will bring your embodied attention to posture in these ways, helping you build a felt vocabulary for the nuance of these changes. If this chapter stands out to you as a doorway you would like to enter, I humbly suggest that you explore yoga.

CHAPTER 36

PASS INTO THE HEART DIRECTLY

On the learning platform I have architected with many hundreds of films and practices, possibly my favorite seven minutes of film in the archive entire, word-for-word, gesture-for-gesture, is of John Stokes, the Founder of The Tracking Project[1], speaking about gratitude.

In the film, he is teaching a permaculture course on the island of Maui, wearing a sleeveless surf shirt, his weathered muscular arms pliant and articulate as he speaks through them. In the background is the yellow of bamboo and the emerald lushness of the island. Periodically, a myna bird calls, or an errant rooster.

His voice, comedically-intoned, is purely reverent. He is being filmed, but this is incidental: he is talking to a group of people gathered there, who you cannot see, but whose laughter you can hear. They are just offstage.

I arrive at gratitude after hearing a man with a knife in his voice. A voice that held–I'm not even talking about the words he was saying–I'm talking about the place in consciousness from which the words were emanating–the kind of sacrificial knife that Abraham used on Isaac. The voice, not even what he said, but the voice itself froze something in my blood. Just slightly, but effectively. *To the quick*, we say. It was like that. My body went cold from the inside: sudden winter. I slept, then awakened. Internally, I was slightly numb, there was an anaesthetic quality to it: a pure shutdown. And even as I knew this, I was simply cold, beginning my morning. Cold inside. A bone-chilling warning, delivered through the voice.

Then accidentally I read something written by the writer Caroline Ross, of the Dark Mountain Project and found myself, like someone who has swooned, nearly passed out on a battlefield, awaken suddenly beside a hearth, wound bandaged, gauzed, the flame-lick of the fire on my face. Blinking in warmth and light, thinking– *how did I get here?* Suddenly home to myself.

1 https://thetrackingproject.org/

This sudden intervention, through words, living words, medicinal words, left me astonished and simply repeating the words, *Thank you.* Gratitude. Thank you. And this caused me to remember what John said.

John begins by speaking about the beauty of life itself. He reminds us that everything we need to live a good life is already here. He notes that sometimes we think we can improve it. He talks about how plants are a perfect solar collector. And then about NASA's decade-long research to figure out how to create oxygen in space, which ended with the recommendation that they bring plants with them. *It's already here*, he says. *It's already perfect, already laid out for us right here.*

And then he says,

"What the old people [the Elders] learned was – what is the proper attitude of a human being in the face of such majesty? And they learned that the human being exists somewhere between the mountain and the ant. And we have to find what's the right attitude between them- and they discovered that it was gratitude. You can just say Thank you, *because it is all so wonderful. And when you say* Thank you, *now the Heartmath guys are telling us that our hearts open up when we are expressing gratitude. And they put electrodes on people, and they test them, and they find that the waves that come off their heart are in a perfect relationship, which we would call 1.618, which is the fibonacci number, which is the Golden mean, which is the way that Nature is constructed. When we say* Thank you, *we become perfectly beautiful. When somebody smiles in happiness their face assumes the ratio of 1.618. And we notice that person is so beautiful because they are smiling all the time, they are happy, they are expressing their gratitude and happiness. And so there is this marvelously simple thing we can all do, and that is to remember to say* Thank you, *a lot."*

And so gratitude is this doorway that can open the heart directly. It is very simple. We can remember to say *thank you.*

Gratitude in the heart overcomes the chaos in the mind. Gratitude is an anchor that we can drop into a stormy sea that reconnects us with who we are. Learning to dwell in, and return to gratitude is a practice. It is a work, an effort. Because so many things around us are broken.

We don't require great insight to see what a mess the world is become. And our biology, as we've noted before tilts in the direction of 'What's wrong,' because this negativity bias has been important to helping us survive. It is easy for me to become righteous in my indignation, to exercise and indulge complaint, to feel that I have been somehow wronged. Even as I tell you this my internet cuts out, this laptop seems to be failing on me, and aggravation rises up in my chest– the feeling hurts my heart. It is like a spearpoint digging into my chest and I have to bring my attention back to what I am telling you. Find gratitude.

I pause, take a deep breath. I survey the kitchen, let my eyes come to rest on a bowl of lemons gathered from the tree in our yard, and, visualizing the tree, I thank it for this winter harvest. There are two tapered candles burned halfway down in geometric holders sitting beside them, and last night, at dinner with my wife and daughter I lit them, so we three would be bathed in softer light, and turned down the overbright kitchen lights. I am grateful for this soft flame that brought us closer together. It is an hour before dawn yet, and I am grateful for the quiet in the house, my family sleeping on the floor above, this cup of green tea that I sip in the stillness. I walk to the back of the house, open the patio doors, and sniff at the weather. The season is changing. Winter is loosing its grip, I can smell the rain coming. There is a bit of wind. I can be grateful for my breath– what a simple thing– and this body I am wearing, that feels so deeply. And there I have picked it up again, the thread of gratitude, returned to it after this agitation. This bringing back the mind, this reining it in, this pointing it back in the direction of something that is useful– this is the province of mind training. In the Brazilian tradition of which I was a part for 15 years, they said, "The Mind is a Hunter." And I like the simplicity of this. It is true.

The mind is relentless in its pursuit of *something*. Anything. But we get to direct this hunter, point him or her, at what we are seeking to pursue. And it is good for me to remind myself that the end goal of pointing myself at indignation is more indignation. The harvest of the object that we pursue is more of that object. It is better to hunt gratitude. It is better to pick the hunter up, when he begins to stalk anger, rage, indignation, and put him back on the trail toward gratitude.

A practice that I find useful in this regard is to recall, at the end of the day, before I go to sleep, all of the things that I am grateful for from the day just completed. I don't write them down anymore, but some people do. This is a way of tracking back through the day and threading, like beads on a necklace, all of the things that I want to open my heart towards, all of the things that I appreciate. Yesterday I taught my first spring class, had a wonderful conversation with a colleague, cleared deadwood in the forest. I received loving advice from a friend. I struck several items off my to-do list. I listened to a marvelous podcast. I watched Federer and Nadal play a set of tennis I had never seen before. I had dinner with my family. And I rested with our dog, who fell asleep with his head on my chest. A pretty long list.

Sometimes it takes me a while to get to gratitude. We returned earlier this month from a trip to Maui that I've come to calling an initiatory anti-vacation. It unfolded as a series of minor catastrophes, the most stressful trip I've possibly ever taken. I went from viewing it as a vacation, to viewing it as a work trip (I had to go back to work three days into the trip, and worked from days three to eight out of ten), to just throwing up my hands and surrendering – I remember the moment I did this. I was sitting outside of a coffeeshop in Kihei in the old Prius we had rented, when I just gave up any sense of control or outcome– realizing we were fastened into a ride that was just going to work me. At that point I just let go trying to classify the trip, which was something I had been doing to recalibrate my disappointment. The trip was, viewed macroscopically, a catalog of near-misses, crises, minor catastrophes. I thanked God the moment our plane touched down at home. And yet, even in this chaos there were many moments– visiting a sacred garden as we descended from upcountry when we found ourselves in an outdoor labyrinth where the quality of sunlight and the immanent touch of the Divine was so present it came over us like a balm. Moments snorkeling with my daughter in a nature reserve off a lava field at the south end of the island where we were free-diving several hundred feet off the coast in the presence of dolphins and a breaching Humpback whale. The moment I tasted the first fresh papaya on the island...that it became worthwhile to string these moments together.

Seed pearls on a string amidst shards of glass.

From a certain distance, I will likely find that the challenges themselves are something to be grateful for: that each one teaches me something about myself. These were wounds that may have been partly self-inflicted, our catalog of problems, and as I get far enough away from them, and they are not so fresh, it will be necessary to return to them to better understand what I need to change in myself. In this way, I suppose, it is possible and necessary to be grateful even for what hurts. Pain, as a signal, is always meaningful. I believe we have gotten into such trouble with ourselves in part because we want to make the pain go away. This is the impulse at the root of the opioid crisis– and it is appropriate for modern civilization, which is the author of such extraordinary harms, that our addiction is to painkillers. We (the collective soma in modernity) don't want to face what we've done. We'd rather dissolve into numbness. Lost in a chaos of our own creation, we'd rather just go to sleep. Yet the cure for the pain *is in the pain*. The way out of the chaos is through the pain. Not losing ourselves in it, not drowning, but traversing it. Moving through.

Pain is always information. Adopting gratitude toward the pain means that we view it as relevant information. That we recognize it is here to tell us something. That it is not without meaning. What hurts needs attention. What hurts directs us to slow down, to notice, to understand, to be compassionate. What hurts tells us what needs to be healed.

Reading through this section, and coming to its conclusion, there are some who might chastise me and say *Gabriel, you've talked about all of these doorways to the connection system, and yet you've left out the doorways that are so obvious: relationships and nature*. And of course, this is true. I found myself reflecting on this the other day. This could have been a book entirely about turning on the Connection System through relationship. I might not be as qualified to write that book, but someone surely is, and should. And for many of us, and for so many historically, this has been the story of their awakening into connection. Healthy relationships, long marriages, great friendships– these are the neurobiology of connection in pure form.

And surely there is another book waiting to be born, or that exists already, that catalogs the neurobiology of connection through our hardwiring into relatedness with the Living World. Through tracking, bird language, meditation in nature, forest bathing, sitting beside still ponds, through mountaineering, rockclimbing, surfing, and sailing the open sea. All the ways that we can build ropes of relating with Nature herself.

Or another book that looks at the neurobiology of connection opening through creativity. Through learning to find and inhabit flow states. The moment your fingers touch the keys of a piano, or your paintbrush hits canvas, or your chisel strikes stone.

All of these and more are the language of connection. Even language itself–there should be a book that addresses the deficits in English, my mother tongue that is filled with lexical gaps, places where she is missing the many words we need to convey our experience of relatedness. Yet this book, gentle reader, holds its focus on the physiology, noting, as it does that this is only one of the layers of context in which we are ensphered. Keeping its gaze pinned to the body we are wearing, to the physiological portals whereby it opens along the origami lines of its intended organic unfolding.

I bring your attention, before we close this chapter, to the gestural shape of gratitude. Look at the photograph that opens this chapter. How beautiful is that? The heart extends out through the arms entire and into the palms. Hands are turned, like radar, in the direction of receiving. We point this listening apparatus of a heart, which fills our chest entire, extends all the way into our palms, in the direction of what we are receiving. We open to experience, through this gesture: the chest is soft, the eyes are closed, the face available: the posture is of receptivity. We embrace it all.

This gesture is of such great openness, such profound availability, that we must be discerning in where we point it. We must reserve this openness for that which is truly life sustaining. But when we find it, something toward which we can fully open–something life-giving, life-sustaining, something worthy of our heart's full acceptance–let us open toward it. And that is grace. It is a prayerful attitude, yet the shape of it, the gesture, is *of being filled*. Somewhere between the ant and mountain we are small enough, and large enough, to be filled by the presence of Spirit, to be blessed by the spark. To have our inward spark fanned by the breath of Life herself. And this is the gesture we stand into. When the medicine woman comes with the Eagle feather fan and the sweetgrass incense, this is the posture we assume to allow her to fan us with Life itself. Brother Hawk calls. Father Sunlight brightens. The heart warms. Practice this gesture, my friends, with emphasis on its inward stance.

DOORWAY OF
INTEROCEPTION

CHAPTER 37

SETTLED INTEROCEPTION

There is some risk, I believe, in describing these processes as though they are linear only. As though safety leads to connection, which leads to interoception, which leads to intuition, which leads to relatedness, which leads to enduring wellbeing– describes some kind of road. These are not merely a set of doorways, they are not merely progressive. This process does not unfold merely in serial fashion, and it is not one time through. Yet there is some elegance in viewing it as a sequence that repeats. By this point, to have made contact with each, or any of the portals to the connection architecture, you will have to have been interocepting.

To make contact with the portals is to feel yourself from inside, or to feel yourself in various ways from inside, or to feel *through* yourself in various ways from inside. But let's give some special attention to this process of interoception *as a process*, because of its singular importance to our wellbeing.

Interoception is, straightforwardly, the collection of inward multi-sensory awarenesses that allow us to answer the question, "How do I feel?"

Interoception allows us to answer the question, "How do I feel?"

When someone asks you this, and you bring your attention inward– to sensation, pain or its absence, emotion, the felt sense[1] in the body, the inward flavor and texture of your experience– interoception is what is providing the answer.
At the simplest level, to be aware that we have to use the bathroom, that we are thirsty, that we are hungry or full: all of this depends on interoception. It depends on being able to feel ourselves accurately from inside.

The neurophysiological substrates of accurate interoception are deep within the domain of the autonomic. In aggregate, this

1 See Eugene Gendlin's pioneering work on the felt sense, a term he coined to indicate the confluence of sensation, emotion, and memory.

domain of inward sensing constitutes a sixth sense[2], yet unlike the other five extero-senses that you know, interoception, tragically, is not ever measured in pediatric medicine.

I'd like to bring our attention to this profound oversight for a moment, because of what it says about how we understand ourselves, how Cartesian we remain. While modern people have developed literally thousands of measurement tools to evaluate cognitive capability–our capacity for thinking–while nearly every single person you know in the world has probably had their IQ tested and been subjected to a battery of cognitive evaluation since their schooling began, the neurophysiological substrate of self-awareness has, for most people, never been tested at all. Not once.

If I cannot feel myself from inside accurately– if the internal rhythmicity of my bodily processes are not coherently synchronized to deliver to me accurate and timely information about the internal state of my body– this foundational deficit cascades forward through my life.

If I cannot feel myself accurately inwardly, I don't know when I have to pee or poop, cannot feel if I am thirsty or too hot or too cold. I don't know if I'm too hungry, or too full. In the absence of this awareness, self-regulation at a physiological level becomes very challenging. I cannot nourish myself if I don't realize I need food, relieve myself if I don't realize I need to use the bathroom, or warm myself if I don't know I'm too cold. My body, which ideally would be the foundation of my awareness, is opaque to me. It can feel mysterious, other, alien.

The neurological systems that transduce pain and temperature also encode pleasure. *Nociceptors*, which is the unfortunate name given to these sensing receptors (unfortunate because it means *pain receptors*) are the same receptors that generate the experience of tactile pleasure in the context of safety and

2 See Stephen Porges, PhD's beautiful article: *The infant's sixth sense: Awareness and regulation of bodily processes.*

connection. So the inability to accurately detect pain makes it hard to detect pleasure. Absent either of these orienting signals, the compass for my navigation toward safety, connection, and pleasure versus danger, disconnection, and distress cannot find its true North.

Social interactions become mystifying. I don't know who to move towards, or who to move away from. The foundation of social cues, which would register in my body as deeper feelings of safety and connection, or feelings of threat, are not clear to me. As social interactions become confused, my own ability to self-regulate and co-regulate with others becomes compromised.

From the substrate of the autonomic, our self-sensing is compromised. This compromise leads to sub-optimal relational sensing, which leads to challenges with self-regulation and co-regulation. Now I am deviated far from the substates of safety, connection, and regulation that are required for optimal cognitive function.

We have created thousands of tools to measure cognitive function, but have failed to measure the geologic substrate on which clear thinking depends.

We've covered the peak of the iceberg with flags, failing completely to register the 90% below the waterline. Those phenomena deep in the Autonomic Nervous System that create the context for clear thinking and self-sensing to emerge.

What trauma, the way that most people think about it, takes away from us is the ability to feel ourselves from inside, or to *enjoy* feeling ourselves from inside. This vacancy of feeling occurs at a neurobiological level because when our bodies get shifted into states of lifethreat we lose interoceptive awareness. Body parts go offline. Moving into states of lifethreat activates the continuum of dissociation. We lose a sense of our boundaries, and awareness migrates out of the body– a process governed in part by the release of endogenous opioids, and we lose this ability to feel ourselves from inside. As we shift into shutdown, move into lifethreat, we lose the firmness of our deepest inward core. The muscles of the transverse abdominus, which form the most inward walls of the cavity between our diaphragm and pelvis lose tone. We stop being able to feel the solid dimensions of our core, delineated at the southern end by the pelvic floor, at the northern end by the diaphragm, and walled on all sides by these deep intrinsic muscles. The chest grows concave, the back collapses, and we fall inward, unable to hold our heads up. These postural adjustments are twinned by a lack of inward feeling awareness. We grow numb, vacant. Interoceptive awareness disappears.

As this interoceptive awareness evacuates, as we cannot feel ourselves from inside, we lose a feeling for who we are. There is an identity component to this. Dr. Ruth Lanius[3], who has spent decades studying post-traumatic stress, teaches us about the way that trauma deforms our sense of self. For people with early developmental trauma, it is common for them to say that they 'do not know who they are.' When people develop post-traumatic stress disorders later in life, it is common to hear them talk about 'losing themselves', or 'losing access to their authentic sense of self' or 'no longer knowing who they are.'

When we get shifted out of the liquid water version of ourselves–or when this liquid water version of us does not get to develop in the first place–we have a hard time knowing who we

---

3 Ruth Lanius, MD, PhD, Professor of Psychiatry is the director of the posttraumatic stress disorder (PTSD) research unit at the University of Western Ontario. She established the Traumatic Stress Service and the Traumatic Stress Service Workplace Program, services that specialized in treatment and research on Posttraumatic Stress Disorder (PTSD) and related co-morbid disorders. She currently holds the Harris-Woodman Chair in Mind-Body Medicine at the Schulich School of Medicine & Dentistry at the University of Western Ontario. Her research interests focus on studying the neurobiology of PTSD and treatment outcome research examining various pharmacological and psychotherapeutic methods. She has authored more than 150 published papers and chapters in the field of traumatic stress and regularly lectures on the topic of PTSD nationally and internationally. She recently published *Healing the Traumatized Self: Consciousness, Neuroscience, Treatment* with Paul Frewen.

are. And this is not a cognitive phenomenon. In the absence of safety and connection scaffolding grounded interoception, we do not *feel* like ourselves.

In recent research, Dr. Lanius found that people with Post-Traumatic Stress are no longer able to light up the Default Mode Network, the part of the brain that activates when the mind wanders at rest, and seems to be connected with our autobiographical sense of self. Deviated from feeling like ourselves by trauma, our brains no longer have access to the passive library of who we are. Interestingly, her research also showed the the network will light up again when a person is experiencing high-level fight-or-flight activation and raw affect–when they are experiencing adrenalized fear, anger, or other very strong emotions. She interviewed people with Post-Traumatic Stress injuries who were aware that they picked fights, or engaged in highly risk-taking behaviors (shoplifting, etc.), in order to feel themselves again. Fascinatingly she discovered that these networks of the recognized sense of self turned on under intense activation by drivers in the PeriAqueductal Gray (PAG), an area of the midbrain associated with raw affect. Yet this sense of self, she noted, is a survival sense of self. A *threatened* sense of self.

If we are shifted by trauma out of the passive experience of who we are, we seek it again in high-intensity activation. We engage in risk-taking, we pick fights, we put ourselves in situations where our safety is compromised. These moments then are the ones when people–removed from themselves by the deviation from a baseline in safety and connection–removed from the liquid water version of themselves–feel alive again.

I invite you to pause and contemplate what this means, in a culture where most modern humans are far sicker than we realize. The multi-generational impacts of trauma are much broader, and more extensive than most of us understand. Is it not possible that this phenomenon, that we only feel ourselves in relationship to an aspect of traumatized identity when we are pushing against it explains in part why we are so ecologically destructive? Unable to experience ourselves as part of the Living World, we only feel our sense of self in relationship to Her when we are destroying. Is it not possible that this explains our hunger for domination? Why we oppress one another so horribly? In the absence of being able to feel ourselves sociologically, we seek domination experiences because they are the only times when we feel alive? Are we not collectively enacting this neurobiological sequelae of undiagnosed collective traumatic stress disorders? I cannot help but think so.

And so this ability to feel ourselves from inside, the experience of what I will call *settled* interoception (for as we are discussing, we can also interocept in high-energy states, we can feel ourselves and even feel alive when we are feeling ourselves in fear, or in anger) is actually a doorway to non-cognitive ways of knowing, ways of knowing that will not open for us if we are accessing interoception through threat states.

What is different about settled interoception, from interoception derived at acute levels of activation and raw affect, is that when we feel ourselves in these high-energy states, it amplifies our distress. To feel our anxiety makes us more anxious, to feel our anger makes us more angry. Our identity gets tangled in with these activation states. We feel ourselves as anxious, as enraged. And these loops entangle with narrative. When I feel myself angry I recycle, in my thinking, in words and pictures, what has made me angry. When I feel myself anxious, I recycle, in my thinking, in words and pictures, what has made me anxious. And in this manner, interoceptive awareness of fight/flight states makes them more acute, and our sense of self more identified with them.[4]

So what I'm saying in this chapter, really, about interoception, is that if we want it to feed us, to nourish us in a manner that will re-inform deeper embodied knowing, deepening access to

4 This is why, at a foundational level, inwardly-oriented mindfulness practice is generally contra-indicated when a person is experiencing fight/flight states.

non-cognitive ways of knowing, and the deeper experience of more relatedness to *All That Is*, the doorway that must precede its opening is of connection.

When we come to interoception through the doorway of connection, what it gifts us is contact with a body we can enjoy being in. What it gifts us is the joy and pleasure of feeling ourselves from inside in a way that feels good.

This is very simple really.

Since attention always potentiates experience– always makes it stronger, deeper– bringing attention inward when our connection system is online is what ensures that the inward tilt of this felt sense moves us in the direction of feeling more fully alive, feeling more fully like who we really are.

Connection states are not simply relaxation states. They include a continuum of settled, calm, and relaxed states where there are elements of stillness as well as higher-energy states where we are still connected to a sense of safety. Interoceptive awareness in the states of safety and stillness is the bodily felt experience of many meditative practices. All body scanning takes place in this domain. Yet connection states, because they are describing a pulse, also include states that bring online the vitality of mobilization without the activation chemistry of danger, and so range into *Play*. Play is mobilized and vital. Play is active and energized. And here too, as we cultivate interoceptive awareness, arises the opportunity to know our bodies in their full vitality. To feel ourselves fully from the inside in the expenditure of breath, the fullness of movement. We can interocept in the midst of great physical effort, and bring a rain of enjoyment to the body. The crucial element here is having the connection circuitry online.

My favorite image of pure connection is two friends having tea or coffee, close-talking animatedly, and leaning in. Here is the classic face/voice expression of sociability. We might be eating

SOCIALITY CHURCH- AKA, THE CAFÉ

or drinking (these are both social, and explain why good house parties end up in the kitchen). We are sitting close enough to note the micro-expressions in our friend's face, and the micro-modulations of their voice. From this distance we can see them, in the *sawubona* sense of seeing. We can really witness them. *Sawubona* is a Zulu word that means – *I see you*. Yet it conveys the notion that this seeing is not just me. My ancestors see you. And in this seeing I acknowledge your humanity, and that I am accountable to you by beholding you thus. Pure connection is this friendship with the gaze of *sawubona*. From within this framework, being seen, we are safe to be who we really are, to bring all of us forward, to leave no parts of ourselves at the door, to tell our deepest and most important and most vulnerable stories, and our friends, the ones we let in to this inmost circle of who we really are, are the ones we trust to see us, and to catch our stories. Take a moment to think about who in your life you allow to see all of you, and who you allow to receive your most sacred stories, and you know who you experience pure connection with. Knowing this is also a template for how people generally want to be received.

Now, imagine that you and your dear friend walk outside, into the forest, or onto the beach. As you begin to move, your Movement System begins to activate. This vitalizing increases your experience of vivid aliveness, which colors into joy. This joy is not the same as the connectedness you were just experiencing. It is more effervescent.

This joy may feed your sense of playfulness. Your desire to run, to move freely, to race, to dart, to soar. Now you are moving into the realm of play. At this point, your Movement System is growing stronger, *yet in service to your Connection System*. It is capacitizing your Connection System, expanding its range. There is no fight-flight chemistry released.

Let's say that you and your friend, on the beach, both take off your shoes. Let's say that you both, out of an abundance of joy, begin to run on the hard-packed sand at water's edge. Now imagine, for just a moment, that you and your friend look at each other, and your friend begins to run faster. You, laughing, accelerate. In this moment, as your exertion increases, you begin to shift into a mode beyond play: *competition*.

Your friend looks back at you, smiles, and accelerates dramatically. Really turns on the afterburners. You do the same. We have now entered the territory of *elite performance*. Both of you smiling, beaming at each other are now running as fast as you can, shooting down the beach, but you are not getting tired. Your legs are blazing but there is no weight. You are in a pure flow state.

At some point you stop, both of you, panting, breathing hard, laughing uproariously. You chortle and snort and guffaw. And then you look back and realize that you've run, barefoot, so far and so fast that you can't even see your shoes. Behind you, on the sand, are two parallel tracks, the imprint of these racing strides, lapped at by the azure.

Okay, let's pause.

This entire transit from coffee to beach racing is happening within the scaffolding embrace of safety and the continuum of connection. This range is the domain of vitality. Every one of these things is good for you. Technically, as mobilization ramps up, your body is experiencing stress. But this is *eustress*. Good stress. And the Movement System that undergirds it, which can be harnessed by fight-or-flight, is actually subservient here to connection and performance. There is not chemistry of adrenaline and cortisol releasing beyond what is required to move. The mobilization, the movement, is pure play, pure freedom. These states of play, competition, and elite performance make you stronger, more resilient, more capable of staying in safety and connection in high-energy and high-stakes situations. We are able to train your resilience because you have a neuroception of safety.

We have moved from CONNECTION to ENJOYMENT to PLAY to COMPETE. Your ventrality has been steady, your neuroeptive center of gravity in safety, as mobilization increases.

You take a few minutes to stretch on the beach, in your visualizing of this, because your body has been exerting heavily. Elite performance happens in this domain of *compete*. Elite performance has connection circuity on board. If it doesn't, it degrades, sometimes astonishingly, into fight energy.

Zenadine Zidane is one the greatest football players of modernity, yet will probably always be remembered for the vicious headbutt he delivered to the chest of Marco Materazzi, an Italian player who was heckling him near the end of the championship game of the 2006 World Cup. France and Italy were locked into a final battle for the championship, and France seemed destined to triumph. Zidane, France's center halfback, had been the dominant force of the entire tournament. He seemed to orchestrate every game. When he got the ball, it looked like it was attached to his foot by a string. Italy, knowing this, spent the entire game harassing him relentlessly, trying to diminish his potency. The entire match– the most important of the four year period, in the world's most popular sport, on the world's largest stage, with hundreds of thousands in attendance, and tens of millions watching–was probably the single most watched sporting event of that entire year.

This incident occurred in the second period of overtime, 110 minutes into the match. The teams were deadlocked. Materazzi entangles Zidane briefly with an arm, as both look in the direction of the ball. They disentangle, Materazzi says something to Zidane, and pats him on the back. Both players begin to move down the field, then Zidane jogs forward so he is in front of Materazzi, turns toward him, and steps in as he lowers his head. Like an enraged bull he strikes Materazzi in the chest with such stunning force that Materazzi drops to the ground, a tree felled in one stroke.

In high stakes competition there is such profound mobilization activation online that if connection withdraws it can surface as naked aggression. This is why fights break out during football, basketball, and hockey games. The neural restraint on all-out war is the connection system and its chemistry. When this withdraws vasopressin, adrenaline, and cortisol flood in and we are in fight. Like a tree felled, Matterazi ends up flat on the ground and Zidane tarnishes his reputation permanently.

DOORWAY OF INTUITION

CHAPTER 38

INTUITION

In May of 2018, I was sitting in Christopher Kuntzsch's backyard, in Wildcat Canyon, trying to figure out whether or not to buy the new townhouse my wife and I had fallen in love with but couldn't afford. And I was trying to figure out why I was not excited about the business I had spent eight years pouring myself into. I was sitting sideways on a reclining lawn chair under the span of several healthy Coastal Live Oaks, and staring up across the bouldered meadow behind his house. I had taken off my shoes to get my feet on the earth, and I must have had a rather agonized expression on my face. Christopher, who was my business/life coach, was sitting obliquely to my left. I was explaining to him that I didn't know how to make the decision before me. Part of me wanted to move forward, and part of me was scared. I felt like I was on the threshold of diving off a cliff, and once you do that it is too late to change your mind.

We had begun the session, probably fifteen minutes prior, with an invocation, and then an awareness practice, and because of Christopher's trustworthiness and extraordinary acuity in tracking my states, I felt both safe and relaxed with him. I felt comfortable naming the dimensions of the problem that I wouldn't have been able to articulate to someone less attuned. At a certain moment, I leaned forward to emphasize a point, and my attention got called to something near the ground. I looked down at my feet. And then something strange happened. My feet awakened and told my head to shut up. Just like that. To my head, my actual feet said, *Be quiet*. Clearly, firmly, in English words. This was odd enough that it startled my thinking into a sort of stunned silence. My feet began to fizzle and hum with energy, and I found that by adjusting my posture, just slightly, I could dial in to the knowing that was coming from them.

Christopher, who was watching, noticed something. *What just happened?* he asked me. Somewhat sheepishly, I told him that my feet had started talking to me. *Ok*, he said, like this was a normal occurrence. *What did they say?*

I told him. *What are they saying now,* he asked me? I closed my eyes.

What came next was very clear. *Regarding your life direction*, my feet told me, *don't listen to your head. It knows nothing about walking. Listen to us. We are the ones who take you where you want to go. Trust us to take you where you want to go.* I opened my eyes, and we looked at each other, because I had said this out loud.

Alright, Christopher said, *Where do your feet want to go?*

Toward the new house, I said. No hesitation whatsoever.

Okay, he said. *Then let's do that.*

I'm sitting in that new house right now, typing these words. That decision was catalytic and initiatory. It scared the shit out of me when I made it, yet I haven't regretted it once. Not a single time. It set in motion a series of events that have led me very clearly to this present moment. The necessity to pay the mortgage led me to transform the business dramatically. The need to transform the business led me to set down what I thought it was, and ask anew. My feet then began walking me in a novel direction, one that I could not have predicted, one that I had to find and feel out as I stepped forward, the way that you feel the shifting stones of the riverbed beneath you as you cross a fast-flowing stream; the way you track an animal that is not here. A direction that was not on my mental radar at all. One that I couldn't see, because it was around a series of corners I didn't know were there.

Several years later, as I interviewed him for a global resilience summit, my friend and mentor Tiokasin Ghosthorse, of the Cheyenne River Lakota, described the Lakota language as a non-mathematical quantum mechanical language of intuition. And he taught me the Lakota word *Anamagoptanye*, which translates roughly as *the earth is constantly uploading information to you.*

When he told me this I remembered, immediately, that day in Wildcat Canyon, behind Christopher's house. Safety is the doorway to connection. Connection is the doorway to interoception. Interoception is the doorway to intuition. All the steps were there. I felt safe. I felt connected. I opened inwardly to sensation, my mind grew quiet (or perhaps more precisely was startled into silence), and then something happened. Was it a coincidence that my feet were bare? It doesn't seem like it. It's harder to feel the earth when there is rubber interposed between you and she. I was sitting barefoot, receptive, in need, when this awareness came online. A visceral signal. And clearly I am not the first person this has happened to, because the Lakota have a name for this way of knowing. I simply for the first time in this life, opened the doorway, accidentally, to a way of knowing that exists, but of which I was unaware.

Like opening our eyes for the first time... or forming words...or hearing a range of frequencies whose warble suddenly becomes detectable out of thin air...I activated something in my innate biology that I didn't know was there, yet exists. This doorway of the soles of the feet, which I did not, at the time, know was a doorway to the Connection System, opened– and suddenly I was in contact with a novel way of receiving information. A new facet of intelligence. An entirely new sense.

If I not been with an attuned guide, someone who reflected back the importance of the non-cognitive way of knowing presenting itself to me, I might not be here writing these words. If Christopher had laughed, stuck in cognition, and dismissed this– *Feet don't talk, you are hallucinating*, my guess is that the doorway would have closed. My ability to open to this doorway, which was outside of the realm of what I understood at the time *could* exist, hinged in part on his taking it seriously. I acknowledge this to honor Christopher, as well as to point out that most of us, when we are little, catch glimpses of these other ways of knowing, yet that if we are not in contexts that validate them, often we do not open these doors.

Returning to the Lakota word for this, *anamagoptanye*, and the recognition that others had accessed this awareness before, when I pressed him on this point, Tiokasin told me that there were many different kinds of knowing that could come up from the earth, depending on who you were asking.

Was it the knowing of the earth herself?
Or of the grandmothers who were your ancestors?

My jaw dropped. There exists a doorway to knowing in a part of my body I was taught was mute: yet the soles of the feet are portals. And through this doorway, this portal to connection, we can communicate not merely with the earth, but with our ancestors. *Specific* ancestors.

I don't like the word *intuition*. I love intuition, the faculty, but I don't like the word. And I don't like the word because it is a blanket term that obscures as much as it reveals. When we say something is intuition, usually that just means that we know something and we don't know how we know it. But I find this lazy, because the knowing of the feet, the way of knowing of the feet that I just described? It is different than the knowing of the hands, or the guts. Intuition is the collapsing of myriad ways of knowing into a single thing, which they are not. There is not *intuition*, singular. It is plural, and it is also differentiable into many distinct yet related ways of knowing, some of which have been practiced and refined in other cultures to a degree we would find unbelievable, conveying information about realms we are taught do not exist and to which we do not have access. Original wisdoms are seeking entry all the time. They can come to and through so many doors, and these doors are differentiated.

Do you ever know something and not know how you know it? Does this happen to you? I certainly hope so. It means you are not dead. That's what it means. Your body knows things all the time that your mind, your thinking, doesn't understand. Knows things your mind can't, or won't, know. It knows things you don't know you know. But then, the next important question. When you know something, and you don't know how you know it, do you listen to the knowing?

This is a provocation, now, that I'm giving you. Because if you can feel the tilt of my words, where I'm leaning, you can tell that I'm going to suggest that you listen when this happens. But most of us don't. Most of us have been trained not to listen. This is one of the ways we are acculturated to be alienated from ourselves.

The most stunning story I have ever heard about this was shared in a class I taught a few years ago on Zoom, during the pandemic. By way of context, I will note that for a period of time I was receiving clients who were leaving cults. I didn't, in any way that I am aware of, put up a shingle that said, "Cult-exiting available here", but for about six months, probably because I had just succeeded in doing it myself, a non-trivial number of people who entered my practice for individual sessions were in the process of deprogramming their neurobiology from a cult. When I say cult, I need to tell you that I'm using this word somewhat loosely. I think modernity is a cult, a death cult, to be precise. I'm not talking about Jim Jones-style situations only.

Most of these were religious cults, which I will define as spiritual systems with increasingly rigid structures of belief that generally involve some sort of messianic component. People were coming to consult with me who had grown up in cults, sometimes in the second or third generation of their family within them, who had left, but were still de-conditioning their nervous systems. It was in the context of this that I heard this story in my class.

Anyway...the woman telling the story was, at that point in her life, traveling through Europe and staying in youth hostels. I'll call her Maxine, because that was definitely not her name. Somewhere in Italy, I believe in the Italian Alps, she become acquainted with another young woman from the United States, who I'll call Lily, who had been raised in an Evangelical church. They became fast friends, and travelled to several hostels in the mountains together.

love with, and was learning and practicing with great enjoyment. I can picture them in the room of their hostel, these two young women, the Italian Alps in the background visible from a window. Hostels in Europe are usually simple, and clean. There is something nearly monastic about the surroundings, in my imaginings. About five minutes into the practice, which consisted of Lily following Maxine's movements, Lily sat down abruptly, curled herself into a ball, and started rocking back-and-forth. Maxine, whose eyes were closed, didn't notice this immediately. When she opened them, Lily was curled up on the floor, holding her knees, a very strange look in her eyes. Maxine could tell that something was wrong.

What is it? she asked.

At first Lily didn't want to say.

Maxine sat down on the ground, quietly reassuring her, and eventually Lily spoke: *It is against my religion*, she said, *to feel the energy moving through my body.*

It is against my religion to feel the energy moving through my body. This sentence, when I heard it, struck me as the clearest expression of someone being conquered by an Empire I had ever heard in my life. What I heard was–*we have outlawed your direct experience of your own interiority.*

The Roman Empire fell long ago. Its buildings have crumbled. But we have rebuilt the empire inwardly. Tiokasin, in a later seminar, taught us how both the Romans and the Greeks forbade mourning. Empire wants to extract and harness your energy. It doesn't want energy it could potentially extract expended in the direction of your sovereignty.

If you cannot feel the energy moving through your body, it means that interoception has been outlawed. If you cannot interocept, you have no intuition. If you have no intuition, you cannot leave the cult. You will always be a citizen of Imperial

Rome.

If you have no intuition, you cannot develop your own morality, and you cannot experience relatedness, or come home to yourself. This makes you the ideal extraction forum for Empire. To be deprived of your inwardness, prohibited from feeling the energy moving through your own body is the clearest interdiction against indigeneity I have come across, in a way.

Remember, *I feel myself, therefore I am?*

In Lily's worldview, this had been outlawed. Authority was totally external. Sovereignty was entirely gone. And yet...when the mind trains us not to act on the things we know but don't know how we know? Is it really so much different?

What we have been acculturated into, most of us, is that cognition holds the reins. Cognition steers, tells us where to go, what to do. If we don't listen to it we are told that we are *irrational.* That our passion is getting the way of our reason. Et cetera. Men, a couple of generations ago, liked to tell this to women. Men, we had some very interesting theories about how this happened. For example, we taught that when a woman's passion got ahead of her reason, this was because her uterus became detached, and began flying around the inside of her body, like a canary in a cage. This was what made women hysterical. You will note that the root of hysterical is the same root as hysterectomy. To have passion outweigh reason was to have an internal migratory ambient uterus. This is, of course, absolute horseshit. Yet it was prevailing medical wisdom for a non-trivial amount of time. Not far from the medical wisdom that had doctor hold the newborn baby upside-down, or inject her immediately on arrival into the earth situation. Where is the best place to hide your domination paradigm? Inside the people, so that they dominate themselves.

To reclaim our intuition is an act of humility. It's a way of saying that possibly we don't know everything. That the Universe itself might be wiser than we are. The thing about cognition, as a master, is that the way it works is to operate as though it knows everything. So to yield to the call of something that we don't understand? To entertain the possibility that there are a plurality of ways of knowing? That our feet might know something our mind does not? That they might be able to lead us into our desired future better than our thinking?

A revolutionary notion to be sure.

DOORWAY OF RELATEDNESS

CHAPTER 39

RELATING AS A VERB

How do we know what we know? One of the differentiators of the human species is our accretion of knowledge, in various forms, and our ability to pass this forward. From the great library of antiquity at Alexandria in Egypt, all the way to ChatGPT, our species differentiates itself by our ability to aggregate and pass forward what we know. Modern humans are literally swimming through information. Drowning in it, at times. And yet there is a fundamental distinction to be made between *knowledge-as-object*, as edifice, as thought architecture that is accreted, and *knowing-through-feeling*.

The heart of this book, in a way, turns on the distinction between *I think, therefore I am*, and *I feel myself, therefore I am*. One is knowledge in thoughtforms, the other is the verb of feeling. The second formulation is, I would propose to you, a more Indigenous view. It regards our purpose in life not as a function of how much knowledge we accumulate (a transactional frame), but in our ability to feel kinship with one another, and all of Life, in ever more refined and beautiful ways (a relational frame). As we experience this kinship ever-deeper, our sense of ourselves as individual, isolated, and disconnected islands dissolves into a deeper experience of being part of something larger.

I remember being alarmed, when I was younger, by people with strong intuition. I remember feeling intruded upon by the knowings of others, about me, at times, without realizing that I was radiating information– state, emotion, thought, intent– from my face, eyes, voice, gesture, posture, stance.

We moderns feature ourselves islands: we don't grow up in a village, don't generally feel part of a band or tribe. Our sense of self is constrained, generally, by the edges of our physical body. And yet this incarnate isolation, this sense of separateness, is at the root of what is killing us and destroying the world. The dial on our individuality is turned all they way up, the boundary of it solid, while our availability to relatedness is turned all the way down. Separate, wounded by this terrible alienation from Source, we endlessly consume to attempt to fill the void created by feeling alone.

The journey we have been walking together, in this book, is a journey back into relatedness. We have been walking it through your Autonomic Nervous System, from safety, to connection, to interoception, to intuition, and now here, to relatedness. To spending more of our moments in feeling. This is the switch at the heart of this book. Most modern people live in their heads– live as heads on sticks, heads that happen to be hauling around a body– and what I am inviting, what we who do this work are inviting, is that you return to the fully embodied landscape of feeling.

The original exile is from this deep landscape of feeling. And most people remain exiled, because we are so traumatized that being in our bodies doesn't feel good. When we are stuck in states of steam and ice, it's no fun wearing a body.

And so my discourse with you, through these chapters, is an inquiry into how to wear a body joyfully again. The answer is through the neurobiology of connection. It is to find the kind of safety that allows you to open into connection in such a way that turning on your inward sensing landscape allows you to experience the groundstates of relatedness, which lead to the joys of village, and in the direction of enduring wellbeing.

This doesn't mean, of course, that you will not feel anxiety and rage and despair. Because to walk through life, probably since always, but particularly amidst the carnage of the modern world, as a feeling person, is to be accosted by such tragedy. In the last 24 hours I have felt anxious, angry, and despairing. Yet I've also had a whole lot of moments of feeling joy, feeling gratitude, and many more moments of simply feeling the nuanced and ever-changing texture of what it is like to wear the body I am wearing. Moments where, to borrow and slightly alter a phrase from Paul Kingsworth, "The fibres of the atmo-

THE FIBERS OF THE ATMOSPHERE HUMMED WITH SOME VITALIZING POWER...

sphere hummed with some vitalizing power."
These feelings are vitalizing because my Connection System is online, and it is prioritizing, and creating a self-reinforcing feedback loop for me that keeps bringing me back into connection and relationship. It is helping me dance into villaging with *All That Is*. It is a kettle of wellbeing, the fires of which I stoke, as a practice.

Because I understand this, because I understand that this is the source of my wellbeing, I make a consistent and moment-to-moment effort to nourish it. When I have a moment where I feel really good, I dwell in it at a sensory level. I sip on it like a cup of tea. And I village all the time.

To village (verb)- to move through the world seeking out timespans (micro-moments, moments, flows) of relatedness with All That Is.

I seek out moments of connection, scan for opportunities to drop-in with humans who are willing to play with me, be seen, be silly. I do this with babies and children and adults. I do it standing in line at the coffee shop and the grocery store. I will drop to the ground to visit with a toddler. I will buy coffee unexpectedly for someone standing next to me. I strike up conversations, like striking up a match, in unlikely moments and with people who do not expect this. I talk to birds. I talk to trees. I talk to stones. I talk to nearly everything.

Much of this talking I do not do with my mouth. I talk with the earth all of the time through my feet. I listen with and through my deep belly. I talk with and listen to trees through my hands. I talk and listen through movement. I talk with people through my smile, through my eyes, through laughter. I talk with them through many of my portals to connection. You can understand, I believe, what I am saying. I do these things because they light up my oxytocin circuits, because they make me feel good, and *I want to feel good*. I do these things because they make me feel less alone.

Every time I break into laughter with a total stranger, or play hide-and-seek with a two-year old, I am injecting wellbeing into my system. Every time I listen to a tree through the radar of my palm I feel my kinship with the earth. When I listen to the language of birds, through the portal of my ears, I understand my environment better and I feel myself related to the birds. I know that to be neuroscientifically true. And so I do it a lot.

People who know this about me sometimes smile when I'm coming. And I feel this. Together we are stoking the fires of connection that keep us all well. We bat this playfulness around like a ball through the village. Together we are tending the hearth of the village, its very fireplace, and whether their brains know this or not, our hearts know it.

I also spend a lot of time outside. As much as I can. If I do not feel well, in any particular given moment, more or less, I go outside. If I feel grumpy, or irritated, or anxious, I go outside. Or I open a window. Usually the moment that I feel the air moving across my face, the moment I step back into contact with the earth, I feel better. I built my office with an open wall, so that I cannot close the Living World out.

There are plenty of things in my life that I cannot control. But where I get to select the inputs my nervous system receives, I am very intentional.

I apply this attention to my relationships rigorously. Like most healthy trees, I prune back my relationships from time to time. I study my relationships through a lens of reciprocity. Who is giving back to me as much as I am giving to them. To whom do I need to give more back?

Where there are long-standing asymmetries, I usually back up. I figure with people I know, if the last five times we've talked is because I've reached out to them, it's time to give them a chance to reach out. I realize that sometimes, for example when someone is sick, or when we are care-giving, there is a long-standing asymmetry. I realize that in the dynamic of relationship, people pass through long spans of time when they need more support. I realize also that some relationships are inherently tilted in one direction or another. I don't look for symmetry with people I'm mentoring. I probably don't give full symmetry to my advisors. But with friends– people I'm horizontal with–I track the call-and-response. It doesn't have to be identical from both sides. But if the direction of flow is asymmetrical *most of the time*, out with the pruning shears.

I pruned the relationships with people who have messed-up boundaries a long time ago.

What this practice does for me is to create the space in my life to feel, because I have been able to build, construct, adjust, arrange, and organize a world that I live in where most of the time I feel safe and seen. I practice keeping shitheads at bay.

My initial goal was 51% of the time. And I did this as an act of both radical self-regard and radical service. Radical in the sense that I saw this rarely modeled. Self-regard in the sense that I did it as an act of love for myself. Service in the sense that if I feel well– if my own cup is full– I can fill the cups of others.

As a healing professional, I cannot give what I do not have. I cannot show up for other people in ways that I do not show up for myself. I contend that if you are able to feel safe enough to have your connection system turned on 51% of the time you will be well. It is that simple. I would reduce this to a definition. Wellbeing is the self-transforming flow of vitality that emerges spontaneously when your Connection System is online more than your defensive systems.

Wellbeing is the self-transforming flow of vitality that emerges spontaneously when your Connection System is online more than your defensive systems.

This is not simply because of the physiology, but because if this is true physiologically, you will create the conditions for wellbeing to flow through your life. You will feel well, and you will create the feedback loops in your life to continue feeding this back to you.

Because of what the Connection System is–your relatedness architecture–when it is online and functioning you see the world differently than when you are locked into the biology of defense. When the Connection System is online, you don't need an enemy. Spend enough time experiencing this, and your worldview will start to indigenize.

Our mentor John Stokes calls this the tracker's syllogism:

What you don't talk to, you don't understand. What you don't understand, you fear. What you fear, you destroy.

Working this in reverse then, he explains: *If we want to stop the destruction, we have to start talking to things.* He recommends the first bird you hear when you wake up in the morning. Yet this applies to everything. Talking to something or someone, greeting them– this is the entrypoint into relationship.

What we experience relatedness with is suddenly no longer *Other*. As soon as we have relationship with someone, they are no longer *Those People*. As soon as we have relationship with place, we no longer drop trash.

What you feel connected to you have a relationship with. What you have a relationship with you take care of. This is the pattern of our lives. Hopefully, the evolutionary process for humans is that as we live more and more in the neurobiology of connection, forging deeper and deeper relationship, we extend our care more and more broadly. This is what it means to be a real human being.

CHAPTER 40

ENDURING WELLBEING

It is said that the Kingdom of Heaven is inside you. But this is nonsense. What is inside you is the doorway that you can open to walk the path in the direction of Heaven. The *way* to the Kingdom of Heaven is inside you.

Can't feel yourself and you'll never find it.

The Kingdom of Heaven– and if we want to get technical about this, the Aramaic word Jesus used was actually closer to Queendom– is the Source context, from which we have fallen out of relationship. The reason that the way to it, the way back to it, is inside you, is because what we have fallen out of, fundamentally, is the experience of safety that opens the dooway to connection, through which we can experience inwardly (interoception) our intuitions of relatedness to *All That Is*. What we have been walking you through, this entire time, is the set of doorways that can lead you back home.

And I mean *Home*, home. Capital H, home. This is not a religious formulation, but it would be hard to deny that it is spiritual, because the sense of belonging we are seeking is indeed Sacred. A sacred sense of belonging. The bonds of relatedness that we long to experience, and that are actually required for us to be Well, are ropes of relatedness. And these ropes are made out of energy. They are woven from energy. We weave them through relating.

I told you, earlier in the book, about a friendship I created with a hummingbird named Flurry, and what happened when we established this thread of relationship. What we build relationship with, and then the way that we tend that garden of relatedness, is what walks us home to ourselves.

Let me distill where we have been heading into a single sentence. This book is about an approach to enduring wellbeing. And enduring wellbeing is yours if you can spend more than half of the moments of your life with your connection system online. It is pretty much that simple. There may be obstacles to getting the system online. But once you have access to it, if you can spend 51% of your waking moments with this neurobiological system, centered in your feet, your pelvic floor, your guts, your heart, your hands, your face, and your voice, you are going to be well.

In order to do this, you are going to have to feed that system. It is likely that, at first, you will find it useful to think of this as a kind of practice. You don't likely forget to eat on any given day, but the odds that you nourish your Connection System with the same diligence that you feed yourself are low. You feed yourself because you are hungry, so generally you don't have to tape up a lot of post-it notes reminding yourself to eat. Your body just tells you– *I'm hungry. Feed me!*

Yet we moderns are so profoundly starved for connection, so de-natured of community, so de-natured of relating, so de-natured of the tribe, of the village, of the hearth, of the original fire, of the ancestral pattern of communion with all, that we are starving and we have no idea. *Wake up! You are starving for connection. Feed thyself.*

Feed thyself connection, and you shall be well.

Nourishing your connection system is as crucial to your wellbeing as eating. I recommend that you eat connection all day long, until you are full. Then rest, and do it again.

We didn't come here to be alone. This doesn't mean that we are not distinct. It doesn't mean that we are not differentiated. It doesn't mean that we cannot have crisp boundaries. But it does mean that wellbeing is social in nature, and that we can and need to learn to open into connection and relatedness where it is safe to do so. There is no such thing as self-care, really. With mammalian nervous systems, it is community care. That's where the magic is. And thankfully, that community doesn't end with the human world. You are related to the entire community of life, which is reassuring, because have you seen the humans these days?

Our hearts are designed, and destined, to come into their fullness in the context of community. And in this way, the evolution of our homecoming journeys is to move from heart to hearth.

First, we have to come down out of our heads back into our bodies, and the fullness of our felt experience. We have to put the guts (the Grounding System), movement (the Movement System), and our hearts (the Connection System) back in their proper places as the fulcrums of our felt experience. We have to re-organize our notional construction of wellness around the reality that the full spectrum of connection states is the doorway to enduring wellbeing, and re-design our lives accordingly.

As we do this, and as our biology begins to settle and re-organize, as we begin to really put attention on feeding and nurturing our daily, felt, embodied experience of connection– of heart-centered, playful, and deep-bellied relating– we start to feel the real hunger for community. Because we are so deeply immersed in the extraction engine of empire, because relationships have become so de-natured and transactional, most communities are actually cults. Most communities, filled with people who have been trained from birth to extract, are not genuinely healthy, any more than most modern people are.

And it is here that we should turn our attention to our cousins– those in the human family who have been, for millennia, living in tribal affiliation– and who have learned to organize their societies more horizontally.

My friend and mentor, the extraordinary trans-disciplinary psychology researcher, Darcia Narvaez, PhD, changed the entire direction of my life with the following sentence: *Small Band Hunter Gatherers are the baseline of human normalcy.* The deviation of modernity from our ancestral baseline could be summarized crisply in the observation that the road we have gone down is one of domination, and this organizes spatially as *power over*. Our modern civilization has, as a result, developed along a vertical axis, with locations of exalted *power over* as we move higher on this ladder, and locations of extraction, sites of ritual and required sacrifice, as we move down. In tandem, we have organized the priority in our bodies to move our awareness up, higher, and higher, until most modern people live in their heads. This movement was accomplished in part by the murder of earth-based people, the repeatedly genocidal actions of empire that sought to destroy earth-based peoples, and chthonic forms of worship, and the subjugation of the feminine, and women more broadly.

God isn't upward. *Heaven* isn't upward. This is a lie promulgated by empire. A literalized mistranslation: propaganda. A lie taught by the same empire seeking to organize power relations vertically, and justify the extraction of life, labor, and resources from the earth, and those closer to the earth.

As we come back home to our hearts and our bellies and our bodies, and home to prioritizing a kinship worldview, we can take great solace in recognizing that this worldview is still present, and being lived by practitioners around the world. These practitioners are not part of some organized network of healers, and they might actually object to me calling them practitioners. I am doing so only to notice that they are engaged in certain kinds of relational practices.

I end this volume by pointing us, in the next logical step of this journey, which is from heart to hearth, from attending to cultivating the neurobiology of connection, to developing healthy neural architectures of the collective, to the Indigenous peoples of the world who are the true lineage carriers of these knowings.

Perhaps my intent, in this book, using the language of neuroscience, is to help you come home to a place where you can understand that the Indigenous worldview is the way forward and the way home.

I now invite you to take just a moment to imagine approaching a campfire. We are in a primal forest, an ancestral forest that has never been de-natured. In a clearing in this forest there is a hearth, and in the center of the hearth a fire: *the original fire.* Around the hearth a village is gathering. In the crisp autumn air, against the edges of night, the immense trees silhouetted, the warmth of the flames speaks to you in the aboriginal language of orange, vermillion; the fire dancing. Approaching the fire you realize you are cold–not freezing, but cold–and as you move closer people step back to make way for you, their faces warm, fire-lit, to welcome you in to the gathering circle.

And curious, yet cautious, you take off your mittens and turn your open palms toward the fire. You are still eight feet from her, but your palms are out now, open, listening for the first shimmer of warmth. And I want you to imagine the moment that you begin to feel it. I want you to put all of your attention in the world into the palms of your hands, to close your eyes, approaching the hearth, and allow yourself to begin to receive, ever so gently, the warmth of the fireplace, cackling at you in her many ancient ancestral tongues.

Think about how gently you can touch someone. Think about how gently you can draw the hair out of your daughter's face. Or how softly you lift the sleeping baby from his crib. Think about the delicate tendrils of electricity that crept through you the first time you held hands with someone you liked in that way. Or the blush that came into your skin on a first kiss with the one you chose. There is a way that we can open into contact with something we want to touch.

Where the ripeness of us rises to the surface, rides at the edge. A way that we can open to receive it that is not merely physical though it makes use of our bodies completely.

I want you to stand at eight feet from the fire and move

into her warmth the way that you enter a bath of the perfect temperature, the temperature where you do not have to pull back at all, the temperature at which you do not shiver at all and yet your whole body shivers– the way you dissolve into the embrace of pure sunlight, the way it awakens every cell of your skin–at each moment of stepping forward letting the warmth move from your hands into your entire body. Letting your body become transparent– *you already know how to do this*. Letting it lap at you like waves. Surrending to the warmth.

This flow of relatedness, passing through the portal of your hands, a portal that is a doorway, a doorway that has opened, that has become transparent, to allow the relating in, a doorway that you control, you able to step forward closer to the Source, the heat, the light, or to step back if it grows too strong– this is the portal that leads you home.

It is so simple, so beautiful, and it is Sacred. My favorite word for Sacred is Aramaic. The word is *Qadash*. It comes from two root words: *qd*, and *ash*. *Qd* is the centerpoint. *Ash* is the circle of light and heat radiating from it.

This marries well with the Tzutuj'il Mayan word for life: *N Kasea*. N Kasea means 'the ember yet glows'. There is another Tzutuj'il word for the bonds of relatedness in the village. It is translated as mutual endebtedness: *Kas Limaal*. But a truer translation is *mutual ensparkedness*.

The entire village is aflame with this: the original fire transmits the primal warmth, which you carry inside you as an inward spark. Look deep enough into your heart and you can find this fireplace, for it is also within you.

Re-kindle the spark, awaken, claim your inheritance. The village has always been here and it calls for your light.

And so I step away with you here, wrapped in blankets, by the side of this great fire, the original fire, this great hearth. I step away from you here, for now, having led you, I pray, through some of the wilderness, the howling night of forgetting, nightmares of the colonial mind, back to the heart of what you already knew, what you came knowing, what you came to remember, and who you came to be.

There are many of us here, gathered around this original fire. Part of my great joy, privilege, and duty is to bring as many of you here as I can.

Behold the beauty of this Creation, these creatures with hands of all colors, shapes, sizes, and ages whose faces are bathed in firelight. Behold *the People.* Many hands that can hold one another close, hands that can make a living bridge to carry us home. Thus concludes *The Neurobiology of Connection: Re-Wilding the Deep Nervous System for Wellbeing.*

NOTES

ACKNOWLEDGEMENTS

This book is a yearning for home, a meditation, and an act of protest, in that order. It is a protest against the hegemony of a brain-centric world, which is far less beautiful than what will be required to call us home. It is a meditation on the wild beauty of the inner landscape of our deep nervous system, far stranger and more wonderful than we could imagine, of which I am grateful to be a tracker and cartographer. And it is most importantly, like all of my work, a calling of us home to the village, where our flourishing resides. All of us have an imaginal ancestral village, a place of homecoming. The fireplace at its center looks strangely like the cardiac ganglion, deep in the heart, which carries the visual imprint of the original fire, and the hearth stones around it, as so:

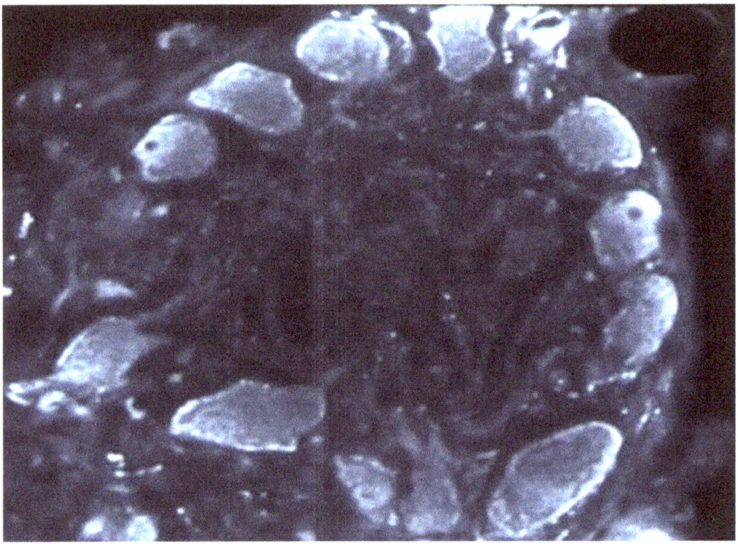

Since everything worthwhile is both universal, and particular, my particular village is filled with my particular fellow villagers, for whose companionship I yearn. When I write it is you for whom I tell fireplace stories. Some of you are still with us, thank God, and some of you have passed on to the spirit realm. As I have written, and been written by this book, I have held you close. To those of my imaginal village: Sigurd Anderson, Stephanie Antonioli, Nic Askew,

Sue Bahnan, Mitchel Berman, Mark Bockley, Alex Briscoe, Jeffrey Bronfman, Rose Brie Code, Deb Dana, JD Daniels, Leah Dawang, Alé Duarte, Ana do Valle, DuShara English, Vincent Felitti, Jimbo Graves, Tiokasin Ghosthorse, Jordhynn Guy, Kristin Griffey, Nadine Burke Harris, Gurucharn Singh Khalsa, Pete Jackson, Carol Jenkins, Alnoor Ladha, Peter Levine, Peter Love, Fleet Maull, Heather MacDuffie, Niloufer Merchant, Ilarion Merculieff, Marcia Miller, Ali Montgomery, Carri Munn, Darcia Narvaez, Tabatha Parker, Stephen Porges, Paula Ramirez, Skeena Rathor, Daniel Rechtschaffen, Jeff Rockwell, Paula Scataloni, Arielle Schwartz, Dan Siegel, Lee Shackleford, Nola Sharp, Earl Simms, John Stokes, Peter Tavernise, Moira Theede, Al Theede, Anthony Valdez, Aubrey Wallace, Twig Wheeler, Jacinta Willems, Marcellus Bearheart Williams, Gloria Yamato, and Shinzen Young.

Which is to say that this book has emerged in the space between us: a document to our relating. In truth, a village has dreamed this book into being. It has been my honor to write it down.

I would like to thank the global Polyvagal Study Group on Facebook, which I stewarded while writing this book, and where I worked out some of these ideas in call-and-response.

I would like to thank all of the participants in the past three semesters of my Post-Doctoral Seminar in Autonomics.

I would like to thank our informal and formal research team at Hearth Science, including specifically Aubrey Wallace ND, Sue Bahnan MA, Heather MacDuffie, PhD, and Samar Singh, PhD. The excitement of working at the edges of the known with such talented colleagues has been an extraordinary joy.

Heather, our conversations at the edge of the neurophysiological known were a fantastic catalyst. I'm so grateful.

Samar, Respiratory Profiling Analysis is a towering achievement. Thank you for your brilliance, and for allowing us to peer into the Autonomic Nervous System through the extroardinary portal of breath.

Aubrey- you are an intuitional genius. It is an honor and a happiness to weave our worlds together.

Sue- your great excitement about the intersection of neuroscience and spirituality, the flame of your passion for it, and the rigor of your intellect make playing with these ideas with you so much fun. I am so appreciative of the collaboration. Looking forward to your podcast. I'll be a guest anytime ;)

I would like to thank Carol Jenkins for long and deeply wonderful conversations about the State of Union. I'm so looking forward to your book on the same.

I would like to thank Jordhynn Ariel Guy for her interoceptive mapping skills and artistic precision in exploring elite performance. For coining the term 'Grounding System', and reminding me that we claim inheritance through the feet.

I would like to thank DuShara English for her pioneering work on the relationship between women's hormonal cycles and the neurobiology of flow states.

I would like to thank Arielle Schwartz, PhD, for embodying and affirming the role of presence and mindful embodied awareness in the teaching of Polyvagal Theory.

I would like to thank Niloufer Merchant, PhD, for exciting conversations situating the state of Appease within the context of cultures both individualist and collective.

I would like to thank Carri Munn for bringing my attention evermore deeply to the role of context in our decision-making and organizational design, for helping me to understand Hearth Science as a living entity not a mechanism, and for her exquisite modeling of how to stay in relationship and connection through difficulty.

I would like to thank Rose Brie Code for her work on world-building in video games based on oxytocin rather than adrenaline/cortisol, her profound morality, and modeling the responsible reduction of bias in her decision-making.

I would like to thank Alnoor Ladha for his distillation of the metapatterns of late stage Capitalism, and his interest in collaboration around the role of the neurobiology of connection in transforming complex systems.

I would like to thank Skeena Rathor for inquiries at the intersection of activism, resilience, and physiology.

I would like to thank Peter Tavernise for his reflections, his tilt towards uncovering deeper meaning in whatever is actually happening, and for introducing me to people I need to know.

I would like to thank Rick Hanson for the friendship that began as this book was concluding, for seeing me and recognizing our work, and for embodying what you are teaching.

I would like to thank Adam Johnson, my creative writing instructor, for teaching me two decades ago that yearning is the great governor of character: this is not just true in fiction ;)

I am grateful to Kevin Barrett Kane for his participation in the design of this book, and to Ash Tyrrell for her proofreading.

I would like to thank Jimbo Graves for marvelous conversations at the intersection of poetics, story, creativity, and transformation as well as your excellent tennis coaching as I learn to strop my patience.

I would like to thank Jed Bickman, from New Harbinger, for variations on the name of the book.

I would like to thank my Autumn 2024 Seminar in General Autonomics, morning section, for brainstorming with me about the name of the book, and Jennie Powe Runde for helping with the sub-title.

I would like to thank Ayssar Arida for helping me with the cover designs.

I would like to thank photographer Lynn Johnson for permission to share a photograph she took of Kapuna Sabra Kauka and her dear friend exchanging the breath of life (*honi honi*) in the spirit of aloha.

I am blessed to be able to explore these ideas that are so close to my heart in beloved community: I am extremely grateful for how meaningful this makes my daily life.

I am grateful to all the readers on Substack who provided comments on chapters of the manuscript. Your enthusiasm was a huge motivation to me.

I am grateful to all participants in the Hearth Science decentralized translation research network in Autonomics.

Finally, I would like to honor and acknowledge the two people who have to deal with me when I am in the grip of the Muse: my wife Lea Marie Wood and my daughter Talullah Azul. Given the sometimes immense intensities that flow through me, and of which you are sometimes downwind, I am grateful to you for putting up with me. I love you both more than I can properly adhere language to. You are the hearth to which I return.

Finally, to the forest that owns me– a bow so deep my lips are pressed to the Earth.

ABOUT THE AUTHOR

Natureza Gabriel (né Gabriel Kram) is the principal neural architect of *Autonomics*, a cutting-edge & ancestral update to our understanding of living autonomic physiology, which he has developed over 30 years of trans-disciplinary study and research with input from well over 5,000 wellness professionals, and 100 mentors and advisors from 25 lineages of healing in 24 cultures. His mind was trained at Yale and Stanford Universities, his heart has been educated in ceremonies and circles. He has spent 30 years studying connection through the lenses of neuroscience, mindful awareness, social justice, deep nature connection, non-cognitive ways of knowing, Indigenous Lifeways, and cultural linguistics.

He is Founder and CEO of Hearth Science: a translation research firm pioneering the union of neurophysiology and ancestral awareness to turn on the deepest drivers of human wellbeing. He is the principal architect of the Autonomic Compass, a proprietary diagnostics and treatment software platform that centralizes autonomic physiology in the diagnosis and treatment of stress-related disorders and the creation of enduring wellbing. He is Host and Executive Producer of *The Restorative Practices Film Series*, *The Connection Masterclass, Evoking Connection States*, and *Lectures on the New Foundation Model in Autonomics*. In autumn 2023 has was asked to lead the global Polyvagal Study Group on Facebook. He has been asked to teach Autonomics to people in 50 countries, executives in Fortune 500 companies, the faculty of medical schools, governments, international NGOs, and tribal leaders. He is the author of the *Connection Phenomenology Series*. This is his tenth book.

He lives with his family on unceded Miwok territory (Bay Area) in South Salmon Nation (California) on western Turtle Island (United States). You can find more of his work, as well as that of the extraordinary faculty of Hearth Science at

HTTP://www.hearthscience.io